Springer Series on Naval Architecture, Marine Engineering, Shipbuilding and Shipping

Volume 18

Series Editor

Nikolas I. Xiros, University of New Orleans, New Orleans, LA, USA

The Naval Architecture, Marine Engineering, Shipbuilding and Shipping (NAMESS) series publishes state-of-art research and applications in the fields of design, construction, maintenance and operation of marine vessels and structures. The series publishes monographs, edited books, as well as selected PhD theses and conference proceedings focusing on all theoretical and technical aspects of naval architecture (including naval hydrodynamics, ship design, shipbuilding, shipyards, traditional and non-motorized vessels), marine engineering (including ship propulsion, electric power shipboard, ancillary machinery, marine engines and gas turbines, control systems, unmanned surface and underwater marine vehicles) and shipping (including transport logistics, route-planning as well as legislative and economical aspects).

The books of the series are submitted for indexing to Web of Science.

All books published in the series are submitted for consideration in Web of Science.

Wentao Huang · Moduo Yu · Hao Li · Nengling Tai

Energy Management of Integrated Energy System in Large Ports

 Springer

Wentao Huang
School of Electronics and Electrical
Engineering
Shanghai Jiao Tong University
Shanghai, China

Moduo Yu
School of Electronics and Electrical
Engineering
Shanghai Jiao Tong University
Shanghai, China

Hao Li
School of Smart Energy
Shanghai Jiao Tong University
Shanghai, China

Nengling Tai
School of Smart Energy
Shanghai Jiao Tong University
Shanghai, China

ISSN 2194-8445   ISSN 2194-8453 (electronic)
Springer Series on Naval Architecture, Marine Engineering, Shipbuilding and Shipping
ISBN 978-981-99-8794-8   ISBN 978-981-99-8795-5 (eBook)
https://doi.org/10.1007/978-981-99-8795-5

This Springer imprint is published by the registered company Springer Nature Singapore Pte Ltd.
The registered company address is: 152 Beach Road, #21-01/04 Gateway East, Singapore 189721, Singapore

Paper in this product is recyclable.

Preface

As strategic points and important hubs for international trade and marine development, large seaports consume a huge amount of energy and are also one of the major sources of carbon emissions. In the context of "carbon peaking and carbon neutrality," it is urgent to prioritize the development and use of new energy, build low-carbon and green ports, and achieve energy saving and emission reduction. Ports are composed of logistics subsystems and energy subsystems. The electrification of logistics and the integration of different energies are key ways to solve the high energy consumption and high emissions in ports. The application of renewable energies like wind and solar, as well as clean fuels such as natural gas and hydrogen, is becoming increasingly widespread. Concurrently, the use of advanced technologies like multiple energy storage systems, shore power, and electric yard bridge cranes is on the rise. This expansion is deepening the interconnectivity of various forms of energy, logistics, and shipping, leading to a complex coupling relationship with heightened mutual influences, thus necessitating a thorough understanding for energy optimization and carbon emissions reduction. Therefore, it is very necessary to study the methods and technologies of low-carbon operation and energy optimization of ports under the coupling of multiple logistics and energy systems.

The port's integrated energy system is a dynamic interplay of logistics and energy subsystems. The energy subsystem, underpinned by the concept of an energy hub, is a tripartite composition involving the energy supply side, the energy conversion side, and the load side. This system works in tandem with the logistics subsystem, which is responsible for both dispatching logistics and serving as an integral part of the energy load. These two subsystems interconnect and form a unified port energy system that delivers an integrated service of energy and logistics to both the port and docked ships. On the supply side, the energy system encompasses the primary power grid, the primary gas grid, and renewable energy sources such as wind turbines and photovoltaic cells. Energy conversion and coupling are facilitated by the energy hub, featuring various types of energy networks—the power grid, natural gas network, and heat network and an array of energy conversion equipment. The load side is equally diverse, comprising port cold and heat loads, civilian electricity loads, and the loads of ships and shore cranes. Through the synchronization of incoming and outgoing ship movements and the allocation of shore cranes, the dispatch center plays a crucial role in maintaining a seamless seaport logistics-energy system. The management of various energy devices' outputs, along with the operational status of ship equipment, further enhances the coordination and efficiency of this complex system. This allows for efficient logistics interaction between the ships and the port. Meanwhile, the shore-based power supply serves as a conduit for energy exchange between the ships and the port, further enhancing the overall system efficiency.

This book aims to analyze the interaction mechanisms of energy and logistics subsystems in green ports, establish models of energy characteristics and carbon emissions for ships, machines, vehicles, multi-energy sources, and port yards, and explore the potential for coupling and complementarity between subsystems on multiple spatiotemporal scales. It proposes methods for logistics-energy multi-timescale coordinated optimization regulation and energy management. Furthermore, it clarifies the complex impact of the multiple uncertainties of logistics and energy systems on energy consumption and emissions and resolves optimization problems related to the coordination of logistics-energy interaction, multiple energy coupling conversion, and graded allocation of production tasks in green ports. This will provide guidance for port economic operations, energy pricing, multi-agent games, and corresponding low-carbon policies.

Chapter 1 provides a comprehensive overview of research status and opportunities on low-carbon technologies for integrated port energy system.

The next five chapters (Chaps. 2–6) present several crucial components of the port electrification approaches. Chapter 2 is about the optimal configuration of renewable energy and energy storage in port microgrid. A novel method for generating typical operational scenarios for port microgrid is proposed, enabling lower computational complexity and more accurate results in port microgrid planning. Chapter 3 aims at shore power variable frequency control in hybrid AC-DC port microgrid and proposes an adaptive bidirectional droop control strategy for the interlinking converter. Chapter 4 systematically studies the basic structure, unified interface, flexible interconnection scheme, and control system and methods of multi-microgrids, thereby expanding the diversity and flexibility of interconnected multi-microgrids. Furthermore, to address a series of issues arising from the grid connection of the ship's power grid with shore power, Chap. 5 develops a smooth control strategy for port-ship islanding/grid-connected mode switching. Chapter 6 proposes a voltage optimization approach for port power supply networks with soft open points, which effectively reduces voltage fluctuations based on real-time model predictive control.

The third part of this book (Chaps. 7–11) addresses comprehensive optimization and management of port energy systems. Chapter 7 proposes a refrigerated container cluster-layered scheduling architecture and a collaborative optimization strategy for multi-agent refrigeration efficiency. This approach addresses the challenges of optimizing the solution and calculation efficiency for large-scale refrigerated container load scheduling in ports. Chapter 8 focuses on the impact of the time-of-use tariff on demand side response (DSR) and proposes a frequency-based DSR strategy, avoiding frequency stability issues related to DSR. Chapter 9 presents an energy cascade utilization method, achieving comprehensive optimization and utilization of various forms of energy in port's electric thermal microgrids. Chapter 10 aims at optimal operation of integrated port energy systems under coordination between logistic and energy systems, so as to minimize total costs including logistic operating and energy service costs. Chapter 11 studies the joint scheduling method of power flow and berth allocation in port microgrid to improve energy efficiency and economic benefits.

The book's last chapter focuses mainly on future prospects of low-carbon management cases in ports under the context of port electrification and integrated energy.

This book is the result of the guidance and support from the Shanghai Engineering Research Center of Intelligent Ship Integrated Power System. We would like to express our profound gratitude to the National Natural Science Foundation of China (5233000026) for the significant support toward our research. Our team members, including Xiaobo Wang, Jie Wang, Yiwen Huang, Congzhe Gao, Li Yang, Pan Wu, and Hao Tian, have all contributed significantly to the research process. Xiao Yu and Hao Li were instrumental in the preparation of the manuscript. We wholeheartedly extend our gratitude to all involved.

Despite our best efforts, we acknowledge that this book may have areas for improvement due to our own limitations and practical experience. We warmly welcome and appreciate any corrections or suggestions from our readers.

Shanghai, China   Wentao Huang
July 2023

Contents

Chapter 1
Overview and Research Opportunities in Energy Management for Port Integrated Energy System

1.1 Introduction

As a strategic pivot and important hub for ocean development and international trade, large ports consume huge amounts of energy and are one of the main sources of global carbon emissions [1]. China has a vast port scale, with seven of the world's top ten ports located in China [2]. The top ten seaports in China based on their annual container throughput as of 2021 are listed in Table 1.1. Various types of ports along the coast, rivers, and inland waterways are densely distributed in China, forming a systematic networked geographical layout with Shenzhen Port, Shanghai Port, and Tianjin Port as the main hub ports for shipping, Ningbo-Zhoushan Port, Guangzhou Port, Qinhuangdao Port, Qingdao Port, Dalian Port, and Rizhao Port as important regional auxiliary ports for shipping, and other medium and small ports as supplementary ports. Against the backdrop of "carbon peak and carbon neutrality", energy-saving and emission reduction in ports have become urgent. Relevant institutions such as the International Maritime Organization have introduced a series of regulations to restrict emissions from the shipping industry, and countries are continuously deepening the adjustment of industrial structure to address the problem of high energy consumption and emissions in the water transport industry. The "14th Five-Year Plan" for Green Transportation Development issued by the Ministry of Transport proposes that by 2025, the proportion of new energy container trucks in international hub ports will reach 60%, and the transformation of existing operational ships into electric power facilities will be accelerated. In 2022, the National Development and Reform Commission and ten other ministries jointly issued the "Guiding Opinions on Further Promoting Electrical Energy Substitution", pointing out the need to deepen the electrification of the transportation sector and promote shore power construction. State Grid Corporation has implemented more than 600 port shore power demonstration projects in the Bohai Rim, southeast coastal large ports, the Yangtze River Delta, Beijing-Hangzhou Grand Canal, and ports along the

© The Author(s) 2023
W. Huang et al., *Energy Management of Integrated Energy System in Large Ports*,
Springer Series on Naval Architecture, Marine Engineering, Shipbuilding and Shipping
18, https://doi.org/10.1007/978-981-99-8795-5_1

Yangtze River. Therefore, building low-carbon and green ports, prioritizing the development and utilization of clean energy, and promoting environmental protection are hotspots that urgently need to be studied in China's energy structure upgrading and ecological green development path [3, 4].

With the development of technology, various renewable energy sources such as solar energy, wind energy, tidal energy, and wave energy have become possible for application in ports [5]. The implementation of projects such as "oil-to-electricity" conversion, shore power, and new energy ships [6, 7] has turned ports into industrial hubs tightly integrated with transportation logistics and energy systems [8]. In this context, the opportunity to replace energy with electricity and construct a low-carbon green port comprehensive energy system not only meets the requirements of the era's development but also represents an inevitable trend in port development. In the future, large-scale operational equipment in green ports will fully adopt electricity substitution technologies [9], resulting in a strong coupling between port logistics and energy systems. Additionally, the port area will encompass various forms of energy, leading to increasingly complex coupling relationships and growing mutual influences between different energy sources and logistics. Under the coupling of multiple systems in logistics and energy, low-carbon methods and technologies for green port comprehensive energy systems have become a current research hotspot [10, 11].

This chapter provides an overview of the development status of low-carbon technologies in ports, focusing on aspects such as electricity substitution, renewable energy generation, and clean fuel applications. It analyzes the operational characteristics of port logistics systems and discusses in detail the coupling mechanisms and modeling methods between energy systems and logistics systems. The article summarizes the typical characteristics of port comprehensive energy systems and existing energy management methods. Finally, based on existing research, it proposes key research directions that urgently need breakthroughs in port comprehensive energy systems.

Table 1.1 The top ten seaports in China based on their annual container throughput as of 2021

Rank	Seaport	Province	Annual Container Throughput (TEUs)
1	Shanghai port	Shanghai	47,030,000
2	Ningbo-Zhoushan port	Zhejiang	31,080,000
3	Shenzhen port	Guangdong	28,770,000
4	Guangzhou port	Guangdong	24,180,000
5	Qingdao port	Shandong	23,710,000
6	Tianjin port	Tianjin	20,270,000
7	Xiamen port	Fujian	12,050,000
8	Beibu gulf port	Guangxi	6,010,000
9	Yingkou port	Liaoning	5,210,000
10	Rizhao port	Shandong	5,170,000

1.2 Low-Carbon Technology in Ports

1.2.1 Electric Energy Substitution

(1) Cold-Ironing Technology

Usually, shortly after a ship arrives at a port and docks, the main generator is shut down while the diesel auxiliary generator is turned on to supply power for communication, lighting, ventilation, cargo handling, and other activities. Generally, the auxiliary generator uses cheap and low-quality fuel, resulting in low fuel efficiency, high losses, and high emissions of pollutants, creating a significant "floating chimney" effect at sea [12]. One important measure to reduce emissions during berthing is to use shore power to supply electricity to the vessel instead of using onboard auxiliary engines. Ship shore power technology refers to the practice of shutting down the ship's own diesel auxiliary generator and using the power system on the dock to provide electricity to the vessel. This way, electrification from shore to ship can be achieved.

Shore power technology has shown significant emission reduction effects and has been gradually applied in multiple ports worldwide, such as Yangshan Port in China [13], Dongguan Port [14], and Jurong Port in Singapore [15]. Studies have shown that the use of shore power technology can reduce global port emissions by 10% [16]. However, the emission reduction effects vary significantly among different ports due to policy differences and variations in charging standards in different countries and regions. According to estimates, the emission reduction proportion of shore power in UK ports is 2% [16], while the emission reduction effect in Kaohsiung Port in Taiwan, China exceeds 57% [17].

The demand for shore power varies among different types of vessels. For bulk carriers, a comparison between shore power and shipboard fuel in literature [18] indicates that shore power has significant economic advantages when the electricity price is below 0.19 USD/kWh, reducing operational costs and energy consumption by up to 75%, providing a win–win solution for shipowners and port authorities. On the other hand, when cruise ships are berthed, a large amount of electricity is required for passenger activities, making the demand for shore power more urgent. A study [19] analyzed cruise ship cases in three different regions and found that the use of shore power can average reduce greenhouse gas emissions by 29.3%. In Norway, France, and Brazil, cruise ports can reduce carbon emissions by 99.5, 84.9, and 85.3%, respectively, through the use of shore power technology.

Although shore power technology has significant advantages in emission reduction, its widespread adoption is still a major challenge. The main obstacles include technological and policy aspects. On the technological side, issues such as power quality, system stability, reliability, safety, and synchronization caused by shore power access need to be addressed [20, 21]. After other industrial or commercial power systems are energized, circuit breakers only trip in maintenance or fault situations, while shore power systems require frequent breaker operations, posing safety hazards to operators. In addition, large-scale port shore power operations often require large substations, with the high-voltage side close to port facilities. In the absence of proper planning, grounding faults on the high-voltage side of substations can cause dangerous contact voltage. At the same time, there are multiple types of ship frequencies. In a country like China where the utility power supply frequency is 50 Hz, frequency conversion is required, which also poses challenges for the synchronization of shore power systems. On the policy side, the degree of marketization of electricity is a key factor restricting the promotion of shore power. Taking China as an example, the electricity regulations clearly state that power companies are the main providers of electricity services, while shore power requires ports to provide power services to vessels, and relevant policies and regulations need to be improved urgently.

(2) **Electrification of Logistics Equipment**

Ports need to provide ample logistics services for docked ships through a variety of different types of logistics equipment, including quay cranes (QC), yard cranes (YC), conveyor belts, and transfer vehicles. QC is used for loading and unloading cargo or containers from the ship's side. YC is divided into rail-mounted gantry (RMG) and rubber-tire gantry (RTG), which are used for container loading, unloading, handling, and stacking in the yard. The main difference is that RMG moves on rails, while RTG moves on rubber tires. Conveyor belts are used to handle bulk cargo, such as coal, fertilizer, and wood. Transfer vehicles include reach stackers (RS) for container lifting, straddle carriers (SC) for container transfer, and lift trucks (LT) for cargo lifting and stacking.

Traditionally, these logistics equipment were almost all manually driven. In recent years, highly automated port equipment, such as automated RTG and RMG, have been used to improve efficiency and reduce labor costs [22]. The energy sources for these devices have also become more diversified. Table 1.2 shows the main energy supply methods for the above equipment [23]. From Table 1.2, it can be seen that electricity is the most commonly used energy supply method in ports, which can provide power for all major equipment in the port, and is energy-saving, easy to control, and easy to automate. This makes the electrification of logistics equipment in large ports an irreversible trend.

QC can recover energy during the lifting process and store it for later use [24]. Therefore, the integration of electrification and energy storage systems can transfer QC's peak load, improving energy utilization efficiency. Literature [25] used supercapacitors to reduce QC's peak load from 1211 to 330 kW. Literature [26] reduced

Table 1.2 Energy source for different equipment

Equipment	QC	RMG	RTG	RS	SC	LT
Diesel	✓	–	✓	✓	✓	✓
Electricity	✓	✓	✓	✓	✓	✓
Natural gas	–	–	✓	✓	✓	–

QC's peak load from 1500 to 150 kW by integrating energy storage systems. The transfer of peak load means higher energy efficiency, achieving energy conservation and emission reduction. In terms of YC, RMG usually has higher energy efficiency than traditional RTG because it is electrically driven, but RTG's advantage is its flexibility in operation, as it is not limited to rails. Therefore, the electrified RTG (E-RTG), which balances the flexibility of RTG operation and the high energy efficiency of RMG, is currently the mainstream trend. Studies have shown that compared with traditional RTGs using diesel as fuel, the energy cost of E-RTGs has decreased by 86.60%, and greenhouse gas emissions have decreased by 67% [27].

For other yard equipment, such as RS, SC, and LT, hybrid diesel-electric engine systems have been integrated. Studies have shown that the fuel efficiency of hybrid SC has increased by 27.1%, and carbon emissions have been reduced by more than 66% [28]. With the development of electrical engineering technology, especially energy storage technology, fully electric RS, SC, and LT will soon become a reality, helping to promote green and low-carbon development in ports [29].

1.2.2 *Renewable Energy*

Renewable energy includes sources such as solar, wind, tidal, wave, and geothermal energy, among others. Unlike fossil fuels, renewable energy has a fast regeneration rate and does not produce gas emissions. It is an important technological tool in achieving low-carbon ports. In recent years, the use of renewable energy in ports has become increasingly widespread. Literature [30, 31] has studied the importance of renewable energy in establishing low-carbon ports. Literature [32] considers "the percentage of energy from renewable resources" as an important Key Performance Index (KPI) for sustainable intelligent ports. Literature [33] has proposed a creative method of covering the roofs of refrigerated areas with photovoltaic cells, generating electricity for lighting, refrigeration, heating, etc. Covering these areas also shields the containers from direct sunlight, reducing the amount of energy needed for cooling.

In practical engineering applications, more and more ports are considering renewable energy as the main measure to reduce carbon emissions. The port of Chennai in India evaluated the possibility of using photovoltaic power to supply energy to the port area from the perspective of open days, capacity utilization factors, and area available for placing photovoltaic cells, and gradually established test engineering projects [34]. Jurong Port in Singapore proposed the concept of a "zero-carbon" port and installed photovoltaic cells on the roofs of warehouses, creating an annual output of 12 million kWh through a leasing model [35]. The German maritime department emphasized the importance of renewable energy, particularly wind, solar, and geothermal energy, for German ports in a report published in 2017. Hamburg Port installed more than 20 wind turbines with a capacity of 25.4 MW [36]. In addition, Hamburg Port has established a photovoltaic power station with an expected annual output of 500 MWh [37]. Other renewable energies such as geothermal energy [38], tidal energy [39], and wave energy [40] are gradually being applied in ports. These examples indicate that there is enormous potential for the use of renewable energy in ports.

However, the planning and layout of renewable energy in ports still needs to be studied. Taking photovoltaic and wind power as the most widely used examples, the installation location of photovoltaic arrays must have sufficient solar radiation. Rooftop photovoltaics are preferred, so the load requirements for distributed photovoltaic power systems on the roofs of large warehouses and other buildings in newly built port areas must be considered. When installing port wind turbines, factors such as changes in sea conditions, reserved selection ranges for axis lines, navigable widths of entry channels, safety distances between anchorages and navigation channels, scale of anchorages, and safety distances between anchorages and wind farm boundaries must be considered [41]. With the adjustment of port scale, the layout of renewable energy power stations in ports also needs to be adjusted, and a safe, efficient, and economical layout plan is urgently needed.

1.2.3 Clean Fuel

Currently, the maritime transportation industry heavily relies on fossil fuels, especially heavy oil, as a source of power. Substituting low-carbon or even zero-carbon fuels for heavy oil is an important way to achieve decarbonization in the shipping industry. Liquefied natural gas, biomass fuels, and hydrogen energy are currently the most promising clean fuels for shipping.

Some ports have already considered using liquefied natural gas for port equipment. In 2008, the Port of Long Beach evaluated liquefied natural gas fueling facilities [42]. As part of the European Union-funded Green Crane project, multiple ports in Europe evaluated terminal tractors, liquefied natural gas or dual fuel RTGs, and liquefied natural gas-dual fuel RS based on liquefied natural gas fuel. The use of liquefied natural gas for terminal tractors is estimated to reduce carbon dioxide emissions by 16%, while nitrogen oxide emissions will also be reduced [27]. Research in [43] shows that using liquefied natural gas can reduce carbon dioxide emissions by 25% compared to fossil fuels. In addition, the International Maritime Organization has set stricter requirements for marine emissions in the latest International Convention for the Prevention of Pollution from Ships, which clearly stipulates that the global marine fuel sulfur content limit will be reduced from 3.5 to 0.5% from 2020 onwards. Therefore, more and more ships will begin to use clean fuels based on liquefied natural gas.

Biodiesel mixed with diesel is also an important clean fuel in ports. The Port of Rotterdam in the Netherlands produces biodiesel mixed fuel by blending biofuels with diesel fuel currently in use at a ratio of 3:7. The port's biomass fuel throughput reached 4.8 million tons in 2016 and has become a major import and export hub [44]. In this process, port waste is used as a renewable resource for producing biomass fuel.

Although liquefied natural gas and biomass fuels are cleaner than traditional fossil fuels, they are not the best long-term solution because they cannot achieve zero emissions. Hydrogen energy, as a zero-carbon fuel, is the best choice for achieving carbon neutrality in ports.

Hydrogen energy is a high-energy density and long-term storage efficient energy storage method. In terms of energy utilization, the high-capacity and long-term storage mode of hydrogen energy is more sufficient for the use of renewable energy; in terms of scale storage economy, the cost of fixed-scale hydrogen storage is one order of magnitude lower than that of battery storage [45]. The basic principle of hydrogen storage is to electrolyze water to obtain hydrogen and oxygen. Taking wind power as an example, when there is sufficient wind power but it cannot be connected to the grid, hydrogen can be produced by using wind power to electrolyze water and store hydrogen; when energy is needed, the stored hydrogen can be converted into electricity through a hydrogen fuel cell. Ports are renewable energy-intensive areas with abundant water resources. Renewable energy can be used to electrolyze water to produce hydrogen, which can be used to achieve pollution-free and zero-emission hydrogen production and utilization, while solving the problems of wind and solar energy waste [46]. Combining hydrogen storage with new energy generation is an excellent choice to reduce hydrogen storage costs and improve new energy generation utilization. Hydrogen energy can be applied to many scenarios in ports, such as commuting, transportation vehicles, loading and unloading machinery, etc. As a clean energy source with no pollutant emissions and no greenhouse gas emissions, hydrogen energy has become the focus of promoting clean energy use in port loading and unloading production equipment. The Shandong Port "14th Five-Year Plan" for scientific and technological innovation clearly stipulates that in order to build a clean energy system in ports and build a "China Hydrogen Port", more than two hydrogen refueling stations for port scenarios should be built, with a total hydrogen refueling capacity exceeding 1000 kg/12 h.

Hydrogen fuel cells are a high-efficiency and clean way to convert hydrogen energy into stable electricity. They have already been applied in ports. For example, the Port of Qingdao in Shandong Province has launched the first hydrogen fuel cell vehicle refueling demonstration and operation project in Chinese ports [47]; The Port of Valencia in Spain tested hydrogen gas produced by renewable energy in fuel cell stackers and analyzed methods to improve energy efficiency, performance, and operational safety using fuel cell terminal equipment [48]. Research in [49] analyzed the factors that need to be considered when using hydrogen fuel cell forklifts in green port construction, including safety, facility investment, and cleanliness, pointing out the necessity of applying hydrogen fuel cell forklifts. The ports of Los Angeles and Long Beach evaluated the feasibility of using commercial fuel cells and hydrogen as clean energy sources for various equipment. The research results show that the large-scale application of hydrogen energy in ports is still restricted by low production efficiency, strict storage conditions, limited maturity level of technology, and low return on investment [50]. This is also the focus of future research.

1.2.4   Low-Carbon Port Management

With the abundance of renewable energy and flexible load resources in ports, more and more ports are introducing digital and intelligent technologies to develop low-carbon integrated management platforms. Figure 1.1 shows a schematic diagram of the integrated carbon reduction work of the low-carbon platform load source network in ports. Generally speaking, low-carbon platforms mainly monitor, control, analyze, and optimize port resources to help promote green and low-carbon development in ports. Figure 1.1 centers around the low-carbon platform and achieves centralized and distributed energy generation, multi-directional flow, and real-time data management. Literature [51] provides a construction roadmap for the port low-carbon platform project, which mainly includes the initial stage and installation stage. The initial stage includes: (1) equipment load analysis; (2) analysis of the intelligent grid scenario in the port; (3) energy balance; (4) benefit analysis. The installation stage includes: (1) analysis of renewable energy and evaluation of daily fluctuations in energy generation; (2) optimization of peak shaving and demand response plans; (3) planning of energy storage; (4) management of electricity prices and costs.

The low-carbon management platform uses key technologies to balance energy supply and demand in an intelligent way. The port low-carbon management platform mainly includes four pillars: (1) energy supply (electricity generation) management, including on-site renewable energy generation, cogeneration, and grid management; (2) battery energy storage capacity; (3) energy demand management, using real-time energy consumption measurement, electrified equipment, and shore power; (4) optimization management and communication of all active resources in the grid through optimization methods, load chart control, peak shaving, and utilization management. Literature [52] proposes an intelligent port low-carbon energy management system

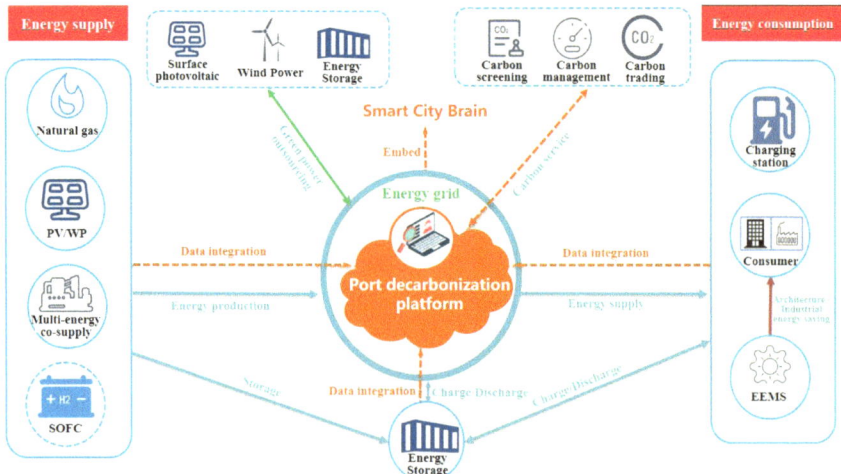

Fig. 1.1 Schematic diagram of port low-carbon platform

composed of microgrids and overall energy planning, and discusses the importance of permanent design, energy efficiency, operational efficiency, and architectural efficiency in overall energy planning. Literature [53] designs a port-grid comprehensive platform composed of sensor technology, advanced smart meters, real-time monitoring systems, control tools, battery technology, and communication technology to optimize port energy demand control and flexibility management. Literature [54] introduces the real-time transmission platform for energy consumption data in the stacking operation of Koper Port in Europe, which achieved a year-round power saving of 281 MWh and fuel saving of 311 tons in the port. Literature [55] proposes an intelligent port energy management system composed of microgrids and overall energy planning platforms, which can reduce port energy consumption and effectively reduce carbon emissions by monitoring the energy efficiency of port equipment facilities, terminal buildings, and other port facilities. In addition, the digital port information platform developed by Singapore Port can provide real-time information for upstream and downstream enterprises in port activities to better coordinate, plan, and allocate port resources [56]. Digital and intelligent technologies will be the backbone of low-carbon management platforms in ports.

1.3 Coupling Mechanism and Modeling for Energy and Logistics

Logistics transportation is the main responsibility of ports, which usually includes operations such as allocating berths and managing container transportation, refrigerated container management, etc. Different logistics operation modes have significant impacts on the distribution of port loads, which in turn affects port energy distribution and system operation. It is necessary to study the coupling relationship between port logistics operations and energy scheduling, as well as the energy modeling method of logistics operations.

1.3.1 Characteristics of Port Logistics Transportation

Terminal operations include allocating berth space and service time for loading and unloading ships, which are usually referred to as berth allocation problems (BAP) [57, 58]. The transfer of goods requires coordination between port equipment such as quay cranes, yard equipment such as stack cranes, and cargo loading and unloading equipment [59]. Therefore, the entire terminal operation is a comprehensive logistics problem, and the operator usually aims to maximize the efficiency and throughput of the port to obtain better economic benefits.

BAP is the focus of port operation scheduling, which determines the berthing position and time of moored ships. Currently, a large amount of research has been conducted on the berth allocation problem in ports. In [60], the impact of uncertain factors such as the arrival and operation time of moored ships on berth allocation was considered, and multi-objective optimization of BAP was achieved based on robust optimization. In [61], the existing BAP formula was modified to address the issue of priority for moored ships. In [62], the problem of dynamic berth allocation for ships was solved in a public berth system. In [63], the risk measure of the given berth timetable was minimized by considering berth productivity while minimizing the total service time of all ships. In [62], the dynamic berth allocation model was first formulated as mixed integer linear programming (MILP), and then extended to the quay crane assignment model in [64]. In [65], the optimal berth allocation model was formulated to minimize the total service time of all-electric ships, and solved using a particle swarm optimization method. In [66], a novel berth allocation model was proposed, which had effective inequalities and rolling horizon heuristics to improve problem-solving efficiency.

In the above berth allocation process, the actual berthing time of the ship depends on the number of allocated quay cranes and the processing speed of quay cranes for containers. Therefore, the berth allocation problem is usually combined with the quay crane allocation problem (QCAP) [66–78]. In [66], a detailed model of crane scheduling considering container groups, non-crossing constraints, and safety distance was proposed for the first time. In [67], an integrated BAP and QCAP was considered, and a berth allocation model based on relative position formula was proposed.

There is also a refrigeration demand in the port transportation process, and the refrigerated area consumes a large amount of electricity. Refrigerated containers must have external power supply, which is also an important energy consumption link. Efficient management of refrigerated containers meets the needs of ships and also reduces related costs. In [79], a spatio-temporal model suitable for relocating refrigerated containers was proposed. In [80], a scheduling method and simulation model of refrigeration-related processes were constructed in a realistic dynamic framework.

Most of the above research is based on the perspective of port logistics management and transportation, aiming to maximize logistics transportation efficiency, and energy problems in logistics transportation are not involved.

1.3.2 Coupling Mechanism of Energy and Logistics

With the increasing electrification level of ports, the coupling between logistics and energy systems is becoming increasingly tight. The close coupling of logistics and energy is an important trend in this special scenario of ports, but it also brings new problems in the port operation process. For example, when a ship docks and uses shore power, the docking time of the ship will affect the time distribution of the shore power load, and the berth (corresponding to the grid node number) determines the spatial distribution of the shore power load. In addition, ship docking brings a series of additional loads, mainly the power load of loading and unloading machinery (such as shore cranes and yard cranes). The docking time and position of the ship further determine the temporal and spatial distribution of the shore cranes, that is, the loading and unloading machinery work when the ship docks, and they are idle when the ship leaves, as shown in Fig. 1.2. Subsequently, new energy container trucks transport containers from the terminal to the yard, and the charging and discharging load of new energy forklifts is closely related to the operation path and time. In the yard, new energy forklifts and yard cranes handle the container transportation, and the charging and discharging load of new energy forklifts is similar to that of new energy container trucks, while the load of yard cranes is related to the frequency of container transportation. It can be seen that the operation mode of the logistics system determines the distribution of power load.

On the other hand, the dispatching of the energy system also has a reverse effect on the operation of the port logistics system. The economic operation of the energy system requires the promotion of renewable energy consumption and the realization of source-load matching. By flexibly adjusting the operational methods of the logistics system, such as berth and gantry allocation, changing the temporal and spatial distribution of port logistics load can achieve the matching between logistics load and renewable energy generation. Thus, more logistics load can be supplied by renewable energy, reducing the energy consumption cost of the port energy system. Therefore, the output of renewable energy and the dispatching of the energy system will also have an impact on the operation of the port logistics system. Based on the above two points, it can be seen that port logistics transportation and energy dispatching are closely coupled and interact with each other, which is a significant characteristic of the comprehensive energy system of the port.

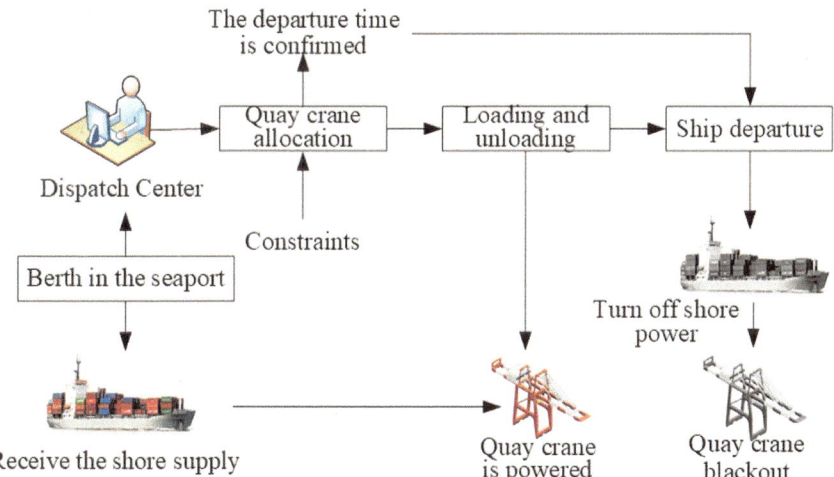

Fig. 1.2 Logistics and power interaction between ship and port during single berthing

1.3.3 Energy-Based Modeling of Logistics

The port logistics system and energy system have different operating characteristics. In order to achieve the coordinated operation of port logistics and energy, it is necessary to establish a quantitative model of port logistics operations.

In current research, in order to associate logistics scheduling with power demand, the common practice is to introduce 0–1 variables to represent the status of ships in port and the operation status of shore cranes and yard cranes. These 0–1 variables are coupled with berth allocation and shore crane scheduling related variables, subject to logistics operation constraints, and can therefore represent the logistics operation process. At the same time, multiplying these 0–1 variables by the rated power of ships and shore cranes can obtain the ship-to-shore power load and shore crane workload related to logistics operations. Therefore, these 0–1 variables can reflect the characteristics of logistics operations and represent the power demand of ships and shore cranes, thereby obtaining a quantitative model of the port logistics system. In [81], the author introduced 0–1 variables to represent the ship's berthing status, associated these variables with berth allocation and shore crane scheduling, and modeled the ship and shore crane power load using these variables, achieving quantitative modeling of ship logistics operations. Similarly, Literatures [82] and [83] both modeled the state of logistics equipment such as ships, shore cranes, and yard cranes by introducing 0–1 state variables, associating them with logistics operation constraints, and using them to represent the power load demand of logistics equipment.

(1) Ships

In the above literature, the state variables of ships are the key to coupling logistics transportation and energy scheduling. Next, we will give an example of how to model the state variables of ships to couple logistics transportation and energy scheduling [81].

Let $X_j(t)$ denote the status of ship j in the port, where $X_j(t) = 0$ before and after the ship berths, and $X_j(t) = 1$ during the ship's stay in the port. Within a single scheduling period, the ship's status in the port can be expressed as shown in Eq. (1.1):

$$X_j(t) = \begin{cases} 0, & t \in [1, t_1] \\ 1, & t \in [t_1, t_{\text{leave}}] \\ 0, & t \in [t_{\text{leave}}, T] \end{cases} \tag{1.1}$$

where t_1 is the time when the ship arrives at the port, and t_{leave} is the time when the ship leaves the port.

Based on (1.1), the energy demand of ships can be expressed as:

$$P_{\text{ship}}(t) = P_{\text{ship}}^{\text{rated}} X_j(t) \tag{1.2}$$

where the ship's power demand changing over time is expressed as the product of rated power and state variable. For a reefer ship, its power demand $P_{\text{ship}}^{\text{rated}}$ should also include the refrigeration power of reefers.

The ship's time window in port is related to the allocation of quay cranes. Meanwhile, the port will be informed in advance of the ship's arrival, so the arrival time of the ship is known. After the ship arrives at the port, it will not immediately berth, but will wait in the anchorage area for port scheduling. The relationship between the time of arrival, time of berthing and time of departure is shown in (1.3). The time of departure t_{leave} is determined by the time of arrival, the amount of cargo loaded and unloaded and the number of quay cranes allocated, as shown in (1.4):

$$t_0 \le t_1 \le t_{leave} \tag{1.3}$$

$$t_{\text{leave}} = t_1 + \left[\frac{N_j}{\eta \times C_j} \right]_{\text{up}} \tag{1.4}$$

where t_0 is the time when the ship arrives at the port. N_j is the quantity of containers loaded and unloaded by the ship, and the unit is TEU. η is the loading and unloading efficiency of the quay crane, and the unit is per TEU/hour. C_j is the allocated number of quay cranes, and $[\cdot]_{\text{up}}$ means rounding up.

Equation (1.4) shows the basic relationship between the ship's time window in port and the number of allocated quay cranes. Ship logistics modeling is closely related to quay crane allocation, which requires further modeling of the quay crane.

(2) Quay Cranes

For quay cranes, their operating status can be represented by introducing new 0–1 variables. Considering the close relationship between quay crane behavior and ship berthing, ship state variables are directly used to model quay cranes to reduce the variable size.

Due to the restrictions on ship length and cargo handling capacity, the number of quay cranes that can be accommodated is limited, and the number of allocated quay cranes cannot exceed the minimum and maximum quay crane demand limits of the ship, as shown in (1.5). At the same time, due to the limited quay cranes, the total number of quay cranes in operation at a certain time cannot exceed the total number of quay crane resources C_{max}, as shown in (1.6).

$$C_j^{\text{min}} \le C_j \le C_j^{\text{max}} \tag{1.5}$$

$$\sum_{j=1}^{n} X_j(t) C_j \le C_{\text{max}} \tag{1.6}$$

The above formulas provide the basic logistics relationship that ships and quay cranes must meet. Based on this, other detailed berth allocation formulas and quay crane scheduling formulas can be expanded on the basis of the above formulas to further model the logistics operation process in a more detailed manner.

Based on ship state variables, the power expression of a single quay crane can be further obtained:

$$P_{\text{crane}}(t) = P_{\text{crane}}^{\text{rated}} C_j X_j(t) \tag{1.7}$$

where the time-varying power demand of quay crane is expressed as the product of rated power and state variable.

Combined with Sects. 1.3.1 and 1.3.2, it can be seen that the state variables of ships and quay cranes are affected by logistics-related constraints. Therefore, the power demand of ships and quay cranes are highly correlated with the characteristics of logistics operations. The energy-based modeling of logistics operations involving ships and quay cranes must be taken into consideration.

The above modeling process indicates the basic idea of energy-based modeling of port logistics operations. By introducing state variables to represent the working status of logistics machinery and associating the working status with logistics-related constraints, the introduced state variables can reflect the logistics operation characteristics of the equipment. On this basis, through expressions similar to Eqs. (1.2) and (1.7), the energy demand of logistics equipment can be achieved by using the introduced state variables.

(3) New Energy Container Trucks

New energy container trucks are one of the main equipment connecting the port and the storage yard. The truck transports cargo between the port and the yard while consuming a certain amount of energy, realizing the coupling between logistics and energy. At present, there is still a lack of corresponding research on the logistics—energy coupling modeling of new energy vehicles. This section presents the preliminary modeling method.

The 0–1 variable $\pi_{i,t}$, $\sigma_{i,t}$ and $\lambda_{i,t}$ are used to indicate that the new energy container truck is in the state of transfer, charging and rest. The 0–1 variable $\beta_{j,t}$ represents whether the truck is on the path j. The state constraint of the new energy container truck can be expressed as:

$$\begin{cases} \pi_{i,t} + \sigma_{i,t} + \lambda_{i,t} = 1 \\ \sum\limits_{t}^{T} \sum\limits_{i=1}^{M} \pi_{i,t} L_i \geq L \end{cases} \tag{1.8}$$

where M is the total number of new energy container trucks. L_i is the transfer capacity of a single new energy container truck within a unit time. L is the total number of transfer tasks within the dispatching cycle. The constraint in the first row represents that each new energy container truck must be in a certain state. The constraint in the second row represents that the transportation volume of all new energy container trucks must meet the total task requirements within the scheduling cycle T.

In addition, there are multiple routes for new energy container trucks to transfer between the wharf and the storage yard. The path selection constraints are as follows:

$$\begin{cases} \sum_{j}^{J} \beta_{i,j,t} \leq 1 \\ \beta_{i,j,t} \leq \pi_{i,t} \\ \sum_{i=1}^{M} \beta_{i,j,t} R_i \leq R_j \end{cases} \tag{1.9}$$

where the first-row constraint means that each new energy container truck can only select one path. The second-row constraint means that only the truck in working state can select the transfer path. The third-row constraint represents the transportation bearing capacity of a certain path. The sum of bearing capacity of all the trucks occupying the path j must be less than the total bearing capacity of the path.

Based on the above model, the energy consumption of the new energy container truck can be further obtained:

$$\begin{cases} \sum_{t}^{T} \pi_{i,t} P_i^{\text{trans}} = \sum_{t}^{T} \sigma_{i,t} P_i^{\text{ch}} \\ P_{i,t} = \sigma_{i,t} P_i^{\text{ch}} \end{cases} \tag{1.10}$$

where P_i^{trans} and P_i^{ch} are respectively the transfer power and charging power of a single truck. $P_{i,t}$ is the charging power of a single truck at every moment. The first-row constraint represents that the total energy consumption of the new energy container truck in the dispatching cycle T is equal to the charging energy, namely, the power balance constraint. The second-row constraint represents the power load of the truck from the perspective of the energy side. The truck will have an impact on the energy side only during the charging period, and will not have an impact on the energy side scheduling in the state of transfer and rest.

Simple modeling can be achieved by introducing 0–1 variables to the logistics and energy consumption constraints of the new energy container truck. Due to the complex port logistics, more detailed modeling of the new energy container truck requires detailed parameters of the logistics transfer model and the charging and discharging model.

(4) Belt Conveyors

The quantitative modeling of belt function is similar to that of ships and quay cranes. It models the transmission speed of the belt conveyor by introducing 0–1 variables, and further associates the transmission power with the speed based on this.

Assuming there are N speed intervals for the belt conveyor, and the transmission speed in each interval is represented by V_i. Introduce 0–1 variables $\alpha_{i,t}$ for each speed interval, representing that the belt conveyor is currently in this speed interval at time t. Then the transmission speed of the belt conveyor at time t can be represented as follows:

$$\begin{cases} V_t = \sum_{i=1}^{N} \alpha_{i,t} V_i \\ \sum_{i=1}^{N} \alpha_{i,t} = 1 \end{cases} \tag{1.11}$$

In (1.11), the first line represents the transmission speed of the belt conveyor at time t, and the second line limits the belt conveyor at time t to only be in a speed range.

Based on (1.11) and using the linearization model pointed out in [84], the transmission power of the belt conveyor at time [84] t can be obtained as follows:

$$P_t = k V_t + b \tag{1.12}$$

The logistics transportation task of belt conveyor is to complete a certain amount of transportation within a specified time T. The modeling is as follows:

$$\sum_{t=1}^{T} V_t L^0 = L^1 \tag{1.13}$$

where L^0 represents the transportation capacity of the belt conveyor in unit time. L^1 represents the total transportation volume that the belt conveyor needs to complete in the specified time T.

From (1.11) and (1.12), it can be seen that the power demand of belt conveyor is coupled with the logistics operation constraint through the transport speed V_i. The quantitative modeling of belt function considering the logistics constraint is realized.

(5) Refrigerated Container Parks

There is a refrigerated area in the port yard, and although the refrigeration power of the refrigerated containers is relatively independent of logistics transportation, it is still an important energy consumption link in the port. The temperature change of the refrigerated containers is coupled with the refrigeration power, which can be compared to the coupling relationship between the state of charge (SOC) of an energy storage battery and the charging power. When the refrigerated container loses power, its temperature rise rate is related to environmental temperature and its own parameters; when the refrigerated container is being charged, its temperature decrease rate is related to the charging power and its own parameters. Therefore, the relationship between the temperature of the refrigerated container and the charging power can be expressed as:

$$T_{t+\Delta t} - T_t = I_{\text{off}} \Delta T_{\text{d}} \left(1 - e^{-c_1 \times \Delta t} \right) - I_{\text{on}} c_2 P_t^{\text{ch}} \Delta t \tag{1.14}$$

where T_t is the temperature of the reefer at time t. ΔT_{d} is the difference between the internal temperature of the reefer and the ambient temperature. I_{on} and I_{off} are 0–1 variables indicating the opening and closing of the reefer respectively. c_1 and [85]

c_2 are the correlation coefficients, whose specific values can be referred to in [85]. P_t^{ch} is the charging power of the reefer at time t.

The above model is a preliminary idea for quantifying the energy consumption in port logistics operations. With the implementation of green and low-carbon projects in ports, ports have become a common node where the three networks of energy, shipping, and logistics intersect. Large-scale equipment in the green port will fully adopt electric energy substitution technology, and the port logistics system and energy system will present strong coupling. However, a large number of discrete variables make it difficult to solve the optimization model, so it is necessary to continue to explore efficient model solving methods.

1.4 Energy Management of Green Port Integrated Energy System

With the development and maturity of low-carbon technologies, the application of electric energy substitution, renewable energy generation, and clean fuels has gradually been introduced into various ports. At the same time, the energy system and logistics system are deeply integrated, forming the Green Port Integrated Energy System. Through efficient energy management of the integrated energy system, the low-carbon development of the port can be achieved.

1.4.1 Port Integrated Energy System

As a typical representative of concentrated energy consumption in industrial production, ports have the characteristics of high energy consumption, high pollution, complex external environment, and resource-intensive, resulting in severe pollution emissions. Building low-carbon and green ports is a key area that urgently needs to be studied for upgrading the energy structure and promoting ecological green development in China. With the development of technology, the application of various renewable energies such as wind energy, solar energy, and tidal energy in ports has become possible [5]. The implementation of projects such as "oil-to-electricity," "shore power," and "low-carbon ships" also makes ports become industrial hubs closely integrated with energy systems and logistics systems. Against this backdrop, the construction of a low-carbon port integrated energy system is not only in line with the requirements of the times but is also an inevitable trend for port development.

The integrated energy system of ports is aimed at the loads of electricity, heat, natural gas, hydrogen, and other loads within the port area. Based on the complementary characteristics of multiple energy types and the principle of energy cascade

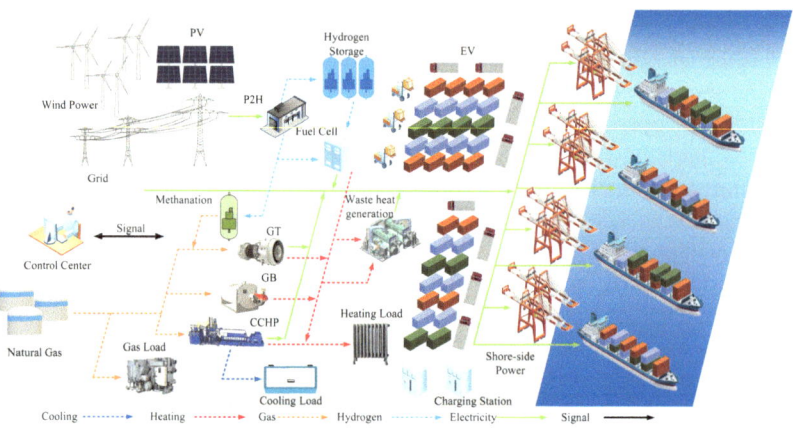

Fig. 1.3 Schematic diagram of integrated port energy system

utilization, the energy system and logistics system are unified in planning and coordinated optimization of operation. This is an important way to improve energy utilization efficiency and reduce carbon emissions in the port area. The schematic diagram of the integrated energy system of ports is shown in Fig. 1.3. Countries with abundant port resources such as Europe, America, and China have already legislated or are in the process of legislating that large-scale port equipment in the future will fully adopt electric energy substitution technology and provide energy through the power grid. At the same time, the integrated energy system of ports will also supply heat, natural gas, and hydrogen to users through infrastructure such as heating networks, gas supply networks, and hydrogen refueling stations to meet various energy needs. In the transmission process of electricity, heat, gas, and hydrogen, the integrated energy system of ports couples electric energy, thermal energy, and cooling energy through combined cooling heating and power (CCHP) units; connects the power grid and hydrogen through power-to-hydrogen technology; realizes the conversion of hydrogen to natural gas through methaneization devices; and supplements heat energy shortage through equipment such as gas boilers and fuel cells. When the energy supply and demand of the system are unbalanced, surplus or insufficient energy can be stored or supplemented through energy storage devices of various energy forms (electricity storage, gas storage, heat storage, etc.) to achieve the energy dynamic balance of the port area.

1.4.2 Flexible Resources of Green Port

The significant difference between the green port integrated energy system and the traditional integrated energy system is the highly coupled energy system and logistics system, which includes various flexible loads. This makes the port integrated energy system have abundant flexibility resources, which can significantly improve the energy management level of the system after reasonable modeling and optimization scheduling. This section will summarize the application of adjustable flexibility resources in green port energy management from four aspects: energy storage devices, vehicle scheduling, berthing ship scheduling, and refrigerated container scheduling.

(1) **Vehicle Scheduling**

Ports require a large number of transportation vehicles in the process of cargo circulation, and the electrification/new energy of transportation vehicles has become a development trend. In Singapore's Jurong Port, electric vehicles have become the main type of transportation vehicles [86]. The Port of Los Angeles in the United States has begun to use hydrogen energy as the power source for container trucks [87]. Reasonably arranging the charging and discharging behavior of transportation vehicles can provide additional flexibility for the port energy system. Literature [88] established a double-layer game model of electric vehicle aggregation merchants integrating electric vehicles to participate in power demand response, which can achieve the multi-subject optimal strategy of electric vehicles and energy suppliers. Literature [89] proposed an ordered charging and discharging scheduling strategy for commercial electric vehicle groups, which can control the number of parked and running vehicles in the electric vehicle group while meeting the operational requirements of commercial vehicles.

(2) **Berthing Ship Sheduling**

Ports need to allocate berths for berthing ships to provide various transportation services. Different types of ships have different energy demand, and different berth services have different service objects [90]. Reasonably deploying all berthing ships in the scheduling cycle and allocating different berths can also balance the energy demand of the port at different time periods, thereby providing additional flexibility for the energy system. Literature [91] established a Nash game model, introduced berth parking duration price incentives, and studied the impact of berth bidding on shore power. Literature [92] studied the energy consumption model of belt conveyor system closely related to the loading and unloading behavior of berthing ships and verified the potential of the model to participate in demand response.

(3) **Refrigerated Container Scheduling**

With the development of society, the demand for fresh food, plants, and flowers from different regions of the world has been increasing. The ocean refrigerated transportation business is constantly developing, and the proportion of refrigerated containers in the port yard is constantly increasing. Maintaining the temperature of goods in refrigerated containers within the allowable range requires a significant amount of energy, which is an important part of port energy consumption. At the same time, the operation of the reefer groups has flexibility in time and power, and optimizing scheduling can significantly improve the efficiency of the port energy system [93, 94]. Literature [95] proposed a multi-subject energy management system for optimizing the operation of large-scale port refrigeration box groups, which can simultaneously schedule a large number of refrigeration boxes in multiple yards to achieve peak shaving and valley filling of port loads. Literature [96] simulated the total power variation of the refrigeration box group under natural conditions, and analyzed the potential for the reefer groups to participate in power balance optimization.

(4) **Energy Storage Devices**

The energy system is a complex system that maintains a balance between power and energy on different time scales, so energy storage devices that can smooth out power and energy fluctuations are an important part of the flexibility resources in the integrated energy system [97]. Thanks to the rich energy sources, ports, especially large seaport integrated energy systems, can apply various energy storage technologies such as electric energy storage, thermal energy storage, natural gas storage, and hydrogen storage. Different types of energy storage have their own advantages, and cooperation can improve the overall economic benefits of the system [98].

Electric energy storage technology is the most mature and has been widely studied in the application of the power system [99–101]. In terms of thermal energy storage, Literature [102] established a detailed mathematical model of thermal energy storage, which can be used to accurately simulate the operating characteristics of thermal storage devices. Literature [103] analyzed the commonalities and differences between thermal energy storage and electric energy storage, and established unified storage models and detailed thermal energy storage models. Literature [104] analyzed the impact of thermal storage equipment on the operation flexibility of combined heat and power units. Regarding natural gas storage and hydrogen storage, Literature [105] established a hydrogen-natural gas hybrid storage system model and proposed corresponding economic dispatch optimization models to analyze its operational benefits. Literature [106] studied the configuration and optimization operation methods of natural gas storage under multiple time scales. Literature [107] analyzed the energy efficiency of hydrogen storage coupled with natural gas combined cycle systems and studied the key factors affecting hydrogen production.

1.4.3 Energy Management Model of Port Integrated Energy System

The green port integrated energy system contains abundant flexible resources and and multiple forms of energy, with great potential for energy optimization management. This section summarizes existing research results on energy management models from two aspects: considering heterogeneous energy characteristics and under uncertainty conditions.

(1) **Energy Management Model Considering Heterogeneous Energy Characteristics**

In port integrated energy systems, different energy types have different time scales and energy transmission characteristics due to different energy transmission media. Fully considering heterogeneous energy characteristics can make the energy management model more accurate and effectively improve the system's energy utilization efficiency. Literature [108, 109] used the node method to establish a thermal system model when managing energy, fully considering the thermal network transmission characteristics and combining transmission heat loss and temperature mixing. Literature [110] used the heat conduction equation to characterize the temperature distribution inside the heat transfer pipeline and calculated the node temperature using the temperature mixing equation. Literature [111] established a dynamic model of thermal systems, considering temperature delay and mass flow rate changes during energy transmission. Regarding natural gas systems, Literature [112] considered natural gas flow rate and pressure to establish an energy optimization scheduling model for port integrated energy systems that consider natural gas flow characteristics.

(2) **Energy Management Model Considering Uncertainty**

In port integrated energy systems, due to the large-scale access of renewable energy and a large number of flexible resources, uncertainty factors are widely present. Currently, scholars often use stochastic optimization models to describe uncertain parameters probabilistically, thereby transforming models containing uncertainty into deterministic mathematical programming problems [113, 114]. Literature [115] used the Monte Carlo method to generate random scenarios to characterize uncertainty factors. Literature [116] considered the propagation of uncertainty between the electricity and gas networks and established a two-level stochastic optimization model and a multi-level stochastic optimization model. However, the accuracy of stochastic optimization models depends on the accuracy of characterizing uncertain factors, and there is a trade-off between solving accuracy and computational difficulty.

Robust optimization methods can also be used for energy management optimization under uncertainty conditions. Robust optimization models are independent of the probability distribution of uncertain parameters and only require setting an uncertainty set to represent the fluctuation range of uncertain parameters. Moreover, under the same conditions, the computational complexity of robust optimization models is often smaller than that of stochastic optimization models [117]. Literature [118] considered the uncertainty of renewable energy and established a virtual power plant stochastic optimization scheduling robust model. Literature [119] established an energy management robust optimization model for multiple microgrids, which considers the uncertainty of renewable energy and loads and can control the conservatism of feasible solutions to a certain extent. However, because robust optimization always needs to meet the worst-case scenario, there are shortcomings such as conservative results and poor economic efficiency.

1.5 Research Directions of Green Port Integrated Energy System

1.5.1 The Current Situation of Typical Ports

(1) **Los Angeles Port**

Los Angeles Port, located at the top of San Pedro Bay on the southwest coast of California, is the second-largest port in the United States. It includes 15 breakbulk/general cargo berths, 36 container berths, and 14 oil berths [120]. Handling equipment includes various types of shore cranes, container cranes, floating cranes, gantry cranes, mobile cranes, loading bridges, and roll-on/roll-off facilities. The container cranes have a maximum lifting capacity of 40 tons, and the floating cranes have a lifting capacity of 350 tons. The container docks can stack up to 25,000 standard containers. The largest oil tanker to dock at the port had a capacity of 220,000 deadweight tons with a tank capacity of 500,000 tons, and the open storage area covers an area of one million square meters.

Los Angeles Port has been developed into a world-leading modern and multifunctional electrified port, with various factories, businesses, and related research institutions within its boundaries. In 2020, Los Angeles Port launched six hydrogen-powered transport trucks, becoming the first port to apply hydrogen fuel cell vehicles [121]. After the massive replacement of electrical and clean energy equipment in the port, Los Angeles Port has become a typical representative of a comprehensive energy supply system that is centered on electricity, has multiple forms of energy, and is coupled with logistics systems.

(2) **Yangshan Deepwater Port**

Opened on December 10, 2005, Shanghai's Yangshan Deepwater Port is located in the rugged archipelago outside the mouth of Hangzhou Bay and consists of Xiaoyangshan Island, Donghai Bridge, and the Yangshan Free Trade Zone, making it the world's largest intelligent container terminal [122, 123]. The main port area of Yangshan Port has completed the replacement of electrical equipment, and some areas have achieved unmanned operations. In 2016, the cargo throughput of Yangshan Port reached 702 million tons, and the container throughput was 37.13 million TEUs, maintaining the world's first place for seven consecutive years since 2010.

Yangshan Port mainly includes the energy operation port area and the container loading and unloading port area. The energy operation port area includes a liquefied natural gas receiving station and a subsea gas pipeline, with an annual import capacity of 3 million tons of liquefied natural gas. At the same time, Yangshan Port is also the largest transfer base for finished oil in the Far East, with a planned and constructed 1900-m-long oil terminal operation area that can store 2.7 million cubic meters of finished oil after the project is completed. The container loading and unloading port area is mainly in the North Harbor area and the West Harbor area, with a planned deepwater coastline of 10 km and more than 30 berths, capable of loading and unloading the world's largest post-Panamax container ships. The variety of goods, rich energy sources, and high level of intelligent operation make Yangshan Deepwater Port a pioneer in China's green port integrated energy system.

(3) **Rizhao Port**

Rizhao Port is the eighth-largest port in China and an important global hub for energy, raw materials, and container transshipment. The main cargo transported by the port includes containers, ores, timber, grain, edible oil, crude oil, natural gas, and coal. The annual electricity consumption of the Port of Rizhao exceeds 100 million kilowatt-hours, which can be converted into more than 40,000 tons of standard coal and 27,000 tons of carbon emissions, with fuel loads, natural gas loads, and cold and heat loads. At the same time, Rizhao Port has abundant sunshine and wind energy, and there is great potential for the electrification transformation of equipment. The various load demands of the port have the conditions for coupling and complementarity, making it a suitable demonstration area for low-carbon energy use in ports.

Currently, Rizhao Port is promoting the "14th Five-Year Plan" for green and low-carbon ports. On the energy supply side, it is developing renewable and low-carbon energy; on the energy consumption side, it is achieving electrification and low-carbonization of port energy consumption; and ultimately, it is constructing a comprehensive energy network with multiple energy coupling and complementarity to form a smart energy solution for the port. By investing in photovoltaics, wind power, combined cooling, heating and power units, and energy storage equipment in stages at Rizhao Port, clean electrical energy replacement can be achieved. By gradually using gas-powered and electric heavy-duty trucks in proportion, electrical energy replacement can be achieved at the port.

1.5.2 Future Research Directions

In recent years, research on integrated energy systems has been flourishing and has achieved relatively complete research results, which can also be applied to the construction and development of port integrated energy systems. Although the port integrated energy system has many similarities with traditional integrated energy systems, its strong coupling requirements for shipping and logistics planning make the low-carbon operation of the port integrated energy system significantly different, and further research by scholars is urgently needed.

(1) **The Accurate Modeling and Application of Liquid Energy Network in Ports**

In the study of traditional integrated energy systems, research on power grids, heat networks, and gas networks has been quite thorough and can be directly applied to the analysis and modeling of integrated energy systems in ports. However, as a transportation hub, ports also contain a large number of liquid networks, such as liquefied natural gas, hydrogen transport networks, and crude oil pipelines. These liquid networks not only transport energy but also interact with the port's energy system to meet its own energy demands. The establishment of an accurate model of the liquid energy network and a clear understanding of its impact on the energy system beyond its transport function is a key difference between integrated energy systems in ports and traditional integrated energy systems, and is also an area that urgently needs further exploration. Based on the existing foundation, it is necessary to establish a refined model that reflects the transport and supply characteristics of liquid networks, and to incorporate this model into the dynamic optimization mathematical model of the port's integrated energy system with multiple network characteristics. The highly nonlinear nature of the above model will significantly increase the difficulty of solving the model. Therefore, the application and rapid solution of the model under the premise of accurate modeling of the liquid energy network is also a key research focus in the future.

(2) **Energy Coupling Between Energy and Logistics Systems**

Currently, research on the coupled operation of energy systems and logistics systems in ports is still in its nascent stage both domestically and internationally. The problem of non-fixed equipment attributes, unstable source-load characteristics, and uncertain network topology after energy coupling urgently needs to be studied. The switching process between ship power stations and shore power, as well as the lifting process of bridge cranes, involve the transformation from load to power source, which has been ignored by existing research and cannot tap the complementary mechanism between logistics and energy. At the same time, the unstable source-load characteristics of ports have been overlooked in existing research, which only considers the time characteristics of renewable energy such as wind and solar power and does not take into account the impact of shock loads such as ships and bridge cranes, as well as logistics scheduling, on the operation characteristics. The uncertain topology of the port network due to the uncertain berthing ship topology and the interconnection node of logistics equipment causes the network topology of the port to be uncertain. Existing research treats the system as a fixed topology and cannot analyze the impact of the spatial transfer of logistics equipment and refrigerated containers on energy, and the analysis of fault energy flow is inaccurate.

Based on the analysis and disclosure of the coupling mechanism between the port logistics system and the energy system, an energy quantification modeling method for the logistics system is proposed. The spatial–temporal correlation characteristics of each logistics operation link and its impact on the energy system operation are revealed, and the two different systems are coordinated and optimized. The dynamic characteristics and dynamic behavior mechanism between each energy supply system and logistics load are currently the bottleneck that urgently needs to be overcome.

(3) **Influence of Extreme Weather on Port Energy System**

In recent years, extreme weather has frequently ravaged various parts of the world. As a hub and convergence point for water and land transportation, ports are significantly affected by extreme weather such as typhoons, heavy rains, and hail. Currently, there is a significant amount of research on the power output fluctuations of wind and solar energy caused by extreme weather and equipment failures. For ports, extreme weather not only directly impacts the energy system but also indirectly affects the logistics system, which in turn affects the energy system. Large seaports contain a large number of temperature control equipment such as refrigerated containers, and drastic changes in external temperature will directly affect the port's load. At the same time, extreme weather in the port area will also affect the berthing plan of ships. More berthing ships and longer berthing times will affect the load demand of the port's energy system. The impact of extreme weather on the efficiency of internal transport equipment in the port may cause a backlog of goods, which will affect the coordination between the port and the external transportation system, disrupt the original logistics plan, and affect the energy system scheduling plan. Therefore, the diverse and complex impact of extreme weather on the integrated energy system in ports is also a key issue that needs to be studied in the future.

1.6 Conclusion

This chapter provides an overview of research on low-carbon technologies for port energy systems both domestically and internationally. Existing methods were analyzed from three perspectives: the current state of low-carbon technology development, coupling of energy-logistics systems in ports, and port energy management. Based on this analysis, future research directions for green port integrated energy systems were explored. The low-carbon development of future ports requires not only advanced technologies in the energy field, but also interdisciplinary development support in areas such as the environment, economy, and management. In the environmental field, continued research is needed on carbon emission measurement standards and development and utilization of ocean carbon sink resources. In the economic field, it is necessary to clarify the distribution of low-carbon development rights and responsibilities among multiple stakeholders in port areas, and further optimize carbon tax schemes. In management, exploration of top-down and bottom-up carbon emission management methods such as carbon quotas and verification of voluntary emission reductions, as well as research into low-carbon emission reduction mechanisms involving the government, ports, and ship owners, is needed.

References

1. Chen, W., Song, B., Zhang, J.: Carbon emission from coastal container ports in China based on AIS data. China Environ. Sci. **42**(07), 3403–3411 (2022)
2. Sina Finance: Seven of the world's top 10 ports are in China (2022). https://baijiahao.baidu.com/s?id=1685110907796769734&wfr=spider&for=pc. Accessed 21 July 2022
3. Zhang, X., Huang, X., Da, Z., et al.: Research on the pathway and policies for china's energy and economy transformation toward carbon neutrality. J. Manage. World **38**(01), 35–66 (2022)

4. Liu, X., Cui, L., Li, B., et al.: Research on the high-quality development path of china's energy industry under the target of carbon neutralization. J. Beijing Inst. Technol. (Soc. Sci. Edit.) **23**(03), 1–8 (2021)
5. Jia, H., Mu, Y., Yu, X., et al.: Thought about the integrated energy system in China. Electr. Power Construct. **36**(01), 16–25 (2015)
6. Mckinlay, C.J., Turnock, S.R., Hudson, D.A.: Route to zero emission shipping: hydrogen, ammonia or methanol? Int. J. Hydro. Energy **46**(55), 28282–28297 (2021)
7. Yuan, Y., Wang, J., Yan, et al.: A design and experimental investigation of a large-scale solar energy/diesel generator powered hybrid ship. Energy **165**(PT.A):965–978 (2018)
8. Yuan, Y., Yuan, C., Xu, H., et al.: Pathway for integrated development of waterway transportation and energy in china. Strategic Study of CAE **24**(03), 184–194 (2022)
9. Corral-vega, P.J., Garcia-trivino, P., Fernandez-ramirez, L.M.: Design, modelling, control and techno-economic evaluation of a fuel cell/supercapacitors powered container crane. Energy **186**(I):115863.1–115863.13 (2019)
10. Fang, S., Zhao, C., Ding, Z., et al.: Port integrated energy systems toward carbon neutrality (I): typical topology and key problems. Proc. CSEE **43**(01), 114–135 (2023)
11. Fang, S., Zhao, C., Ding, Z., et al.: Port integrated energy systems toward carbon neutrality (II): flexible resources and key technologies in energy-transportation integration. Proc. CSEE **43**(03), 950–969 (2023)
12. Ericsson, P., Fazlagic, I.A.: Feasibility study and a technical solution for an on-shore electrical infrastructure to supply vessels with electric power while in port. Doctoral thesis, Chalmers University of Technology, Gothenburg (2008)
13. Top China Travel: Yangshan Deep Water Port (2019). https://www.topchinatravel.com/china-attractions/yangshan-deep-waterport.html. Accessed 01 June 2019
14. FleetMon: Dongguan port (2019). https://www.fleetmon.com/ports/dongguan_cndgg_16434/?language=zh. Accessed 01 June 2019
15. Next Gen Multipurpose Port. https://www.jp.com.sg/. Accessed 06 September 2017
16. Zis, T., North, R., Angloudis, P., et al.: Evaluation of cold ironing and speed reduction policies to reduce ship emissions near and at ports. Maritime Econom. Logist. **16**(4), 371–398 (2014)
17. Chan, C., Wang, C.: Evaluating the effects of green port policy: a case study of Kaohsiung harbor in Taiwan. Transport. Res. Part D Transport. Environ. **17**(3), 185–189 (2012)
18. Yigit, K., Kokkulunk, G., Parlak, A., et al.: Energy cost assessment of shoreside power supply considering the smart grid concept: a case study for a bulk carrier ship. Maritime Policy Manage. **43**(3/4), 469–482 (2016)
19. Hall, W.J.: Assessment of CO_2 and priority pollutant reduction by installation of shoreside power. Resour. Conserv. Recycl. **54**(7), 462–467 (2010)
20. Iris, C., Lam, S.: A review of energy efficiency in ports: operational strategies, technologies and energy management systems. Renew. Sustain. Energy Rev. **118**, 170–182 (2019)
21. Fang, S., Wang, Y., Gou, B., et al.: Toward future green maritime transportation: an overview of seaport microgrids and all-electric ships. IEEE Trans. Vehic. Technol. **69**(1), 207–219 (2019)
22. Hossein, G.A., Debjit, R., Rene, D.K.: Sea container terminals: new technologies, or models, and emerging research areas. Soc. Sci. Electron. Publish. **18**(2), 103–140 (2016)
23. Wilmsmeier, G., Spengler, T.: Energy consumption and container terminal efficiency. FAL Bull. **350**(6), 1–10 (2016)
24. Zhao, N., Schofield, N., Niu, W., et al.: Hybrid power-train for port crane energy recovery. In: IEEE Conference and Expo Transportation Electrification Asia-Pacific (ITEC Asia-Pacific), Beijing (2014)
25. Tran, T.K.: Tudy of Electrical Usage and Demand at the Container Terminal. Doctoral thesis, Deakin University, Victoria (2012)
26. Parise, G., Honorati, A.: Port cranes with energy balanced drive. In: AEIT Annual Conference—From Research to Industry: The Need for a More Effective Technology Transfer, Trieste (2014)

27. Greencrans: Green Technologies and Eco-Efficient Alternatives for Cranes and Operations at Port Container Terminals. Greencranes Project (2012)
28. Hangga, P., Shinoda, T.: Mostion-based energy analysis methodology for hybrid straddle carrier towards eco-friendly container handling system. J. East. Asia Soc. Transport. Stud. **11**, 2412–2431 (2015)
29. Seddiek, I.A.: Application of renewable energy technologies for eco-friendly sea ports. Ships Offshore Struct. **15**(9), 953–962 (2020)
30. Sadiq, M., Ali, S., Terriche, Y., et al.: Future greener seaports: a review of new infrastructure, challenges, and energy efficiency measures. IEEE Access **9**, 75568–75587 (2021)
31. STP Smart: Energy Efficient and Adaptive Port Terminals (Sea Terminals). Sea Terminals Project Technical Report (2015)
32. Buiza, G., Cepolina, S., Dobrijevic, A., et al.: Current situation of the Mediterranean container ports regarding the operational, energy and environment areas. In: International Conference on Industrial Engineering and Systems Management (IESM), Seville (2015)
33. Werner, B.R.: Eduction of the CO_2 Footprints of Container Terminals by Photovoltaics. Green Efforts Project (2014)
34. Misra, A., Venkataramani, G., Gowrishankar, S., et al.: Renewable energy based smart microgrids—a pathway to green port development. Strat. Plann. Energy Environ. **37**(2), 17–32 (2017)
35. Song, S., Poh, K.: Solar PV leasing in Singapore: enhancing return on investments with options. In: IOP Conference Series: Earth and Environmental Science, IOP Publishing, vol. 67, issue 1, pp. 012–020 (2017)
36. Hafen, H.: Energy Cooperation Port of Hamburg. Port of Hamburg (2015)
37. Acciaro, M., Ghiara, H., Cusano, M.: Energy management in seaports: a new role for port authorities. Energy Policy **71**, 4–12 (2014)
38. Hao, H., Zhang, P.: An overview of U.S. new energy development. Power Syst. Technol. **35**(07):48–53 (2011)
39. Tang, H., Qu, K., Chen, G., et al.: Potential sites for tidal power generation: a thorough search at coast of New Jersey, USA. Renew. Sustain. Energy Rev. **39**, 412–425 (2014)
40. Wang, Q., Qing, C., Ju, P., et al.: A coordinated control strategy for hybrid offshore renewable energy power generation considering state of charge of battery energy storage system. Power Syst. Technol. **39**, 412–425 (2014)
41. Mang, L., Jiang, J.: Challenges and countermeasures of wind power industry development in China under low-carbon energy transition. China Energy **43**(11), 8 (2021)
42. PLB: Yard Hostler Demonstration and Commercialization Project Final Report of Liquefied Natural Gas (LNG). Port of Long Beach Technical Report (2008)
43. Na, J., Choi, A., Ji, J., et al.: Environmental efficiency analysis of Chinese container ports with CO_2 emissions: an inseparable input-output SBM model. J. Transport. Geogr. **65**, 13–24 (2017)
44. Geerlings, H., Van, D.R.: A new method for assessing CO_2-emissions from container terminals: a promising approach applied in Rotterdam. J. Clean. Product. **19**(6–7), 657–666 (2011)
45. Ren, Z., Luo, X., Qin, H.: A mid/long-term optimal operation method of regional integrated energy systems considering hydrogen physical characteristics. Power Syst. Technol. **191**(2), 116860 (2022)
46. Duan, J., Xie, J., Feng, L., et al.: Gain allocation strategy for wind-light-water-hydrogen multi-agent energy system based on cooperative game theory. Power Syst. Technol. **46**(05), 1703–1712 (2022)
47. Zhang, M.: The first hydrogen fuel cell vehicle charging demonstration project in China's ports was launched. China Plant Eng. **03**, 2 (2022)
48. Viviana, C.: He's Role of Hydrogen in European Port Ecosystems. Energia Ambiente E Innovazione, Italy (2021)
49. Li, X.: Application of hydrogen fuel forklift in green port construction. China Ports **09**, 60–61 (2020)

50. PLA: Hydrogen and Fuel Cells in the Ports Workshop Report. Port of Los Angeles Technical Report (2016)
51. Liang, T., Guo, H., Moser, J., et al.: A roadmap towards smart grid enabled harbour terminals. CIRED Workshop-Rome **25**, 528–542 (2014)
52. Parise, G., Parise, L., Pepe, F., et al.: Innovations in a container terminal area and electrical power distribution for the service continuity. In: 2016 IEEE/IAS 52nd Industrial and Commercial Power Systems Technical Conference (I&CPS), Detroit, pp. 1–6 (2016)
53. Parise, G., Parise, L., Martirano, L., et al.: Wise port and business energy management: port facilities, electrical power distribution. IEEE Trans. Indust. Appl. **52**(1), 18–24 (2016)
54. Reencranes: Real-Time Energy Consumption Monitoring System. Greencranes Project Technical Report (2014)
55. Sharma, K., Saini, L.M.: Formance analysis of smart metering for smart grid: an overview. Renew. Sustain. Energy Rev. **49**, 720–735 (2015)
56. International Maritime Information: How does the port create a "low-carbon" new name card? (2022). http://www.simic.net.cn/news-show.php?id=257820. Accessed 01 May 2022
57. Bierwirth, C., Meisel, F.: A survey of berth allocation and quay crane scheduling problems in container terminals. Europ. J. Oper. Res. **202**(3), 615–627 (2010)
58. Chang, D., Jiang, Z., Yan, W., et al.: Integrating berth allocation and quay crane assignments. Transport. Res. Part E Logist. Transport. Rev. **46**(6), 975–990 (2010)
59. Liang, C., Huang, Y., Yang, Y.: A quay crane dynamic scheduling problem by hybrid evolutionary algorithm for berth allocation planning. Comp. Indust. Eng. **56**(3), 1021–1028 (2009)
60. Xi, X., Liu, C., Miao, L.: A bi-objective robust model for berth allocation scheduling under uncertainty. Transport. Res. Part E Logist. Transport. Rev. **106**, 294–319 (2017)
61. Imai, A., Nishimura, E., Papadimitriou, S.: Berth allocation with service priority. Transport. Res. Part B Methodol. **37**(5), 437–457 (2003)
62. Imai, A., Nishimura, E., Papadimitriou, S.: The dynamic berth allocation problem for a container port. Transport. Res. Part B Methodol. **35**(4), 401–417 (2001)
63. Bierwirth, C., Meisel, F.: A follow-up survey of berth allocation and quay crane scheduling problems in container terminals. Europ. J. Oper. Res. **244**(3), 675–689 (2015)
64. Imai, A., Chen, H.C., Nishimura, E., et al.: The simultaneous berth and quay crane allocation problem. Transport. Res. Part E Logist. Transport. Rev. **44**(5), 900–920 (2008)
65. Ting, C.J., Wu, K.C., Chou, H.: Particle swarm optimization algorithm for the berth allocation problem. Expert Syst. Appl. **41**(4), 1543–1550 (2014)
66. Agra, A., Oliveira, M.: MIP approaches for the integrated berth allocation and quay crane assignment and scheduling problem. Europ. J. Oper. Res. **264**(1), 138–148 (2018)
67. Chao, S., Chen, C.: Applying a time-space network to reposition reefer containers among major Asian ports. Res. Transport. Business Manage. **17**, 65–72 (2015)
68. Agra, A., Constantino, M.: Description of 2-integer continuous knapsack polyhedral. Disc. Optim. **3**(2), 95–110 (2006)
69. Al-Dhaheri, N., Diabat, A., Diabat, A.: The quay crane scheduling problem with nonzero crane repositioning time and vessel stability constraints. J. Manuf. Syst. **36**, 87–94 (2015)
70. Beens, M.A., Ursavas, E.: Scheduling cranes at an indented berth. Europ. J. Oper. Res. **253**, 298–313 (2016)
71. Bierwirth, C., Meisel, F.: A survey of berth allocation and quay crane scheduling problems in container terminals. Euro. J. Oper. Res. **202**, 615–627 (2010)
72. Bierwirth, C., Meisel, F.: A follow-up survey of berth allocation and quay crane scheduling problems in container terminals. Europ. J. Oper. Res. **244**, 675–689 (2015)
73. Boysen, N., Briskorn, D., Meisel, F.: A generalized classification scheme for crane scheduling with interference. Europ. J. Oper. Res. **258**, 343–357 (2010)
74. Chang, D., Jiang, Z., Yan, W.: Integrating berth allocation and quay crane assignments. Transport. Res. Part E **46**, 975–990 (2010)
75. Chen, J., Lee, D., Cao, J.: A combinatorial benders cuts algorithm for the quayside operation problem at container terminals. Transport. Res. Part E Logist. Transport. Rev. **48**, 266–275 (2012)

76. Constantino, M., Gouveia, L.R.: Equationtion by discretization: application to economic lot sizing. Oper. Res. Lett. **35**, 645–650 (2007)
77. Daganzo, C.: The crane scheduling problem. Transport. Res. Part B **23**, 159–175 (1998)
78. Diabat, A., Theodorou, E.: An integrated quay crane assignment and scheduling problem. Comp. Indust. Eng. **73**, 115–123 (2014)
79. Fang, S., Wang, Y., Gou, B., et al.: Toward future green maritime transportation: an overview of seaport microgrids and all–electric ships. IEEE Trans. Vehic. Technol. **69**(1), 207–219 (2020)
80. Hartmann, S.: Cheduling reefer mechanics at container terminals. Transport. Res. Part E Logist. Transport. Rev. **51**, 17–27 (2013)
81. Huang, Y., Huang, W., Wei, W., et al.: Logistics–energy collaborative optimization scheduling method for large seaport integrated energy system. Proceedings of CSEE **42**(17):1–12 (2022)
82. Iris, A., Lam, J.S.L.: Optimal energy management and operations planning in seaports with smart grid while harnessing renewable energy under uncertainty. Omega **103**(3), 102445 (2021)
83. Mao, A., Yu, T., Ding, Z., et al.: Optimal scheduling for seaport integrated energy system considering flexible berth allocation. Appl. Energy **308**, 118386 (2022)
84. He, D., Pang, Y., Lodewijks, G.: Green operations of belt conveyors by means of speed control. Appl. Energy **188**, 330–341 (2017)
85. Kanellos, F.D.: Multiagent-system-based operation scheduling of large ports' power systems with emissions limitation. IEEE Syst. J. **13**(2), 1831–1840 (2019)
86. Iris, C., Lam, S.: A review of energy efficiency in ports: operational strategies, technologies and energy management systems. Renew. Sustain. Energy Rev. **112**, 170–182 (2019)
87. The Port of Los Angeles: Port of Los Angeles rolls out hydrogen fuel cell electric freight demonstration (2021). https://www.portoflosangeles.org/references/2021-news-releases/news_060721_zazeff. Accessed 07 June 2021
88. Cai, G., Jiang, Y., Huang, N., et al.: Large-scale electric vehicles charging and discharging optimization scheduling based on multi-agent two-level game under electricity demand response mechanism. Proc. CSEE **43**(01), 85–99 (2023)
89. Huang, J., Zhu, J., Hou, D., et al.: Group charging and discharging optimization strategy for electric vehicles considering real-time power trading. Southern Power Grid Technol. **15**(03), 113–120 (2021)
90. Xiang, X., Liu, C., Miao, L.: A bi-objective robust model for berth allocation scheduling under uncertainty. Transport. Res. Part E Logist. Transport. Rev. **106**, 294–319 (2017)
91. Fang, S., Wang, H., Shang, C., et al.: A decision-making method for berthed electric–ships based on generalized Nash game. In: IET Renewable Power Generation, Shanghai (2019)
92. Mu, Y., Yao, T., Jia, H., et al.: Optimal scheduling method for belt conveyor system in coal mine considering silo virtual energy storage. Appl. Energy **275**, 115368 (2020)
93. Heij, R.: Opportunities for Peak Shaving Electricity Consumption at Container Terminals. Master's thesis, Delft University, Delft (2015)
94. Tran, T.K.: Tudy of Electrical Usage and Demand at the Container Terminal. Ph.D. Dissertation. Deakin University, Melbourne (2012)
95. Kanellos, F.D., Volanis, E.M., Hatziargyriou, N.D.: Power management method for large ports with multi-agent systems. IEEE Trans. Smart Grid **10**(2), 1259–1268 (2019)
96. Deng, S., Xiao, J., Liu, J., et al.: Prospects analysis on power balancing optimization of reefer container ship. Guangdong Shipbuild. **37**(04), 36–38 (2018)
97. Xie, X., Ma, N., Liu, W., et al.: Functions of energy storage in renewable energy dominated power systems: review and prospect. Proc. CSEE **43**(01), 158–169 (2023)
98. Nadeem, F., Hussain, S.M.S., Tiwari, P.K., et al.: Comparative review of energy storage systems, their roles, and impacts on future power systems. IEEE Access **7**, 4555–4585 (2018)
99. Li, J., Tian, L., Lai, X.: Outlook of electrical energy storage technologies under energy internet background. Autom. Electr. Power Syst. **39**(23), 15–25 (2015)
100. Wang, Y., Tian, J., Sun, Z., et al.: A comprehensive review of battery modeling and state estimation approaches for advanced battery management. Renew. Sustain. Energy Rev. **131**, 110015 (2020)

101. Sun, Y., Yang, M., Shi, C., et al.: Analysis of application status and development trend of energy storage. High Volt. Eng. **46**(1), 80–89 (2020)
102. Zhang, Y., Weu, Z., Wang, C.: Hierarchical optimal dispatch for integrated energy system with thermal storage device. Proc. CSEE **39**(S1), 36–43 (2019)
103. Diao, H., Li, P., Lyu, X., et al.: Coordinated optimal allocation of energy storage in regional integrated energy system considering the diversity of multi-energy storage. Trans. China Electrotech. Soc. **36**(01), 151–165 (2021)
104. Lyu, Q., Wang, H., Chen, T., et al.: Operation strategies of heat accumulator in combined heat and power plant with uncertain wind power. Autom. Electr. Power Syst. **39**(14), 23–29 (2015)
105. Liu, J., Zhou, C., Gao, H., et al.: A day-ahead economic dispatch optimization model of integrated electricity-natural gas system considering hydrogen-gas energy storage system in microgrid. Power Syst. Technol. **42**(01), 170–179 (2018)
106. Qiao, Z., Guo, Q., Sun, H., et al.: Multi-time period optimized configuration and scheduling of gas storage in gas-fired power plants. Appl. Energy **226**, 924–934 (2018)
107. Zhou, Z.: Energy efficiency analysis of hydrogen storage coupled gas-steam combined cycle. Acta Energiae Solaris Sinica **42**(01), 170–179 (2018)
108. Dong, S., Wang, C., Xu, S., et al.: Day-ahead optimal scheduling of electricity-gas-heat integrated energy system considering dynamic characteristics. Autom. Electr. Power Syst. **42**(13), 12–19 (2018)
109. Zhou, H., Li, Z., Zheng, J., et al.: Robust scheduling of integrated electricity and heating system hedging heating network uncertainties. IEEE Trans. Smart Grid **11**(2), 234–246 (2019)
110. Wang, D., Zhi, Y., Jia, H., et al.: Optimal scheduling strategy of district integrated heat and power system with wind power and multiple energy stations considering thermal inertia of buildings under different heating regulation modes. Appl. Energy **240**, 341–358 (2019)
111. Sartor, K., Dewalef, P.: Experimental validation of heat transport modelling in district heating networks. Energy **137**, 961–968 (2017)
112. Qin, X., Shen, X., Sun, H., et al.: A quasi-dynamic model and corresponding calculation method for integrated energy system with electricity and heat. Energy Proc. **158**, 6413–6418 (2019)
113. Alabdulwahab, A., Abusorrah, A., Zhang, X.: Comordination of interdependent natural gas and electricity infrastructures for firming the variability of wind energy in stochastic day ahead scheduling. IEEE Trans. Sustain. Energy **6**(2), 606–615 (2015)
114. Alipour, M., Zare, K., Seyedi, H.: A multi-follower bilevel stochastic programming approach for energy management of combined heat and power microgrids. Energy **149**, 135–146 (2018)
115. Bazmohammadi, N., Tahsiri, A., Anvari-Moghaddam, A., et al.: Optimal operation management of a regional network of microgrids based on chance-constrained model predictive control. IET Gen. Transmis. Distrib. **12**(15), 3772–3779 (2018)
116. Qadrdan, M., Wu, J., Jenkins, N., et al.: Operating strategies for a GB integrated gas and electricity network considering the uncertainty in wind power forecasts. IEEE Trans. Sustain. Energy **5**(1), 128–138 (2014)
117. Yu, D., Yang, M., Zhai, H., et al.: An overview of robust optimization used for power system dispatch and decision-making. Autom. Electr. Power Syst. **40**(7), 134–143 (2016)
118. Tan, Z., Fan, W., Li, H., et al.: Dispatching optimization model of gas-electricity virtual power plant considering uncertainty based on robust stochastic optimization theory. J. Clean. Product. **247**, 119106 (2019)
119. Hussain, A., Bui, V.H., Kim, H.M.: Ruobust optimization based scheduling of multi-microgrids considering uncertainties. Energies **9**(4), 278 (2016)
120. Sou, H.: Port of Los Angeles (2022). https://www.sofreight.com/ports/us/uslsa. Accessed 01 June 2022
121. Port of Los Angeles rolls out hydrogen fuel cell electric freight demonstration. https://www.portoflosangeles.org/references/2021-news-rleases/news_060721_zazeff. Accessed 06 June 2021
122. Shanghai Deep water Port comprehensive information service network. https://www.yangsh antermnal.com/https://www.sofreight.com/ports/CN/CNYGS.html. Accessed 06 June 2022

123. Sou, H.: Shanghai Yangshan Port (2022). https://www.sofreight.com/ports/us/uslsa. Accessed 01 June 2022

Chapter 2
Optimization Configuration of Renewable Energies and Energy Storages in Port Microgrids

2.1 Introduction

Ports are important hubs for maritime activities and trade transportation. The large amount of handling machinery, vehicles, and berthed ships make ports a major energy consumer and emitter. Driven by the goals of "carbon peaking" and "carbon neutrality," it is urgent to explore the carrying capacity of port power grids for the electrification transformation of high-energy consumption and high-emission loads, and to develop, utilize, and absorb high-proportion new energy on site. To make the most of the new energy for transportation, handling, and ship power supply, optimizing the configuration of distributed power sources and energy storage has become the key to energy conservation and emission reduction in green port areas.

Unlike conventional microgrids, port microgrids include not only micro power sources, conventional power loads, energy storage systems, and control devices, but also a large number of electric loading and unloading machinery and shore power equipment. The load characteristics of the port area are closely related to logistics and transportation, as well as production operation plans, especially affected by the number of ships, tonnage, and types of cargo, without obvious natural periodic characteristics. On the other hand, new energy sources such as photovoltaics in the port area show power generation characteristics strongly correlated with natural periods such as days, weeks, and months. The inconsistency of randomness and periodic fluctuations between sources and loads poses challenges to the depiction of port operation scenarios, and also poses severe challenges to the optimization and configuration methods based on typical periods such as daily and weekly cycles [1, 2].

The scenario method can represent uncertain variables through multiple deterministic typical scenarios to achieve optimal planning of high-stochastic systems [3]. Increasing the quantity of typical scenario data can effectively improve the accuracy of configuration, but too many operating points not only create redundancy but also result in enormous computational complexity, which can even make the problem

W. Huang et al., *Energy Management of Integrated Energy System in Large Ports*, Springer Series on Naval Architecture, Marine Engineering, Shipbuilding and Shipping 18, https://doi.org/10.1007/978-981-99-8795-5_2

unsolvable. Therefore, how to utilize massive data such as load, wind speed, and light intensity, extract key information to generate typical scenarios containing limited operating points, and accurately reflect the system's operating conditions is crucial for the optimization of renewable energy and hybrid storage configuration in a port microgrid.

Selecting a series of typical days or weeks as typical scenarios is a common method in microgrid planning. Literature [4] selects typical days from each month included in the planning period for capacity expansion optimization. Literature [5] selects operating days with the highest load, lowest load, and maximum load fluctuation as typical days. These two methods are easy to implement and operate but difficult to ensure the representativeness of the selected typical scenarios. The above literature uses days as the length of the scenario, which cannot reflect the fluctuations of load and distributed generation over a longer period of time and is not suitable for port microgrids with unclear natural cycles. Literature [6] uses planning scenarios with a length of 5 days to optimize the configuration of large-capacity energy storage, and literature [7] selects typical week data for optimization. The above methods can effectively reduce the computational complexity of microgrid optimization configuration but are difficult to reflect the correlation between different regions and data types. For example, when the load in the port production area is high, the load in the living area is usually low, and on sunny days, there is often more air convection and wind compared to cloudy days.

In order to extract potential correlation information from massive data, some literature has proposed data compression algorithms to generate typical scenarios. K-means clustering algorithm [8] and hierarchical clustering algorithm [9] are often used to filter typical information from large data sets. Literature [10] integrates the above two algorithms for selecting typical scenarios, which not only retains the advantages of fast calculation speed of K-means algorithm, but also the sensitivity to outlier values of hierarchical clustering. Literature [11] analyzes the spatiotemporal correlation of wind power using multivariate normal distribution and Copula function, and combines Monte Carlo sampling to generate a set of wind power output scenarios with spatiotemporal correlation. Literature [12] applies vector K-medoids clustering method to classify loads, generates typical loads, and then applies K-means clustering method to compress typical loads and distributed generation data to obtain typical scenarios. Literature [13] constructs spatial and temporal feature extraction units for time-series generative adversarial networks to generate daily/monthly wind and solar power time series. The above literature studies the selection method of typical scenarios in planning and achieves good application effects in conventional microgrids. However, the selected scenarios are divided into time scales according to natural cycles, and the effect is average when applied to data with weak periodicity such as port load.

This chapter proposes a high-fidelity compression and reconstruction method for operating scenarios of port microgrid, and applies it to the optimization configuration of port source and storage. In view of the characteristic that the natural cycle of port microgrid operation data is not obvious, hierarchical clustering algorithm and continuous hierarchical clustering algorithm are proposed. According to the dynamic adjustment of the time scale corresponding to each operating point based on the operation characteristics of the port microgrid, the high-fidelity compression and reconstruction of operating scenarios are realized through two data processing steps, and the typical operating scenarios of the port are obtained. Then, the source and storage optimization configuration model is established, and the time scale of the model is dynamically adjusted based on the obtained typical scenarios to obtain the optimal configuration of the port microgrid source and storage. Finally, taking Rizhao Port microgrid as an example, the representativeness of the selected typical scenarios and the accuracy of the optimization configuration results are verified.

2.2 Scenario-Based Depiction of the Fluctuating Characteristics of Port Microgrid

In green port microgrids with rich distributed generation resources and concentrated loads, the penetration rate of renewable energy will increase to over 80%. Therefore, port microgrids should also be equipped with various types of energy storage devices to improve system regulation capabilities and promote renewable energy consumption. Compared to traditional power grids, renewable energy in port microgrids does not have scale effect, and its randomness and volatility are stronger. At the same time, the load in port microgrids includes a high proportion of shore power, belt conveyors, shore cranes, and other equipment, and the load curve is closely related to port production operations. The load operation of conventional microgrids has a clear intra-day rule between days, and the load curve basically coincides with the average value of the two-day load. As shown in Fig. 2.1, the load in port microgrids does not have an intra-day rule and has the characteristics of long-lasting and large fluctuations in a short period of time. Therefore, accurately characterizing the fluctuation characteristics of renewable energy and load in the system is the key and difficulty in the optimization configuration process of source and storage in the microgrid of the port.

Fig. 2.1 The load curve of port microgrids

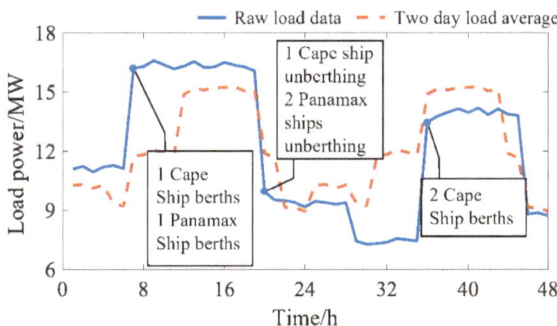

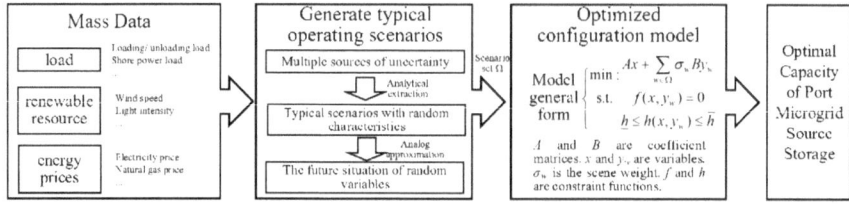

Fig. 2.2 The optimal generation and storage configuration of port microgrids

This chapter uses a scenario-based optimization method to address the impact of source load fluctuations on system optimization configuration. The operating scenarios contain multiple operating points, each of which includes system operating information such as load, renewable energy generation, and energy prices at a port within the same period. In existing methods, the time scale corresponding to the operating point is determined based on the scenario's purpose. In most literature focused on optimization configuration, an operating point corresponds to one hour. This chapter focuses on the characteristics of a port microgrid and proposes a high-fidelity compression and reconstruction method for operating scenarios with dynamically adjustable time scales for each operating point. The time scale corresponding to each operating point in the scenario is determined by the system's operating state, ranging from one hour to several hours, thereby increasing the density of data information in typical scenarios. The overall process of source-storage optimization configuration for a port microgrid is shown in Fig. 2.2. First, the fluctuation characteristics of uncertain sources, such as load, wind speed, and solar irradiance, are extracted from a massive amount of operating data to construct a series of typical operating scenarios for the port microgrid. Through simulation and approximation of the distribution of random variables in scenarios, the random variables' future situations are characterized. Then, using the typical scenarios as input, the scenario-based stochastic planning problem is transformed into a deterministic model corresponding to the scenario set Ω. Finally, the model is solved to obtain the optimal capacity of renewable energy and energy storage devices.

2.3 High-Fidelity Compression and Reconstruction Method

The accurate characterization of operational characteristics of microgrids in ports through typical scenarios is of great significance for improving the accuracy and feasibility of optimizing source and storage configurations in green ports. This chapter uses a vector autoregression model to generate microgrid operational scenarios based on massive historical data inputs, simulating the operational situation of microgrids in ports over the next few years. The more operational points included in the scenario, the more accurate the simulation of the random variable, but a large number of operational points can make the model difficult to solve. Selecting typical scenarios with an appropriate number of operational points and accurately describing the random variables is crucial to determine the accuracy of the optimization results.

This section proposes a high-fidelity compression and reconstruction method for generating typical scenarios of microgrids in ports. Based on existing system operational scenarios, this method can dynamically adjust the temporal scale of operational points according to operational characteristics, as shown in Fig. 2.3. The method reduces the data size by utilizing the similarities between operational weeks and adjacent operational points in ports microgrids, while retaining the properties of the original data and the correlation between different data types, thereby increasing the density of data information.

2.3.1 Port Data Extraction and Integration

The data contained in the operating scenarios of a port microgrid may exhibit differences in both the temporal scale and the data types. Therefore, prior to further processing of data, various types of data contained in the scenarios need to be extracted and integrated. Firstly, the entire temporal cycle data is processed into hourly units using systematic sampling or interpolation methods. Then, a two-dimensional data matrix is obtained by using time as one dimension and different nodes or data types as the other. This matrix serves as the input matrix for data processing. For instance, when the initial operating scenario covers a one-year period, including data on solar irradiance, wind speed, load level, and electricity price of three nodes, the input matrix format should be 8760×12. A year corresponds to 8760 h, and the four types of data for three nodes correspond to $3 \times 4 = 12$.

Due to the potential large differences in the values of data among different types and nodes, it is necessary to calculate the distance between corresponding data of different operating points during data processing. To ensure that various types of data have equal weight, data standardization is required. This chapter employs the Min–max normalization method, as shown in Eq. (2.1).

$$\alpha_{nor} = \frac{\alpha - \min}{\max - \min} \tag{2.1}$$

where α_{nor} represents the standardized data; α represents the original data; min represents the minimum value of the data set; max represents the maximum value of the data set.

2.3.2 High-Fidelity Compression Based on Operational Week

In the operation data of the port microgrid, the data such as load and wind speed do not exhibit obvious diurnal patterns. In order to capture the fluctuation characteristics of the data over a longer period of time, the number of operating points is reduced

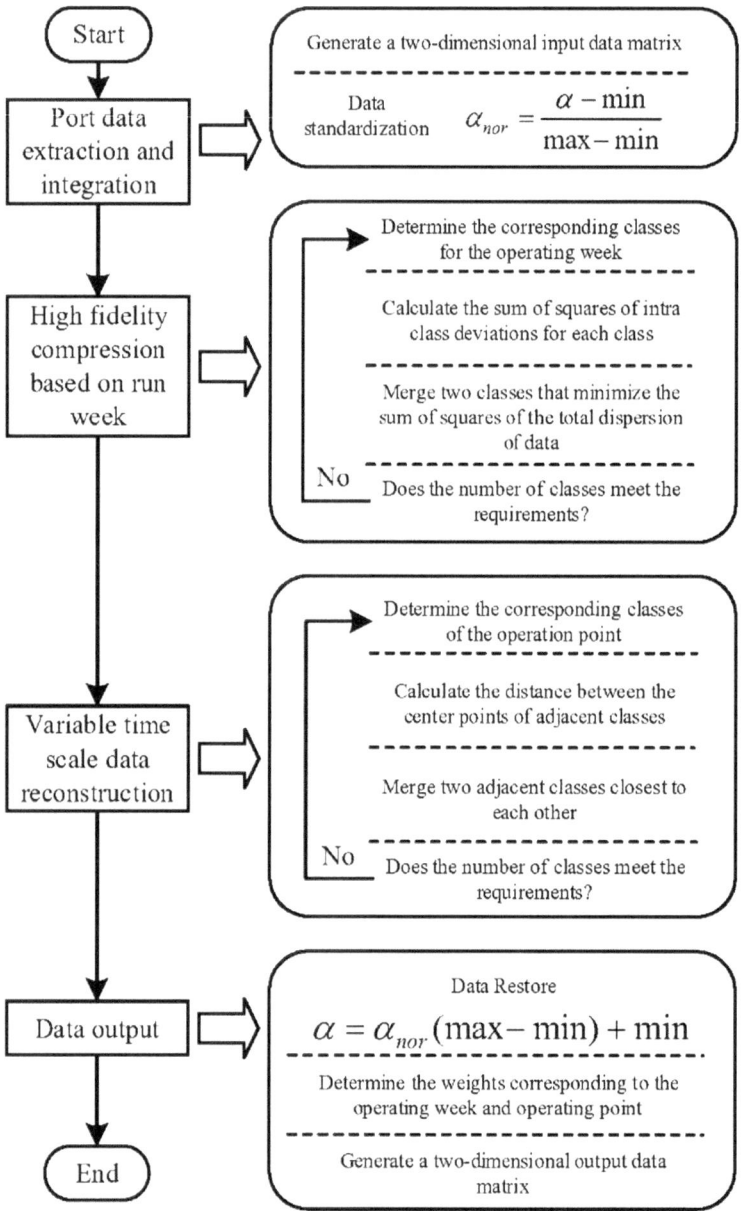

Fig. 2.3 High-fidelity compression and refactoring method of port microgrid operation scenarios

by utilizing the similarity between operating weeks in the port. Firstly, all operating weeks included in the operation scenario are treated as an initial cluster, and then the number of clusters is reduced using hierarchical clustering until the remaining number of clusters meets the expected target.

The Euclidean distance is used to measure the difference between the data of each week, as shown in Eq. (2.2).

$$E(A, B) = \sqrt{(x_{A1} - x_{B1})^2 + (x_{A2} - x_{B2})^2 + \cdots + (x_{An} - x_{Bn})^2} \qquad (2.2)$$

where $E(A, B)$ is the European distance between Week A and Week B; x_{A1} to x_{An} are the data in week A; x_{B1} to x_{Bn} are the data in week B.

The Ward's minimum variance method is used as the criterion for merging clusters in the hierarchical clustering process. This method requires calculating the centroid of each cluster, using the formula shown in Eq. (2.3).

$$\overline{I} = \frac{1}{|I|} \sum_{A \in I} A \qquad (2.3)$$

where $\overline{I}$ is the center point of class I; A represents all operating weeks included in Class I; $|I|$ is the number of running weeks in class I.

Then, calculate the within-group sum of squares for each class, which is the sum of the squared Euclidean distances between all observations and the class centroid. The calculation formula is shown in Eq. (2.4).

$$DSS_I = \sum_{A \in I} \left[E\left(A, \overline{I}\right) \right]^2 \qquad (2.4)$$

where DSS_I is the sum of squares of the intra class dispersion of class I.

The within-class sum of squares can characterize the dispersion among the running weeks within each class and measure the clustering of elements within each class. As the number of running weeks in a class increases, its within-class sum of squares will be greater than or equal to the original within-class sum of squares. During the clustering process, the total sum of squares of the data is iteratively calculated, as shown in Eq. (2.5).

$$SDSS = \sum_{I \in data} DSS_I \qquad (2.5)$$

where $SDSS$ is the sum of squares of the total dispersion of the data; $data$ represents all data during data processing.

When two classes are merged, the total sum of squares of the data increases. Since a smaller sum of squares indicates a higher degree of clustering among the elements, the two classes with the minimum total sum of squares are selected for merging. After the clustering is completed, this chapter uses the running week closest to the centroid of each class as the representative week for that class, rather than the centroid of each class. This avoids weakening the fluctuations between adjacent running points within the representative week caused by taking the average of multiple weeks.

In the data processing process, the clustering algorithm reduces one running week every time it is run. When the entire running scenario contains i running weeks and needs to be reduced to j representative weeks, the clustering algorithm needs to be run i-j times. The number of running weeks contained in each class in the clustering result is the weight of the representative week.

2.3.3 Variable-Time-Scale Data Reconstruction

The operation of a microgrid at a port is closely related to the production and operational behavior of the port. Under unchanged operational tasks, the same operational state can persist for a long time. Therefore, to further reduce the data scale of typical scenarios, the time scale of each operational point can be dynamically adjusted based on the similarity between operational points, in order to reconstruct the operational scenarios. For example, when the operational state at the port changes little, the time scale of an operational point can be increased to represent multiple hours, thereby reducing the data scale. Conversely, when the operational state changes rapidly, the time scale of an operational point should be decreased to represent one hour, in order to retain key operational information.

Based on the data processing results from the previous stage, the continuous hierarchical clustering method is used to cluster operational points that represent adjacent weeks. Each operational point included in each representative week is used as an initial class, and the classes with the smallest difference and adjacent time series will be merged. Classes from different representative weeks cannot be merged. The difference between classes is measured by the distance between their center points, which are calculated using the same formula as Eq. (2.3). Weighted Euclidean distance, as shown in Eq. (2.6), is used to calculate the distance.

$$WE(X, Y) = \frac{2\sqrt{\sigma^A}}{{}^1\!/_{|X|} + {}^1\!/_{|Y|}} E(X, Y), \forall X, Y \in A \tag{2.6}$$

where $WE(X, Y)$ is the weighted Euclidean distance between class X and class Y; σ^A is the weight representing week A; $|X|$ and $|Y|$ are the number of elements in class X and class Y, respectively.

From Eq. (2.6), it can be seen that the distance calculation process incorporates weights representing inter-cluster proximity and intra-cluster element count. This approach amplifies the distances between different classes in high-weight inter-cluster proximity, making it more cautious when reducing running points in high-weight clusters. At the same time, it avoids having too many elements included in each class.

After the distance calculation between each class is completed, the two adjacent classes with the closest distance in time sequence are merged, and the centroid of each class is taken as the representative running point of that class. This process is repeated until the desired target number of representative running points is reached, completing the reconstruction of running points. The number of running points contained in each class is the weight of the representative running point for that class.

2.3.4 Data Output

In the preprocessing stage of port data, the data in the running scenario is standardized, so the processing results need to be restored to their original form in order to represent the correct meaning. The method of data restoration corresponds to the min–max normalization method, as shown in Eq. (2.7).

$$\alpha = \alpha_{nor}(\max - \min) + \min \tag{2.7}$$

where α is the restored data. α_{nor} is the standardized data obtained after data processing.

The high-fidelity compression and reconstruction of data are based on representative operating points, which are the basic unit of composition. Multiple adjacent representative operating points on different time sequences constitute a representative week. Each operating point corresponds not only to multiple types of input data but also to corresponding week weights and operating point weights. The typical output scenario is a two-dimensional data matrix. For example, when three nodes with four types of data are reduced to 672 representative operating points, the output matrix format is 672×14. The 672 corresponds to different operating points, the first 12 columns correspond to the four types of data from three nodes $3 \times 4 = 12$, the 13th column corresponds to the representative week weight, and the 14th column corresponds to the representative operating point weight.

2.4 Optimal Configuration Model for Port's Renewable Energies and Energy Storages

In order to show the advantages of the proposed high-fidelity compression and reconstruction methods for the microgrid's operating scenario, an optimization configuration model for source and storage in microgrid for ports is established. The model's decision variables include the installation capacity of each renewable energy source and energy storage device, and the objective function is to minimize the total system cost. The model parameters include representative week weights and operating point weights, which dynamically adjust the time scale of the operating points in the model.

2.4.1 Objective Function

The objective function includes three parts: investment cost (C^{inv}), operation cost (C^{ope}), and equipment residual value (C^{rem}). The investment cost includes the investment costs of each renewable energy source and each energy storage device. The operation cost includes the operating costs of each energy storage device and the system's purchased electricity costs. The equipment residual value refers to the residual value that can be recovered when the equipment is retired or scrapped during the planning period.

$$\min : C^{inv} + C^{ope} - C^{rem} \tag{2.8}$$

$$C^{inv} = \sum_{n \in N} \sum_{g \in G} A_g^{gen} \cdot C_g^{gen} \cdot \overline{P}_{gn}^{gen} + \sum_{n \in N} \sum_{s \in S} A_s^{sto} \cdot C_s^{sto} \cdot \overline{S}_{sn} \tag{2.9}$$

$$C^{ope} = \sum_{w \in \Omega} \sigma_w \sum_{n \in N} \sum_{t \in w} \sigma_t \left(\sum_{s \in S} O_s^{sto} \cdot \overline{S}_{sn} + O_{wt}^{grid} \cdot P_{wnt}^{grid} \right) \tag{2.10}$$

$$C^{rem} = \sum_{n \in N} \sum_{g \in G} B_g^{gen} \cdot R_g^{gen} \cdot \overline{P}_{gn}^{gen} + \sum_{n \in N} \sum_{s \in S} B_s^{sto} \cdot R_s^{sto} \cdot \overline{S}_{sn} \tag{2.11}$$

where N is a collection of system nodes; G is the collection of renewable energy types; S is the set of energy storage devices; Ω represents the weekly set; A_g^{gen} and A_s^{sto} are the fund recovery rates of renewable energy g and energy storage equipment n within the planning cycle, respectively; C_g^{gen} is the power cost coefficient of renewable energy g; $\overline{P}_{gn}^{gen}$ is the rated power of renewable energy g at the node n; C_s^{sto} is the capacity cost coefficient of the energy storage equipment s; $\overline{S}_{sn}$ is the installed capacity of the energy storage s device at the node n; σ_w and σ_t are weights representing week w and operating point t, respectively; O_s^{sto} is the operating cost per unit capacity of the energy storage equipment s; O_{wt}^{grid} is the electricity price at the representative week w and operating point t; P_{wnt}^{grid} is the active power transmitted from the grid to the node n at the representative week w and operating point t; B_g^{gen} and B_s^{sto} are the residual value discount rates of renewable energy g and energy storage equipment s within the planning cycle, respectively; R_g^{gen} is the equipment residual value coefficient of renewable energy g; R_s^{sto} is the equipment residual value coefficient of the energy storage equipment s.

2.4.2 Constraints

(1) Hybrid Energy Storage Constraints

The port microgrid system comprises various energy storage devices, and this model assumes a fixed ratio of capacity to power for each type of energy storage device. The time scale of operating points in the high-fidelity compression and reconstruction results of the port microgrid operating scenario depends on their weights. Therefore, when calculating the energy level of the energy storage device, this model takes into account the corresponding weights and matches different time scales accordingly.

$$0 \leq P_{wsnt}^+ \leq \frac{\overline{S}_{sn}}{\phi_s} \tag{2.12}$$

$$0 \leq P_{wsnt}^- \leq \frac{\overline{S}_{sn}}{\phi_s} \tag{2.13}$$

$$E_{wsnt} = E_{wsnt-1} + \sigma_t \left(\eta_s^+ P_{wsnt}^+ - \frac{P_{wsnt}^-}{\eta_s^-} \right) \tag{2.14}$$

$$\alpha_s^{min} \overline{S}_{sn} \leq E_{wsnt} \leq \alpha_s^{max} \overline{S}_{sn} \tag{2.15}$$

$$E_{wsnt0} = E_{wsnt\ max} = \tfrac{1}{2} \overline{S}_{sn} \tag{2.16}$$

where P_{wsnt}^+ and P_{wsnt}^- are respectively the charging and discharging power of the energy storage device; ϕ_s is the full power charging duration of the energy storage device; E_{wsnt} is the energy level of the energy storage device; η_s^+ and η_s^- are respectively the charging and discharging efficiency of the energy storage device; α_s^{min} and α_s^{max} are the minimum and maximum charge levels of the energy storage device, respectively.

(2) Renewable Energy Generation Constraints

Given the significant trend towards increasing penetration of renewable energy generation under the "Dual Carbon" goal, this model imposes a minimum share for renewable energy generation. The calculation of renewable energy generation takes into account the dual effects of representative weekly weights and operational point weights.

$$\kappa \cdot \left[\sum_{w \in \Omega} \sigma_w \sum_{n \in N} \sum_{t \in w} \sigma_t \left(P_{wnt}^{grid} + \sum_{g \in G} P_{wgnt}^{gen} \right) \right]$$
$$\leq \sum_{w \in \Omega} \sigma_w \sum_{n \in N} \sum_{t \in w} \sigma_t \sum_{g \in G} P_{wgnt}^{gen} \tag{2.17}$$

where κ is the proportion of renewable energy power generation to the total power supply of the system; P_{wgnt}^{gen} refers to active power generation from renewable energy.

(3) Other Constraints

The model's constraints include branch flow constraints, node injection power balance constraints, and system operating constraints, as shown in Appendix 1. Among them, the branch flow constraint contains nonlinear terms. To facilitate solving, the nonlinear model is transformed into a second-order cone programming model, as shown in Appendix 2. Literature [14, 15] has proven that the second-order cone transformation is strictly accurate for radial distribution networks when the objective function is a convex function and a strictly increasing function. The optimized configuration model for the source and storage of the transformed port microgrid can be quickly solved using the commercial software Gurobi.

2.5 Case Studies

2.5.1 Case Description

This chapter utilizes the topology structure and historical data of the Shijiu Port microgrid in Rizhao Port to verify the accuracy and effectiveness of the proposed method. The microgrid at this port is planned to include wind power and photovoltaic power, connected to the superior power grid via a public connection point (PCC point). To increase the proportion of clean energy in the total energy supply, the amount of wind and photovoltaic power generation is limited to more than half of the total system power generation. Additionally, to improve the overall performance of the energy storage system, including longer cycle life and excellent discharge performance, the microgrid contains two types of energy storage devices: lithium-ion batteries (S_1) and lead-acid batteries (S_2). The topology and location of the installed equipment in the port microgrid are shown in Fig. 2.4. The planning period for the calculation example is four years. The time-of-use electricity price is based on the general industrial and commercial electricity price announced by the National Development and Reform Commission. The parameters for renewable energy and energy storage are shown in Table 2.1.

This chapter employs a Vector Autoregression (VAR) model to generate future data on wind speed, irradiance, and load for a microgrid operating in a port based on 6 years of historical data from the Rizhao Port. To simplify the calculation process, the wind speed and irradiance are converted into generating equipment capacity factors during the data extraction and integration stage.

The case study also includes the results of typical day scenario generation method (Method 1) and typical week scenario generation method (Method 2) to compare and analyze the effectiveness of the high-fidelity compression and reconstruction method for the microgrid operating scenarios in the port proposed in this chapter (Method 3). The three methods generate typical scenarios with the same data compression rate based on the operating scenarios. Based on the elbow rule, the compression rate of the data is determined to be 1.92% ($672 \div 4 \div 8760 \approx 1.92\%$) by finding the "elbow" point according to the degree of data distortion, which reduces 4 years of data to 672 operating points. Method 1 generates 28 ($28 \times 24 = 672$) typical operating days; Method 2 generates 4 ($4 \times 7 \times 24 = 672$) typical operating weeks; Method 3 generates 16 representative weeks during the high-fidelity compression

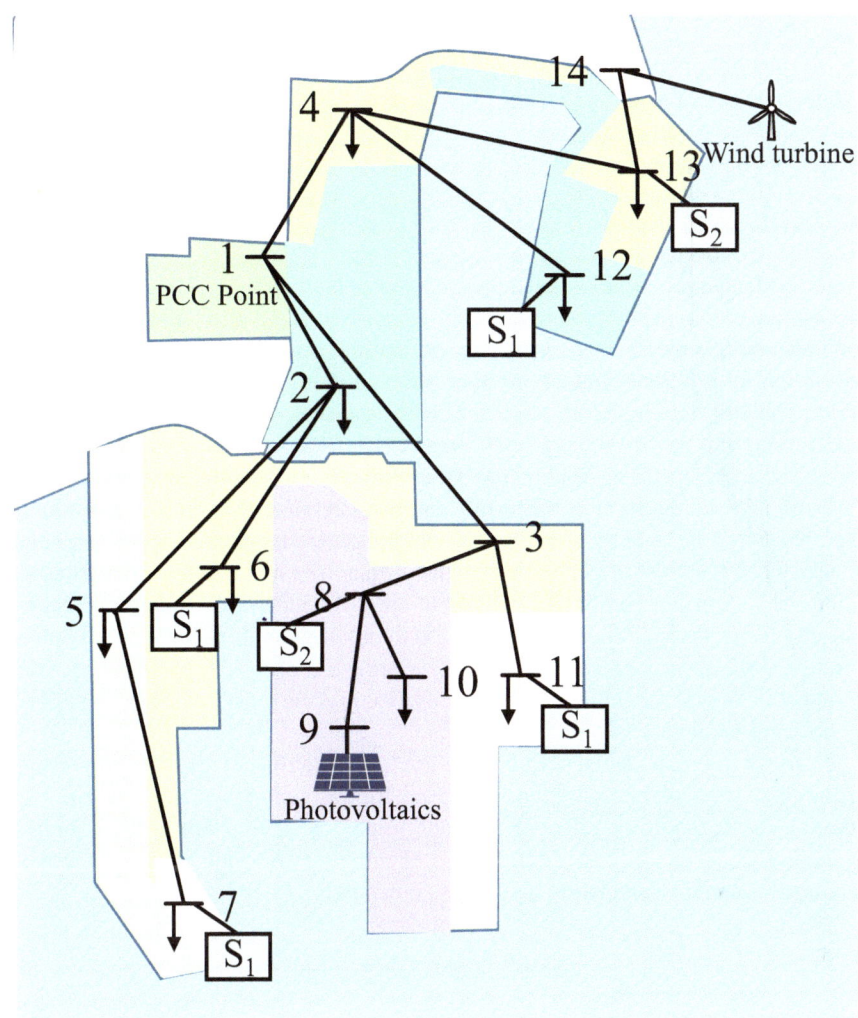

Fig. 2.4 The topology of Rizhao Port microgrid system

Table 2.1 Device parameters

Power generation equipment		Investment cost (yuan/kW)		Equipment service life (year)
WT (wind turbine)		15,900		25
PV (photovoltaic)		13,800		25
Energy storage equipment	Investment cost (yuan/ kW)	Full power charging duration (h)		Equipment service life (year)
Lithium cell	5600	4		10
Lead accumulator	11,800	8		15

stage based on the operating weeks, and further reduces the 16 representative weeks to 672 operating points during the data reconstruction stage based on the operating points.

2.5.2 Analysis of Typical Scenario Results

The representative scenarios generated by the operating scenarios directly affect the optimization results of microgrid configuration in green ports. Load is the most fundamental parameter in typical scenarios, and the ability to retain the load distribution during different methods of data processing is measured by load duration curve. The load duration curve is not arranged according to the chronological order, but according to the load changes within a statistical period, and is rearranged and composed according to the accumulated duration of each different load value within the period. When the level of renewable energy generation is high, the system may need to charge the energy storage devices and generate curtailed wind and solar power; when the level of generation is low, it may be supplied by grid purchasing and energy storage. Therefore, the distribution of the renewable energy generation capacity factor is also important. Analogous to the load duration curve, wind power duration curve and solar power duration curve are established. The comparison of the duration curves for the initial operating scenarios and the typical scenarios obtained by different methods is shown in Fig. 2.5.

From the figure, it can be seen that for the load duration curve, the fitting results of the three methods are all good. However, for the wind power and solar power duration curve, the fitting effect of methods 1 and 3 is significantly better than that of method 2. To quantify the fitting effect of the curve, a fitting accuracy index is established, and the calculation method is shown in Eq. (2.18).

$$RMSD(X, Y) = \frac{\sqrt{\frac{1}{n} \sum_{i=1}^{n} (X_i - Y_i)^2}}{\max(Y) - \min(Y)} \tag{2.18}$$

where X is the duration curve corresponding to a typical scenario; Y is the duration curve corresponding to the original data. The smaller the value of RMSD, the closer the two curves are.

Table 2.2 presents a comparison of the duration curves of three methods. It can be observed that method 1 provides the closest fit to the load, and also performs well in fitting the duration curves of wind and photovoltaic power. Method 2 exhibits significant errors in fitting the duration curve of wind power, indicating that the four typical operating weeks selected by this method do not fully represent the wind power generation scenarios. Considering the fitting results of the duration curves of load, wind, and photovoltaic power, method 3 yields the highest accuracy in fitting.

Another measure of the accuracy of typical scene is the correlation between different types of data. The correlation between load and renewable energy generation capacity coefficient reflects the degree of synchronization between demand and

Fig. 2.5 The comparison of duration curves

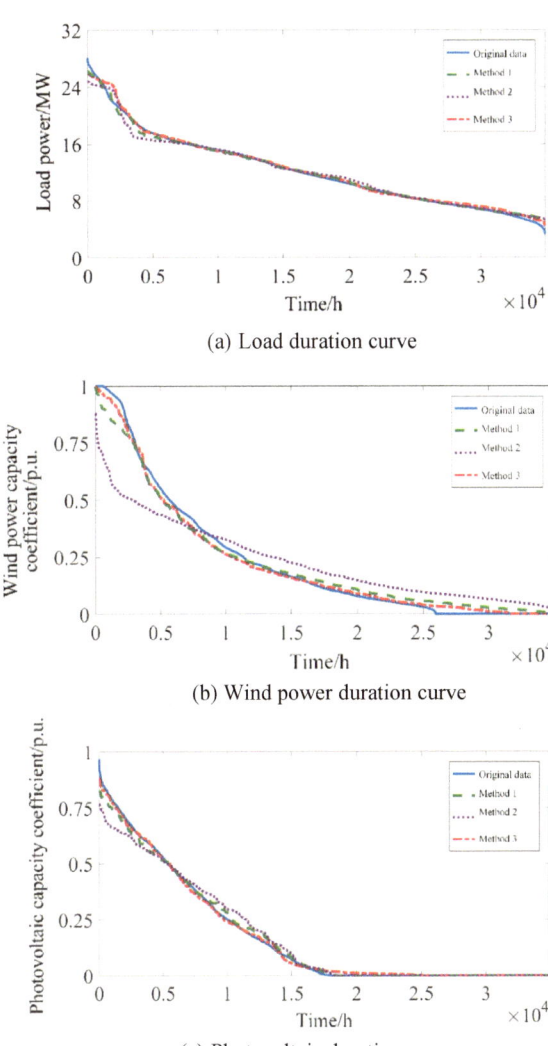

(a) Load duration curve

(b) Wind power duration curve

(c) Photovoltaic duration curve

Table 2.2 The RMSD of duration curves

	Method 1	Method 2	Method 3 (method of this article) (%)
Load	1.53	2.61	1.60
WT	3.76	12.27	2.22
PV	1.84	3.88	1.00

supply, which determines the net load of the port microgrid. The correlation between different renewable energy generation coefficients reflects the degree of complementarity between two generation technologies. The correlation between electricity price and load, and the renewable energy generation coefficient directly affects the system's purchasing cost. The Pearson correlation coefficient is used to measure the correlation between different types of data, and the calculation method is shown in Eq. (2.19).

$$PCC(X, Y) = \frac{\sum_{i=1}^{n} (X_i - \overline{X}) \cdot (Y_i - \overline{Y})}{\sqrt{\sum_{i=1}^{n} (X_i - \overline{X})^2} \cdot \sqrt{\sum_{i=1}^{n} (Y_i - \overline{Y})^2}} \tag{2.19}$$

where $\overline{X}$ and $\overline{Y}$ are the average values of data X and Y. The calculated result of PCC is between -1 and 1. "1" represents a complete positive correlation between the two types of data, "-1" represents a complete negative correlation, and "0" represents no correlation.

The correlation coefficients between different data types in the typical and initial scenarios obtained by different methods are shown in Fig. 2.6. Table 2.3 lists the errors between the data correlations in the typical scenario and the original data correlations. It can be seen that, except for the poor performance in preserving the correlations between "wind power-photovoltaics" and "wind power-electrovalence", the correlation coefficients obtained by method 1 are generally close to those of the original data. The correlation coefficients obtained by method 2 are relatively poor, indicating that the data characteristics of the initial scenario cannot be well preserved when there are 672 operating points in the typical scenario. Among the three methods, method 3 preserves the data correlations best, except for the weaker correlations between "load-wind power" and "photovoltaics-electrovalence" compared to method 1, the correlations between other data types are significantly better than those of the other methods.

Fig. 2.6 Comparison of correlation coefficient

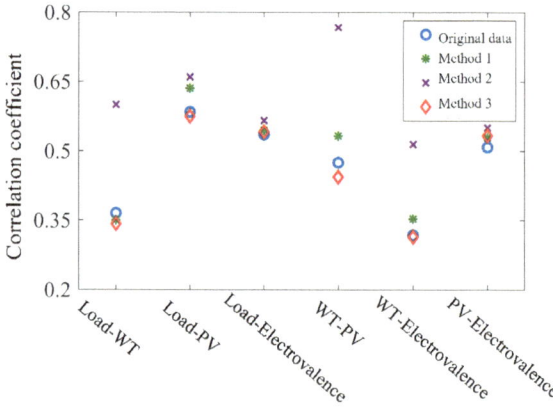

Table 2.3 Correlation coefficient error between raw data and processed data

Type	Method 1 (%)	Method 2 (%)	Method 3 (method of this article) (%)
Load-WT	−4.17	64.36	−6.40
Load-PV	8.70	12.91	−1.61
Load-electrovalence	1.66	5.56	1.02
WT-PV	12.16	61.49	−6.68
WT-electrovalence	11.12	61.99	−1.15
PV-electrovalence	4.05	8.26	4.85

2.5.3 Analysis of Optimization Configuration Results of Renewable Energy and Energy Storage

Firstly, the optimization configuration was performed using original data spanning 4 years as the baseline to evaluate the accuracy of other methods. Then, the optimized configuration model was solved using the typical scenarios obtained from the three methods as input to obtain the optimized configuration results of the port microgrid source and storage.

Table 2.4 shows the results obtained by each method and their errors relative to the baseline. It can be seen from the table that the wind power capacity obtained by the optimization configuration of method 1 is closest to the baseline among the three methods, and the error of the configuration result's total cost is also relatively small. However, the storage capacity configured by this method is too large, especially the lead-acid battery capacity is much larger than the baseline. This is because the typical scenario obtained by method 1 fits poorly with the photovoltaic data, resulting in a small photovoltaic capacity in the configuration result, which requires the installation of more storage equipment to supply peak loads. At the same time, the larger storage capacity also enables method 1 to fully utilize the time-of-use electricity pricing to reduce costs, and the purchased electricity cost is significantly lower than the baseline case. In addition, method 1 uses the typical day as the basic unit of scenario construction, and the storage equipment operates on a daily basis, making it difficult to balance energy over longer time intervals, which also increases the difference between the storage configuration results and the baseline.

The overall optimization configuration results of method 2 are the worst, except for the total cost and purchased electricity cost, which have some reference value, the configuration capacity of the power generation and storage equipment has large errors compared with the baseline. The typical scenario constructed by this method poorly characterizes the light and wind conditions, resulting in an overestimation of the wind power capacity and an underestimation of the photovoltaic capacity in the optimization configuration results. Meanwhile, wind power does not have an obvious intraday pattern, and its output fluctuation cycle is longer. Therefore, compared with the baseline case, the lead-acid battery capacity with a large capacity-to-power ratio

Table 2.4 Optimal sizing results of port microgrid

	Datum	Method 1	Method 1 error (%)	Method 2	Method 2 error (%)	Method 3 (method of this chapter)	Method 3 error (%)
Total cost (million yuan)	277.62	258.41	−6.92	252.98	−8.88	279.29	0.60
Electricity purchase cost (Million yuan)	128.45	114.97	−10.49	120.12	−6.49	127.94	−0.40
Wind power capacity (MW)	13.20	13.98	5.91	20.04	51.82	12.41	−5.98
Photovoltaic capacity (MW)	19.02	15.20	−20.08	7.02	−63.09	20.22	6.31
Lithium battery capacity (MWh)	16.04	20.86	30.05	11.58	−27.81	17.06	6.36
Lead acid battery capacity (MWh)	1.81	5.61	209.94	6.11	237.57	2.25	24.31

is overestimated, and the lithium battery capacity with a small capacity-to-power ratio is underestimated in the configuration results of method 2.

The optimized configuration results obtained by method 3 are the closest to the baseline, and the error of the total cost is only 0.60%. The maximum capacity error is observed for the lead-acid battery, which is 24.31%, but compared with the other two methods, this result still has a significant advantage. This is because method 3 preserves the system operating characteristics more accurately with the same amount of data by dynamically adjusting the time scale of the operating points, which makes the configuration results of wind and photovoltaic more accurate and conducive to storage capacity configuration. At the same time, the operating cycle of this method is one week, which allows the storage to play a role over a longer operating cycle compared to method 1, making the storage capacity configuration more accurate. Therefore, it can be seen that the method proposed in this chapter can be effectively applied to the optimization configuration of renewable energy and hybrid storage in port microgrids.

Fig. 2.7 Comparation of calculation time

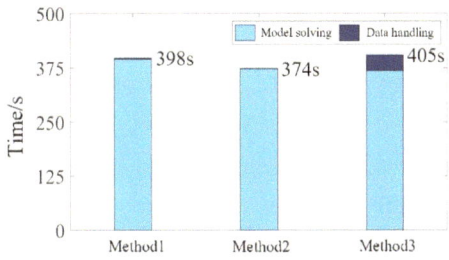

2.5.4 Calculation Time Analysis

The purpose of using typical scenarios for optimization configuration is to reduce the difficulty of problem solving and improve the speed of model solution. The results of the numerical examples were obtained using a computer with 8 cores, a main frequency of 3.6 GHz, and 64 GB of memory. Figure 2.7 compares the total computation time for optimizing source and storage using the three methods, including both model solving time and data processing time. Method 3 has a longer data processing time compared to Methods 1 and 2, but considering the model solving time, the total time for all three methods remains at the same level. This indicates that the proposed method in this chapter does not increase the difficulty of problem solving while improving the accuracy of the solution. In addition, when using the initial running scenario for optimization configuration, the solving model took 25 h, which fully demonstrates the effect of applying typical scenarios in accelerating problem solving speed.

2.6 Conclusion

The operational status of a port microgrid is closely related to the production and berthing of ships. Describing the system's operating characteristics directly through natural periods such as days and weeks would generate redundant information. This chapter proposes a high-fidelity compression and reconstruction method for port microgrid operating scenarios, which effectively increases the data density of the system's typical operating scenarios. The method combines weeks with similar operating characteristics using the Ward minimum variance method, and then proposes a hierarchical clustering method that dynamically adjusts the time scale of each operating point in the representative week to further reduce redundant information. With the high-fidelity compression and reconstruction method, an optimization model for the source and storage configuration of the port microgrid is established. An example analysis is conducted using Rizhao Port to demonstrate that the proposed method better preserves the fluctuation characteristics of the port's source and load and the correlation between different types of data. In addition, the proposed method provides more accurate source and storage optimization configuration results without increasing the computational resources required for problem-solving, with a total cost error of only 0.60%, which is better than existing methods.

Appendix 1

The equations for the branch flow constraint, nodal power injection balance constraint, and system operation constraint in the optimization configuration model for source and storage in microgrids of ports are as follows.

Branch flow constraints

$$P_{wnt} = \sum_{n \to o} P_{wnot}^{\text{line}} - \sum_{m \to n} \left(P_{wmnt}^{\text{line}} - r_{mn} |I_{wmnt}|^2 \right) + g_n |V_{wnt}|^2 \tag{2.20}$$

$$Q_{wnt} = \sum_{n \to o} Q_{wnot}^{\text{line}} - \sum_{m \to n} \left(Q_{wmnt}^{\text{line}} - x_{mn} |I_{wmnt}|^2 \right) + b_n |V_{wnt}|^2 \tag{2.21}$$

$$V_{wnt} = V_{wmt} - (r_{mn} + jx_{mn}) I_{wmnt} \tag{2.22}$$

where P_{wnt} and Q_{wnt} are the active injection power and reactive injection power of node n, respectively; P_{wnot}^{line} and Q_{wnot}^{line} are the active and reactive power from node n to node o, respectively; r_{mn} and x_{mn} are the resistance and reactance of line mn, respectively; I_{wmnt} is the current flowing through the line mn; g_n and b_n are the conductivity and admittance of node n, respectively; V_{wnt} is the voltage of the node n.

Node injection power balance constraint

$$P_{wnt} = P_{wnt}^{\text{grid}} + \sum_{g \in G} \left(P_{wgnt}^{\text{gen}} - P_{wgnt}^{\text{gen,cur}} \right) + \sum_{s \in S} \left(P_{wsnt}^{-} - P_{wsnt}^{+} \right) - P_{wnt}^{\text{load}} \tag{2.23}$$

$$Q_{wnt} = Q_{wnt}^{\text{grid}} + \sum_{g \in G} Q_{wgnt}^{\text{gen}} - Q_{wnt}^{\text{load}} \tag{2.24}$$

where Q_{wnt}^{grid} is the reactive power transmitted by the power grid to node n; Q_{wgnt}^{gen} is the reactive power generation of renewable energy g; $P_{wgnt}^{\text{gen,cur}}$ is the renewable energy power that has not been consumed; P_{wnt}^{load} and Q_{wnt}^{load} are the active and reactive loads of node n, respectively.

System operation constraints

$$\overline{V}_n \le V_{wnt} \le \overline{V}_n \tag{2.25}$$

$$0 \le P_{wmnt}^{\text{line}} \le \overline{P}_{mn}^{\text{line}} \tag{2.26}$$

$$0 \leq P_{wnt}^{\text{grid}} \leq \overline{P}_n^{\text{grid}} \tag{2.27}$$

$$0 \leq Q_{wnt}^{\text{grid}} \leq \overline{Q}_n^{\text{grid}} \tag{2.28}$$

$$0 \leq P_{wgnt}^{\text{gen}} \leq \overline{P}_{gn}^{\text{gen}} \tag{2.29}$$

$$0 \leq P_{wgnt}^{\text{gen,cur}} \leq P_{wgnt}^{\text{gen}} \tag{2.30}$$

where $\overline{V}_n$ and $\overline{V}_n$ are the maximum and minimum values of node voltage, respectively; $\overline{P}_{mn}^{line}$ is the maximum transportable power of line mn; $\overline{P}_n^{grid}$ and $\overline{Q}_n^{grid}$ are the maximum transferable active power and maximum transferable reactive power from the power grid to node n.

Appendix 2

The nonlinear second-order cone transformation process is as follows. In the above model, Eqs. (2.20) and (2.21) contain nonlinear terms, which are replaced and transformed to transform the nonlinear model into a second-order cone programming model. The definitions are as follows:

$$l_{wmnt} = \frac{P_{wmnt}^{\text{line}\ 2} + Q_{wmnt}^{\text{line}\ 2}}{|V_{wmt}|^2} \tag{2.31}$$

$$v_{wnt} = |V_{wnt}|^2 \tag{2.32}$$

Convert Eqs. (2.20) and (2.21) into Eqs. (2.33) and (2.34) through Eqs. (2.31) and (2.32).

$$P_{wnt} = \sum_{n \to o} P_{wnot}^{\text{line}} - \sum_{m \to n} \left(P_{wmnt}^{\text{line}} - r_{mn} l_{wmnt} \right) + g_n v_{wnt} \tag{2.33}$$

$$Q_{wnt} = \sum_{n \to o} Q_{wnot}^{\text{line}} - \sum_{m \to n} \left(Q_{wmnt}^{\text{line}} - x_{mn} l_{wmnt} \right) + b_n v_{wnt} \tag{2.34}$$

Using phase angle relaxation and retaining only modulus constraints, convert Eq. (2.22)–(2.35).

$$v_{wnt} = v_{wmt} - 2 \left(r_{mn} P_{wmnt}^{\text{line}} + x_{mn} Q_{wmnt}^{\text{line}} \right) + \left(r_{mn}^2 + x_{mn}^2 \right) l_{wmnt} \tag{2.35}$$

By using second-order cone relaxation transformation, Eq. (2.31) is relaxed to Eq. (2.36).

$$\left\| \begin{array}{c} 2P_{wmnt}^{\text{line}} \\ 2Q_{wmnt}^{\text{line}} \\ l_{wmnt} - v_{wmt} \end{array} \right\|_2 \le l_{wmnt} + v_{wmt} \tag{2.36}$$

The Eq. (2.25) in the system operation constraint can be replaced by Eq. (2.37).

$$\underline{V}_n^2 \le v_{wnt} \le \overline{V}_n^2 \tag{2.37}$$

At this point, all nonlinear constraints in the optimized configuration model are transformed into linear constraints and second-order cone constraints.

References

1. Wang, X., Huang, W., Tai, N., et al.: Two-stage full-data processing for microgrid planning with high penetrations of renewable energy sources. IEEE Trans. Sustain. Energy **12**(4), 2042–2052 (2021)
2. Morstyn, T., Hredzak, B., Agelidis, V.G.: Control strategies for microgrids with distributed energy storage systems: an overview. IEEE Trans. Smart Grid **9**(4), 3652–3666 (2018)
3. Quevedo, P.M., Munoz-Delgado, G., Contreras, J.: Impact of electric vehicles on the expansion planning of distribution systems considering renewable energy, storage and charging stations. IEEE Trans. Smart Grid **10**(1), 794–804 (2019)
4. Koltsaklis, N.E., Georgiadis, M.C.: A multi-period, multi-regional generation expansion planning model incorporating unit commitment constraints. Appl. Energy **158**, 310–331 (2015)
5. Belderbos, A., Delarue, E.: Accounting for flexibility in power system planning with renewables. Int. J. Electr. Power Energy Syst. **71**, 33–41 (2015)
6. Yang, P., Nehorai, A.: Joint optimization of hybrid energy storage and generation capacity with renewable energy. IEEE Trans. Smart Grid **5**(4), 1566–1574 (2014)
7. Kiptoo, M.K., Lotfy, M.E., Adewuyi, O.B., et al.: Integrated approach for optimal techno-economic planning for high renewable energy-based isolated microgrid considering cost of energy storage and demand response strategies. Energy Convers. Manage. **215**, 112917 (2020)
8. Liu, L., Da, C., Luo, N., et al.: A planning method of renewable distributed generation based on polymerization of output curves and tolerance evaluation of typical scene in interval power flow. Proc. CSEE **40**(14), 4400–4410 (2020)
9. Alvarez, R., Moser, A., Rahmann, C.A.: Novel methodology for selecting representative operating points for the TNEP. IEEE Trans. Power Syst. **32**(99), 2234–2242 (2017)
10. Liu, Y., Sioshansi, R., Conejo, A.J.: Hierarchical clustering to find representative operating periods for capacity-expansion modeling. IEEE Trans. Power Syst. **33**(3), 3029–3039 (2017)
11. Zhao, S., Jin, T., Li, Z., et al.: Wind power scenario generation for multiple wind farms considering temporal and spatial correlations. Power Syst. Technol. **43**(11), 3997–4004 (2019)
12. Li, B., Sun, J., Yu, P., et al.: A multi-dimensional typical scenarios generation algorithm for distribution network based on load clustering and network structure equivalence. Proc. CSEE **41**(08), 2661–2671 (2021)
13. Li, H., Ren, Z., Hu, B., et al.: A sequential generative adversarial network based monthly scenario analysis method for wind and photovoltaic power. Proc. CSEE **42**(02), 537–548 (2022)
14. Steven, H.L.: Convex relaxation of optimal power flow—part I: formulations and equivalence. IEEE Trans. Control Netw. Syst. **1**(1), 15–17 (2014)
15. Steven, H.L.: Convex relaxation of optimal power flow—part II: exactness. IEEE Trans. Control Netw. Syst. **1**(2), 177–189 (2014)

Chapter 3
Adaptive Bidirectional Droop Control Strategy for Hybrid AC-DC Port Microgrids

3.1 Introduction

Port Electric-thermal microgrid is one of the typical applications of integrated energy systems. Its integrates the supply, conversion, and storage equipment in electric and thermal energy flows based on users' electrical and thermal demands, and to coordinate and optimize protection and control methods to achieve economical and reliable operation [1–4]. With the increasingly diverse energy needs of industrial, commercial, and residential users supplied by microgrids, there exist more complex multi-energy couplings, such as energy cascade utilization, which poses difficulties for energy optimization management of electric thermal microgrids [4, 5]. Energy cascade utilization is an effective method to improve energy utilization efficiency and supply quality. It is an important direction in current research on energy optimization management of electric-thermal microgrids [6–8].

AC-DC hybrid microgrid mainly consists of AC microgrid, DC microgrid and microgrids interlinking converter (MIC). The MIC is the core device to balance the power of both microgrids, which can coordinate the control of AC and DC microgrids to achieve bi-directional power mutual aid and optimal stability. MIC not only ensures the stable and efficient operation of hybrid microgrid under different operating conditions, but also further improves the capacity of renewable energy consumption. As a bridge for AC-DC power exchange, the interconverter plays a key role in maintaining the stability of the system frequency and voltage [9–12], and the design of suitable interconverter control strategy becomes an important research content in the development of hybrid AC-DC microgrid.

AC-DC hybrid microgrid interconverters need to regulate AC frequency and DC voltage simultaneously, and the traditional active droop method, which only considers single-side quantities, is no longer applicable. By improving the traditional droop control equations or studying the coupling relationship between AC frequency and DC voltage, scholars have proposed bidirectional power control strategies that can regulate AC frequency and DC voltage simultaneously [13–16]. The literature [17,

W. Huang et al., *Energy Management of Integrated Energy System in Large Ports*, Springer Series on Naval Architecture, Marine Engineering, Shipbuilding and Shipping 18, https://doi.org/10.1007/978-981-99-8795-5_3

18] unifies the droop characteristics of AC side and DC side for bidirectional droop control of microgrid interconnection converter based on typical hybrid microgrid structure through the scalarization process. In [19], the droop relationship between frequency and voltage squared is introduced for control through the energy conservation relationship on the AC and DC sides. The literature [20] analyzed the linear coupling relationship between AC frequency and DC voltage by virtual inertia and virtual capacitance characteristics, and also obtained the droop relationship between AC frequency and DC voltage squared. To further improve the stability of hybrid AC-DC microgrid, literature [21] proposed an improved inner-loop control strategy for microgrid interconnection converter based on disturbance observation link, and literature [22, 23] provided inertia support for frequency and voltage based on virtual synchronous motor control strategy to improve the system anti-disturbance capability. Literature [24] proposed a control strategy for microgrid interconnection converter with AC-DC microgrid internal energy storage device to improve the overall energy management capability of hybrid microgrid. In the literature [25], an AC-DC microgrid two-layer optimization method is proposed, which incorporates secondary regulation and DG generation cost optimization control strategies based on bidirectional droop control of the interconnection converter. In the literature [17–25], a unified droop control based on the scalarization method or linking the AC frequency and DC voltage from the energy balance perspective is performed for the control of microgrid interconnection converters to realize the bidirectional control of power between AC and DC microgrids. However, in the process of power regulation of hybrid microgrid with islanded operation, there is often a conflict between AC frequency optimization and DC voltage optimization due to the limitation of adjustable DG power regulation capability, and the aforementioned literature is based on fixed droop coefficient or proportional coefficient to regulate the power between AC and DC sub-microgrids without considering the regulation priority between AC frequency and DC voltage. Due to the different amount of DG access, the strength of the AC-DC sub-micro-network's ability to stabilize frequency and voltage will also change. The two-way droop control strategy based on fixed coefficients will lead to larger frequency or voltage deviations when load changes occur on the weaker side of the AC-DC sub-micro-network, which cannot give full play to the support role of the strong micro-network to the weak micro-network, and the frequency and voltage equalization effect is poor.

To solve the above problems, this chapter proposes an interconverter adaptive bidirectional droop control strategy with the control objective of balancing and compensating the AC frequency and DC voltage deviations of the hybrid microgrid. In order to reduce the combined deviation of AC frequency and DC voltage, the adaptive value of the weight coefficient is proposed to ensure that the MIC gives priority to the quantity with larger deviation in the active dynamic regulation process. Set the start-up deadband to avoid the non-essential operation of the interconverter. Establish the small-signal model of the interconnection converter and analyze the influence of the control parameters on the system stability. A typical AC-DC hybrid microgrid system is established in the PSCAD/EMTDC simulation platform and RT-LAB experimental platform to verify the effectiveness and reliability of the proposed control strategy.

3.2 Hybrid Microgrid and Its Interlinking Converter

3.2.1 AC-DC Hybrid Microgrid

The typical topology of a hybrid AC/DC microgrid is shown in Fig. 3.1. The AC microgrid and DC microgrid are interconnected at the bus through an interlinking converter, allowing for flexible power flow between the two grids. By controlling the interlinking converter, power can flow bidirectionally between the AC microgrid and DC microgrid. The entire hybrid microgrid system is connected to the public AC grid through a point of common coupling (PCC), enabling seamless switching between grid-connected and islanded modes.

The hybrid microgrid with both AC and DC components adopts a peer-to-peer control mode, which does not require communication and can meet the requirement of plug-and-play for both distributed generators (DGs) and loads. The controllable DGs within the AC microgrid jointly maintain the stability of AC bus frequency and voltage according to the $P - f$ and $Q - U_{\mathrm{ac}}$ droop characteristics outlined in Eq. (3.1).

$$\begin{cases} f = f_0 - K_{fi}(P_i - P_{0i}) \\ U_{\mathrm{ac}} = U_{\mathrm{ac0}} - K_{qi}(Q_i - Q_{0i}) \end{cases} \tag{3.1}$$

Among them, the subscript i represents the index of the AC DG, P_i represents the active power output of DG i, P_{0i} represents the active power output of DG i under rated conditions, f_0 is the power frequency, f is the current operating frequency of the AC microgrid, K_{fi} is the frequency droop coefficient, Q_i represents the reactive power output of DG i, Q_{0i} represents the reactive power output of DG i under rated conditions, U_{ac0} is the rated AC voltage, U_{ac} is the current voltage of the AC micro-grid, and K_{qi} is the voltage droop coefficient. In order to achieve load sharing, the droop coefficients K_{fi} and K_{qi} are inversely proportional to the capacity of DG i.

The controllable distributed generators (DGs) in the DC microgrid collectively maintain the stability of the DC bus voltage according to the droop relationship of Eq. (3.2) for P-U_{dc}.

$$U_{\mathrm{dc}} = U_{\mathrm{dc0}} - K_{uj}\left(P_j - P_{0j}\right) \tag{3.2}$$

where the subscript j represents the index of the DC distributed generation (DG), P_j represents the active power output of DG_j, P_{0j} represents the output power of DG_j under rated conditions, U_{dc0} represents the rated DC voltage, U_{dc} represents

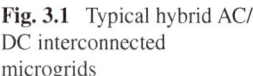

Fig. 3.1 Typical hybrid AC/DC interconnected microgrids

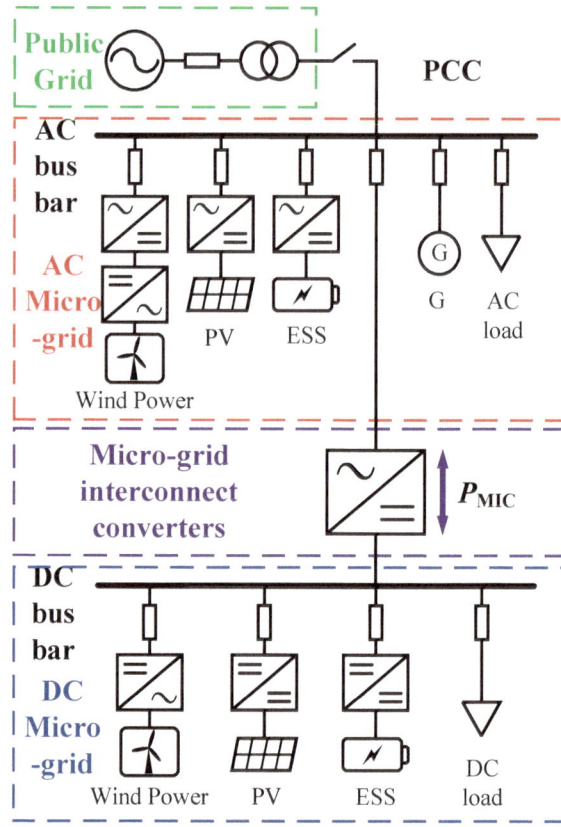

the current operating voltage of the DC microgrid, and K_{uj} represents the droop coefficient of the DC voltage. In order to achieve load sharing, the droop coefficient K_{uj} is inversely proportional to the capacity of DG_j.

3.2.2 The Structure of Hybrid Microgrid Interlinking Converter

The microgrid interlinking converter adopts voltage source converter (VSC) structure, and the topology is shown in Fig. 3.2, where the meanings of the variables are as follows: U_{abc} and i_{abc} are the three-phase voltage and current on the AC side of the converter, E_{abc} is the three-phase voltage of the AC bus, L_{ac} is the total inductance of the filter and line between the AC side of the converter and the AC bus, C_{dc} is the DC filter capacitor, U_{dc} and i_{dc} are the voltage and current of the DC side of the converter, respectively, and P_{MIC} is the active power exchanged between the AC and

Fig. 3.2 Topological structure of the MIC

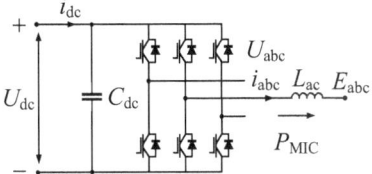

DC microgrids, and the analysis below takes the inverting direction as the positive power direction.

The microgrid interlinking converter of the microgrid adopts a voltage source converter (VSC) structure, as shown in Fig. 3.2. The meanings of the variables are as follows: U_{abc} and i_{abc} are the three-phase voltage and current on the converter's AC side, E_{abc} is the three-phase voltage of the AC bus, L_{ac} is the total inductance of the filter and line between the converter's AC side and the AC bus, C_{dc} is the DC filter capacitor, U_{dc} and i_{dc} are the voltage and current on the converter's DC side, P_{MIC} is the active power exchanged between the AC and DC microgrids. The analysis in the following text is based on the inverter direction as the positive power direction.

There is no reactive power balance issue in a DC microgrid, and the interlinking converter does not need to control the reactive power on the DC side. This chapter only discusses the impact of the active power control strategy of the interlinking converter on the frequency stability of the AC microgrid and the voltage stability of the DC microgrid under islanding operation mode.

3.3 Bidirectional Adaptive Droop Control Strategy for Interlinking Converter

3.3.1 Bidirectional Droop Control Targets

The interlinking converter, in the operation of an AC/DC hybrid microgrid island, coordinates the magnitude of the AC frequency deviation and the DC voltage deviation by controlling the active power between the AC and DC microgrids. To simplify the analysis, line impedance is ignored, and the relationship between power variation in the AC/DC sub-microgrids with load changes can be expressed as Eq. (3.3).

$$\begin{cases} \sum_{i=1}^{m} P_i = \sum_{i=1}^{m} P_{0i} + \Delta P_{\text{load\_AC}} - P_{\text{MIC}} \\ \sum_{j=1}^{n} P_j = \sum_{j=1}^{n} P_{0j} + \Delta P_{\text{load\_DC}} + P_{\text{MIC}} \end{cases} \tag{3.3}$$

where m and n are the number of controllable DGs in the AC microgrid and DC microgrid, respectively. ΔP_{load_AC} and ΔP_{load_DC} are the incremental loads of the AC load and DC load respectively. Substituting Eqs. (3.1) and (3.2) into Eq. (3.3) yields the variations in AC frequency and DC voltage, as shown in Eq. (3.4).

$$
\begin{cases}
\Delta f = f - f_0 = -\dfrac{\Delta P_{load\_AC} - P_{MIC}}{\sum_{i=1}^{m} \frac{1}{K_{fi}}} \\
\Delta U_{dc} = U_{dc} - U_{dc0} = -\dfrac{\Delta P_{load\_DC} + P_{MIC}}{\sum_{j=1}^{n} \frac{1}{K_{uj}}}
\end{cases}
\tag{3.4}
$$

From Eq. (3.4), it can be seen that the deviation between the AC frequency and the DC voltage is related to the magnitude of the load change, the power transmitted by the interlinking converter, and the load-carrying capacity of the AC/DC hybrid microgrid itself. If the interlinking converter does not participate in power regulation, different load changes or the strength of the microgrid can cause a large deviation in either AC frequency or DC voltage, while the other quantity has a small deviation. When the load change is large or the microgrid is weak, the quantity with a large deviation in frequency and voltage will exceed the allowable range, affecting the safe and stable operation of the hybrid microgrid system.

The power control strategy for interlinking converters in a hybrid microgrid can provide power support from the side with smaller AC frequency or DC voltage deviation to the other side, ensuring that both AC frequency and DC voltage stay within limits. In order to quantitatively describe the impact of frequency and voltage deviations on the stability of the hybrid microgrid, a comprehensive deviation index called the Global Variation Index (G_{vi}) is defined as shown in Eq. (3.5) [24].

$$
G_{vi} = \left(\frac{\Delta f}{f_{max} - f_{min}}\right)^2 + \left(\frac{\Delta U_{dc}}{U_{dcmax} - U_{dcmin}}\right)^2
\tag{3.5}
$$

where f_{max} and f_{min} are the maximum and minimum frequencies allowed for AC sub-microgrid, U_{dcmax} and U_{dcmin} are the maximum and minimum voltages allowed for DC sub-microgrid, respectively. The objective is to obtain the active reference value of the interlinking converter based on the current AC frequency and DC voltage of the hybrid microgrid system, and to make the hybrid microgrid to achieve a smaller active reference value under different load changes, DG switching, and AC and DC microgrid strengths and weaknesses by adaptive parameter droop control without communication. The adaptive parameter droop control enables the hybrid microgrid to achieve a small integrated deviation under different conditions of load variation, DG switching and AC/DC microgrid strength.

3.3.2 Adaptive Bidirectional Droop Control Strategy

To standardize the dimensions and simplify control, normalization is first applied to the AC frequency and DC voltage [18], as shown in Eq. (3.6).

$$\begin{cases} f_{pu} = \dfrac{f-0.5(f_{max}+f_{min})}{0.5(f_{max}-f_{min})} \\ U_{dc.pu} = \dfrac{U_{dc}-0.5(U_{dcmax}+U_{dcmin})}{0.5(U_{dcmax}-U_{dcmin})} \end{cases} \tag{3.6}$$

where f_{pu} is the normalized AC frequency and $U_{dc.pu}$ is the normalized DC voltage. After the normalization process, the frequency and voltage of the hybrid microgrid are unified to a range of $[-1, 1]$ when normally allowed.

The power response of the interlinking converter to the AC/DC sub-microgrid can be obtained based on the "active power-frequency" droop relationship and the "active power-DC voltage" droop relationship. The difference power is the power reference value of the interlinking converter, with the inverter direction taken as the reference of positive direction, as shown in Eq. (3.7).

$$P = P_0 - K \cdot f_{pu} + K \cdot U_{dc.pu} \tag{3.7}$$

where K is the droop coefficient, which can be set based on the maximum power P_{max} of the interlinking converter [17], or adjusted according to actual operational requirements [23].

The control strategy of Eq. (3.7) implements bidirectional control of active power for the interlinking converter in the microgrid based on the droop relationship between AC and DC. However, this strategy cannot effectively adjust the priority of AC frequency and DC voltage response in different operating states due to the fixed droop coefficient. To address this, Eq. (3.8) is derived by introducing an adjustable weighting coefficient l into the equation.

$$P = P_0 - \lambda \cdot K \cdot f_{pu} + (1 - \lambda) \cdot K \cdot U_{dc.pu} \tag{3.8}$$

The range of variation for the weighting coefficient λ is (0, 1). By adjusting λ, the priority of power support between the AC and DC sides can be coordinated, ensuring a balanced compensation for AC frequency and DC voltage during disturbances.

To ensure the safe and reliable operation of an independent AC/DC microgrid, priority should be given to provide power support to the side with larger AC frequency or DC voltage deviations. By optimizing the power exchange between the two sides of the interlinking converter, the balance and compensation for AC frequency and DC voltage deviations can be achieved, thereby better ensuring the overall performance of the system. When the frequency deviation (f_{pu}) is greater than the DC voltage deviation ($U_{dc.pu}$), power is supported from the DC side to the AC side, ensuring that the AC frequency is adjusted according to the $P-f_{pu}$ droop relationship. Conversely, when the power deviation (f_{pu}) is smaller than the DC voltage deviation ($U_{dc.pu}$), power is supported from the AC side to the DC side, ensuring that the DC voltage is adjusted according to the $P - U_{dc.pu}$ droop relationship. Therefore, the weighting coefficient λ is positively correlated with the AC frequency deviation and negatively correlated with the DC voltage deviation.

To adapt to different scenarios, λ is set to have sufficient sensitivity to both AC frequency and DC voltage deviations.

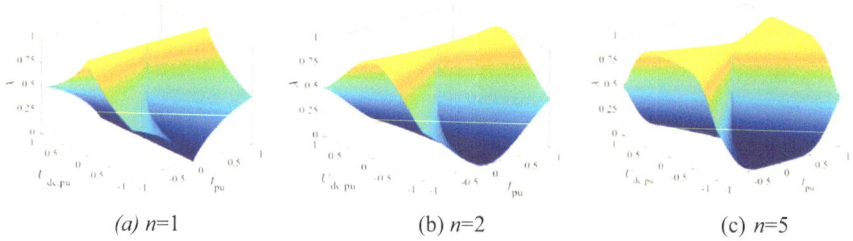

(a) n=1   *(b) n=2*   *(c) n=5*

Fig. 3.3 Diagram of the variation of λ with *n*

$$\lambda = \begin{cases} \dfrac{\left|f_{pu}\right|^n}{\left|f_{pu}\right|^n + \left|U_{dc.pu}\right|^n}, & \left|f_{pu}\right| > \varepsilon \, or \left|U_{dc.pu}\right| > \varepsilon \\ 0.5, & \left|f_{pu}\right| < \varepsilon \, and \left|U_{dc.pu}\right| < \varepsilon \end{cases} \quad (3.9)$$

In the equation, ε represents a small value for optimal selection according to the microgrid operation requirements. In this text, e is set to 0.05. When the normalized AC frequency deviation and DC voltage deviation are within 5%, the weighting factor for the droop control on both sides is not adjusted, and in this case, λ is set to 0.5. The variation of λ with different values of n is shown in Fig. 3.3.

From Fig. 3.3, it can be seen that when n is set to 1, the change in λ is too gradual, making it difficult to achieve optimal control of support power when f_{pu} and $U_{\text{dc.pu}}$ have significant deviations. When n has a large value, the change in λ is too rapid, and the support power becomes too sensitive to fluctuations in f_{pu} and $U_{\text{dc.pu}}$, which can easily lead to control instability.

In order to give λ a certain adaptive ability to the deviation of f_{pu} and $U_{\text{dc.pu}}$, while avoiding the instability caused by λ being too sensitive to the changes of f_{pu} and $U_{\text{dc.pu}}$, by setting $n = 2$ in Eq. (3.9), the method for determining the adaptive weighting coefficients of λ is obtained as shown in Eq. (3.10).

$$\lambda = \begin{cases} \dfrac{f_{\text{pu}}^2}{f_{\text{pu}}^2 + U_{\text{dc.pu}}^2}, & \left|f_{\text{pu}}\right| > \varepsilon \text{ or } \left|U_{\text{dc.pu}}\right| > \varepsilon \\ 0.5, & \left|f_{\text{pu}}\right| \le \varepsilon \text{ and } \left|U_{\text{dc.pu}}\right| \le \varepsilon \end{cases} \quad (3.10)$$

Substituting Eq. (3.10) into Eq. (3.8), the adaptive bidirectional droop control equation can be obtained as shown in Eq. (3.11).

$$P = \begin{cases} P_0 - K \cdot \dfrac{f_{\text{pu}}^3 - U_{\text{dc.pu}}^3}{f_{\text{pu}}^2 + U_{\text{dc.pu}}^2}, & \left|f_{\text{pu}}\right| > \varepsilon \text{ or } \left|U_{\text{dc.pu}}\right| > \varepsilon \\ P_0 - 0.5 \cdot K \cdot \left(f_{\text{pu}} - U_{\text{dc.pu}}\right), & \left|f_{\text{pu}}\right| \le \varepsilon \text{ and } \left|U_{\text{dc.pu}}\right| \le \varepsilon \end{cases} \quad (3.11)$$

The control equation can achieve the control effect shown in Table 3.1. The adaptive bidirectional droop control strategy prioritizes the side with larger deviations in AC frequency and DC voltage, and smoothly transitions between $P - f_{\text{pu}}$ droop control and $P - U_{\text{dc.pu}}$ droop control, without causing any impact during the control strategy switching process. Since both AC frequency and DC voltage can be directly

Table 3.1 Adaptive bidirectional droop control characteristics

Operation status	Control characteristics
$f_{pu} \gg U_{dc.pu}$	$P = P_0 - K \cdot f_{pu}$
$U_{dc.pu} \gg f_{pu}$	$P = P_0 + K \cdot U_{dc.pu}$
$f_{pu} \to 0,\ U_{dc.pu} \to 0$	$P = P_0$

Fig. 3.4 Reference power of MIC in different situations

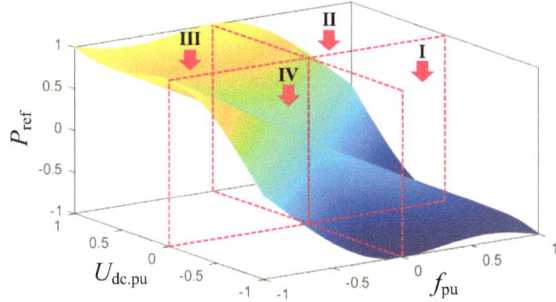

measured at the interface of the microgrid interlinking converter, this method does not require communication and is suitable for the plug-and-play requirements of microgrids.

When Δf is large, priority is given to maintaining f within the normal operating range through the $P - f$ droop relationship. When ΔU_{dc} is large, priority is given to maintaining U_{dc} within the normal operating range through the $P - U_{dc}$ droop relationship. This achieves adaptive adjustment of the deviation.

The active reference values of the MIC under different conditions are shown in Fig. 3.4. The equation for adaptive bidirectional droop control is a binary function, and its function graph is the characteristic surface of P_{ref} in the spatial coordinate system with respect to f_{pu} and $U_{dc.pu}$. Based on the relative size relationship between normalized frequency and voltage, the operating range can be divided into four quadrants.

(1) Quadrant I: $f_{pu} > 0$ and $U_{dc.pu} < 0$. In this case, there is an excess of power in the AC microgrid and a shortage of power in the DC microgrid. The interlinking converter operates in rectifier mode, providing power support from the AC microgrid to the DC microgrid.

(2) Quadrant II: $f_{pu} > 0$, $U_{dc.pu} > 0$. In this case, both the AC and DC sub-microgrids have power surplus. The interlinking converter operates in standby, rectifier, or inverter mode based on the relative magnitudes of the AC frequency and DC voltage, reducing overall deviation.

(3) Quadrant III: $f_{pu} < 0$, $U_{dc.pu} > 0$. In this case, there is a shortage of power in the AC sub-microgrid, while the power in the DC sub-microgrid is surplus. The interlinking converter operates in the inverter mode, providing power support from the DC sub-microgrid to the AC sub-microgrid.

(4) Quadrant IV: $f_{pu} < 0$, $U_{dc.pu} < 0$. In this case, both AC and DC microgrids have a power deficit. Interlinking converters operate in standby, rectifier, or inverter mode based on the relative sizes of the AC frequency and DC voltage, reducing the overall deviation.

The above analysis demonstrates that the proposed adaptive bidirectional droop control strategy can achieve bidirectional power flow based on the magnitude of the AC frequency and DC voltage deviation. Furthermore, when there is a conflict between AC frequency control and DC voltage control, it can prioritize compensating the side with larger deviation in order to reduce the overall deviation.

3.3.3 Start-Up Conditions

The real-time variation of DG and load within the hybrid microgrid can lead to small-scale fluctuations in active power, even when the system is operating stably. In order to prevent frequent operation of the interlinking converter during steady-state operation, a deadband is required as a start-up condition in practical applications.

By factoring Eq. (3.11), we obtain Eq. (3.12):

$$P = P_0 - K'\left(f_{pu} - U_{dc.pu}\right) \tag{3.12}$$

where K' is the adaptive bidirectional droop coefficient, which can be expressed as:

$$K' = \begin{cases} \frac{f_{pu}^2 + f_{pu} \cdot U_{dc.pu} + U_{dc.pu}^2}{f_{pu}^2 + U_{dc.pu}^2} \cdot K, & \left|f_{pu}\right| > \varepsilon \text{ or } \left|U_{dc.pu}\right| > \varepsilon \\ 0.5 \cdot K, & \left|f_{pu}\right| \leq \varepsilon \text{ and } \left|U_{dc.pu}\right| \leq \varepsilon \end{cases} \tag{3.13}$$

In Eq. (3.12), $(f_{pu} - U_{dc.pu})$ can be regarded as a new bidirectional droop control input variable, with a range of $[-2, 2]$. The magnitude of $(f_{pu} - U_{dc.pu})$ determines the active power regulation direction of the interlinking converter. The interlinking converter should remain in idle state under the following two conditions: (1) both the AC frequency and DC voltage are near their rated values. (2) there is power surplus or deficit in both the AC subgrid and DC subgrid (i.e., quadrant II or quadrant IV in Fig. 3.3), and the relative deviation of AC frequency and DC voltage is close in magnitude. These two conditions can be unified as $(f_{pu} - U_{dc.pu}) \approx 0$. Based on the above analysis, Eq. (3.11) can be improved as Eq. (3.14).

$$P = \begin{cases} -K' \cdot \frac{f_{pu} - U_{dc.pu} - D}{2 - D}, & f_{pu} - U_{dc.pu} > D \\ 0, & -D < f_{pu} - U_{dc.pu} < D \\ -K' \cdot \frac{f_{pu} - U_{dc.pu} + D}{2 - D}, & f_{pu} - U_{dc.pu} < -D \end{cases} \tag{3.14}$$

where D is the action deadband, and in this text, $D = \varepsilon = 0.05$. The characteristics of droop control before and after improvement are shown in Fig. 3.5.

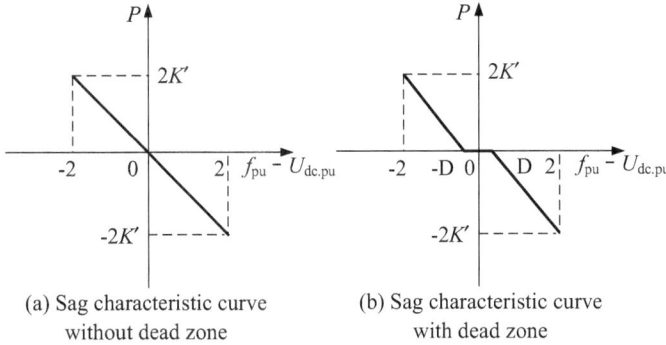

(a) Sag characteristic curve
without dead zone

(b) Sag characteristic curve
with dead zone

Fig. 3.5 Curve of droop control characteristic

The interlinking converter does not transfer power in the deadband and remains in standby mode. It only starts when the normalized deviation between the AC frequency and the DC voltage ($f_{pu} - U_{dc.pu}$) exceeds the deadband range, thus avoiding frequent device actions.

3.4 Small-Signal Modeling and Stability Analysis of Interlinking Converter

The adaptive bidirectional droop control strategy is based on the control of the inter-linking converter in a hybrid microgrid using AC frequency and DC voltage. Different droop coefficients K and weight coefficients λ may have an impact on the stability of the hybrid microgrid system, and only when the coefficients are within a reasonable range can the stable operation of the hybrid microgrid system be ensured. In this section, the influence of changes in the droop coefficient and weight coefficient values on the stability of the system is studied using eigenvalue analysis.

The adaptive bidirectional droop control strategy takes the normalized AC frequency (f_{pu}) and the DC voltage ($U_{dc.pu}$) as inputs, and the active power (P) of the interlinking converter as the output. The active power reference value is obtained by linearizing Eq. (3.8) as shown in Eq. (3.15).

$$\Delta P_{ref} = -\lambda \cdot K \cdot \Delta f_{pu} + (1 - \lambda) \cdot K \cdot \Delta U_{dc.pu} \qquad (3.15)$$

The power decoupling double closed-loop control of the microgrid interlinking converter is implemented in the dq coordinate system. The power outer loop adopts PI control, while the current inner loop has a much wider bandwidth than the power outer loop and can be simplified as an inertial element [26]. Its transfer function is shown in Eq. (3.16).

$$\begin{cases} G_{\text{PI}}(s) = k_p + \frac{k_i}{s} \\ G_c(s) = \frac{1}{1+T_s \cdot s} \end{cases} \tag{3.16}$$

where k_p and k_i are the proportional and integral coefficients respectively, and T_s is the switching period of the interlinking converter.

The active power angle relationship on the AC side of an interlinking converter is written as:

$$P = \frac{U_{\text{ac}} \cdot E}{X_L} \delta \tag{3.17}$$

where X_L is the total impedance between the AC output of the interlinking converter and the AC bus, and δ is the phase angle difference between the converter and the AC bus.

The relationship between the DC side power of the interlinking converter is:

$$P = U_{\text{dc}} \left(I_{\text{dc}} - C_{\text{dc}} \frac{dU_{\text{dc}}}{dt} \right) \tag{3.18}$$

The transfer functions for active power of the interlinking converter with respect to AC frequency and DC voltage, obtained by linearizing Eqs. (3.17) and (3.18), are shown in Eq. (3.19).

$$\begin{cases} G_f(s) = \frac{X_L \cdot s}{2\pi \cdot U_{\text{ac}} \cdot E} \\ G_{\text{Udc}}(s) = \frac{1}{I_{\text{dc}} - C_{\text{dc}} \cdot U_{\text{dc}} \cdot s} \end{cases} \tag{3.19}$$

According to Eqs. (3.15)–(3.19), the small signal model of the interlinking converter closed-loop system can be obtained, as shown in Fig. 3.6. The influence of the droop coefficient K and the weight coefficient λ on the distribution of system eigenvalues is studied separately, and the root locus is plotted as shown in Fig. 3.7.

Figure 3.7a shows the root locus plot as the droop coefficient K increases from 0 to 5, and Fig. 3.7b shows the root locus plot as the weighting coefficient λ increases from 0 to 1. The characteristic roots of the closed-loop system are all located in the left half-plane of the imaginary axis, indicating system stability under small disturbances.

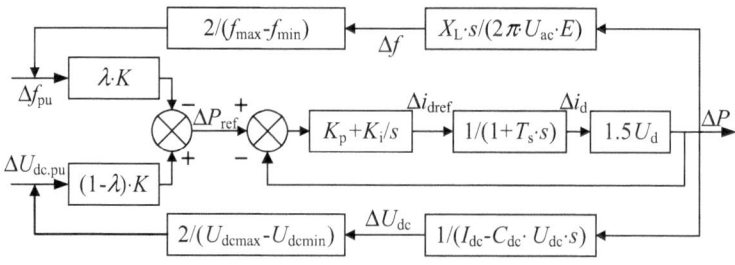

Fig. 3.6 Small-signal model of close-loop system

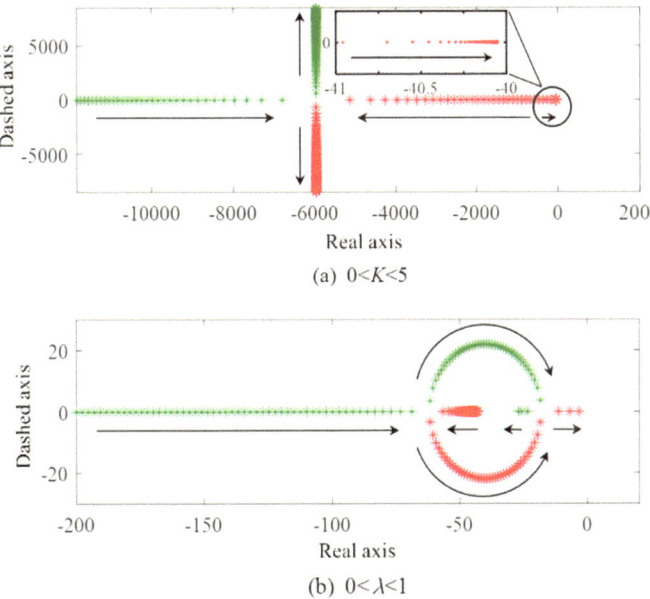

Fig. 3.7 Locus of eigenvalues for close-loop system

The distribution of dominant poles in the complex plane changes with variations in K and λ, thereby affecting the dynamic response characteristics. Based on the stability analysis results in this chapter and existing stability analysis conclusions for AC/DC microgrids [12], it can be concluded that within the specified range, the droop coefficient K and weighting coefficient λ can satisfy the stability requirements of the system during normal operation.

The weight coefficient λ of the adaptive bidirectional droop control strategy varies actively within the range of (0, 1), aiming to reduce the comprehensive deviation index and achieve better AC frequency and DC voltage balance while ensuring system stability. To prevent the interlinking converter from overloading, the value of the droop coefficient K should not exceed the maximum value of the interlinking converter capacity. At the same time, to fully utilize the bidirectional power regulation capability, the droop coefficient K can be chosen as the interlinking converter capacity S_{MIC}.

3.5 Simulations

A hybrid microgrid simulation model in PSCAD/EMTDC software is built as shown in Fig. 3.1. The AC microgrid consists of two controllable power sources DG1 and DG2 with *P-f* control, and one uncontrollable power source DG3 with *PQ* control.

Table 3.2 System parameters

Parameters	Numerical value
Rated AC voltage U_{ac0} (kV)	0.38
Rated AC frequency f_0 (Hz)	50
AC DG capacity S_{DG1}, S_{DG2}, S_{DG3} (MW)	0.55, 0.55, 0.2
AC load P_{load_AC} (MW)	0.6
Rated DC voltage U_{dc0} (kV)	0.7
DC DG capacity S_{DG4}, S_{DG5}, S_{DG6} (MW)	0.55, 0.55, 0.2
DC load P_{load_DC} (MW)	0.6
Interconverter capacity S_{MIC} (MW)	1.2
DC Capacitance C_{dc} (μF)	7800
Filter Inductors L (mL)	0.25
AC frequency upper and lower limits f_{max}, f_{min} (Hz)	50.5, 49.5
DC voltage upper and lower limits U_{dcmax}, U_{dcmin} (kV)	0.735, 0.665
Droop coefficient K (MW)	1.2
Motion deadband D (pu)	0.05

The DC microgrid consists of two controllable power sources DG4 and DG5 with P-U_{dc} control, and one uncontrollable power source DG6 with PQ control. The system parameters are shown in Table 3.2. The rated value of the AC and DC loads is 0.6 MW, and the initial active power for each DG is 0.3 MW with an initial output of 0.1 MW. The power is balanced within the AC and DC microgrids, and the power transfer through the interlinking converter is 0.

3.5.1 Scenario 1: Load Variation

To verify the effectiveness of the proposed adaptive bidirectional droop control strategy in responding to changes in AC and DC loads and maintaining stable AC frequency and DC voltage, different scenarios were set up including unilateral load changes, simultaneous load changes on both sides, etc. At the initial moment, DG1 ~ 6 were all put into operation, and both AC and DC loads are 0.3 MW.

(1) Scenario 1.1: Load Variation in One Side

The AC load increases to 0.6 MW at 2 s, the DC load also increases to 0.6 MW at 4 s, and the AC load returns to 0.3 MW at 6 s. The simulation results are shown in Fig. 3.8.

As shown in Fig. 3.8, at the beginning, the AC frequency and DC voltage are both close to their rated values, and the interlinking converter is in standby mode. At 2 s, the AC load increases, causing the AC frequency to drop, and the interlinking converter operates in the inverter mode. At 4 s, the DC load increases, causing the

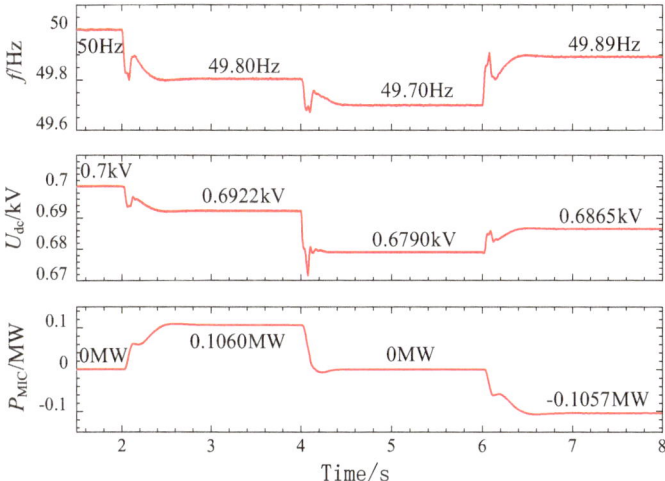

Fig. 3.8 Simulation results of scenario 1.1

DC voltage to drop. At this point, the power shortage in the AC and DC microgrid is equal, and adjusting the power on either side will increase the overall deviation value. The interlinking converter is in standby mode. At 6 s, the AC load recovers, but there is a power shortage on the DC side, and the interlinking converter operates in rectifier mode. The above results show that the proposed strategy can effectively respond to changes in AC and DC loads, balance the power between AC and DC, avoid excessive deviation in AC frequency or DC voltage, and keep them within the allowable range.

(2) Scenario 1.2: Bilateral Load Variation

The mixed microgrid experiences changes in both AC and DC loads, which may result in complementary or repulsive (reducing the frequency or voltage deviation on one side, while increasing the deviation on the other side) AC and DC power. Therefore, an interlinking converter is required to coordinate the bidirectional power based on the deviation values of AC frequency and DC voltage. Different scenarios of changes in AC and DC loads were simulated, and the steady-state values of AC frequency, DC voltage, and active power of the interlinking converter were recorded. The simulation results are shown in Table 3.3.

According to Table 3.3, an increase in load is positive and a decrease in load is negative. The active power of the interlinking converter is positive for inversion and negative for rectification. Therefore, the following conclusions can be drawn: (1) When the load of AC and DC changes in a complementary manner, the interlinking converter can provide support from the surplus side to the deficient side in controlling power. (2) When the load of AC and DC increases or decreases simultaneously, there is a contradiction between optimizing AC frequency and optimizing DC voltage. If the power change on both sides is equal, adjusting the power on either side will

Table 3.3 Simulation results of scenario 1-2

Operation status	$\Delta P_{\text{load\_AC}}$ (MW)	$\Delta P_{\text{load\_DC}}$ (MW)	f (Hz)	U_{dc} (kV)	P_{MIC} (MW)
Cross-straight balance	0	0	50.00	0.7000	0
Cross-direct complementary	+0.2	−0.2	49.90	0.7071	0.1067
	−0.2	+0.2	50.10	0.6931	−0.1032
Intersection and direct repulsion (ΔP_{load} equal)	+0.2	+0.2	49.80	0.6857	0
	−0.2	−0.2	50.20	0.7153	0
Intersection and direct repulsion (ΔP_{load} is not equal)	+0.3	+0.1	49.77	0.6879	0.0672
	+0.1	+0.3	49.83	0.6838	−0.0674
	−0.3	−0.1	50.24	0.7106	−0.0703
	−0.1	−0.3	50.18	0.7172	0.0793

increase the overall deviation, and the interlinking converter will operate in standby mode. (3) If the power change on AC and DC is not equal, the side with larger deviation is more likely to exceed the limit. In this case, the interlinking converter adaptively provides power support to the side with larger deviation to reduce the overall deviation. The above analysis shows that the proposed method can adaptively adjust the power and direction of the interlinking converter based on the relative deviation of AC frequency and DC voltage, ensuring that both AC frequency and DC voltage remain stable within the optimal range.

3.5.2   Scenario 2: DG Output Change or on/off Switch

The changes in DG output or the switching of DGs may cause power imbalance and changes in load capacity in a hybrid microgrid. In such scenarios, the strong microgrid should provide power support to the weak microgrid. To validate the effectiveness of the proposed method in this scenario, simulations were conducted for both uncontrollable DG output changes and controllable DG switching.

(1)   Scenario 2.1: Uncontrollable Output Variation of DG

At the initial moment, both AC and DC loads are at 0.3 MW. DG1 ~ 6 are all in operation, with non-controllable DG3 and DG6 generating 0.1 MW each. At 2 s, the output of AC DG3 increases to 0.2 MW, and at 4 s, the output of DC DG6 also increases to 0.2 MW. At 6 s, the output of AC DG3 returns to 0.1 MW. The simulation results are shown in Fig. 3.9.

As shown in Fig. 3.9, the response characteristics of uncontrollable DG are similar to load variations. However, since uncontrollable DGs are generally renewable energy sources, their output has fluctuating characteristics, leading to fluctuations in frequency and voltage. The adaptive bidirectional droop control strategy

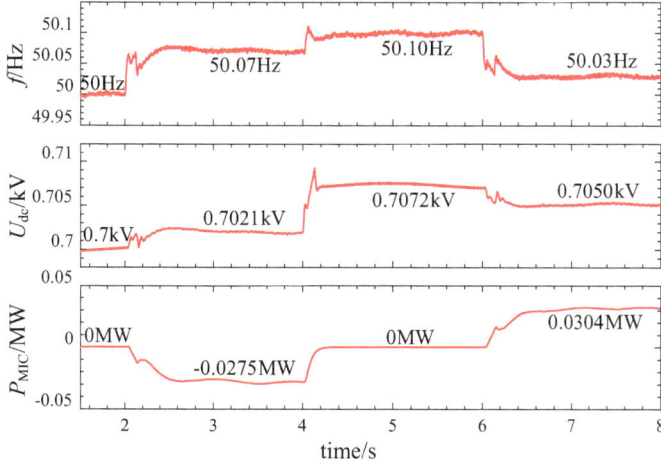

Fig. 3.9 Simulation results of scenario 2.1

can balance the uncontrollable DG power between the AC and DC sides, which is beneficial for improving the capacity of renewable energy integration.

(2)   Scenario 2.2: Adjustable DG Switching

At the initial moment, both the AC and DC loads are 0.3 MW, and controllable DG1, DG2, DG4, and DG5 are put into operation. At 2 s, DC DG5 is disconnected; at 4 s, AC DG2 is disconnected; at 6 s, DC DG5 is reconnected to the grid. The simulation results are shown in Fig. 3.10.

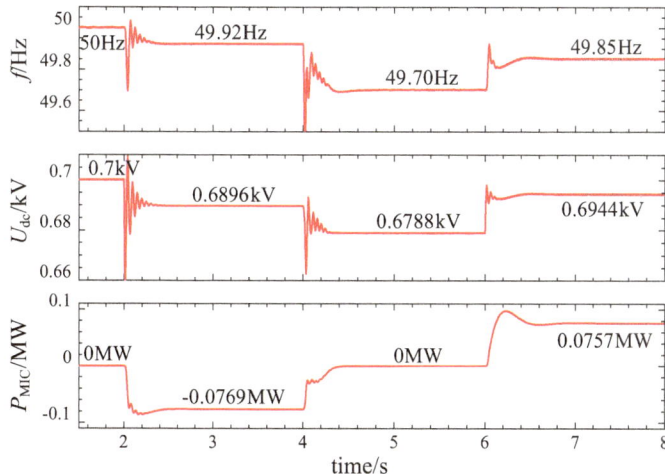

Fig. 3.10 Simulation results of scenario 2.2

As shown in Fig. 3.10, at 2 s, the voltage drops and the DC microgrid weakens due to the disconnection of DG5. At this time, the interlinking converter operates in rectifier mode. At 4 s, the AC DG2 is also disconnected, and the strength of the AC and DC microgrids decreases, as well as the power deficit. The interlinking converter switches to standby mode. At 6 s, the DC DG5 is reconnected, with a weak AC grid and a strong DC grid, and the interlinking converter operates in inverter mode. The above results indicate that the interlinking converter, using adaptive bidirectional droop control strategy, can meet the requirements of flexible integration of DG, improve the mutual support capability of AC and DC DG, and fully exert the power support function of AC and DC DG in hybrid microgrid.

The simulation results of scenario 1 and scenario 2 demonstrate that the proposed adaptive bidirectional droop control strategy can effectively regulate the bidirectional power between AC and DC microgrids in various operating scenarios, such as load variations, changes in distributed generation (DG) output, or switching. It balances the AC frequency and DC voltage deviation, and enhances the power regulation capability of the hybrid AC/DC microgrid.

3.6 RT-LAB Simulation Verification

The RT-LAB semi-physical simulation platform is built, as shown in Fig. 3.11, to verify the performance of the proposed adaptive bidirectional droop control strategy. The platform includes the RT-LAB OP7000 hardware simulated real-time simulator, the FPGA controller of XILINX FPGA ZYNQ7020, the AN706 controller sampling module with a maximum sampling frequency of 200 K, and the waveform measurement using Tektronix MSO44 oscilloscope.

A real-time simulator has been built in which a hybrid AC/DC microgrid and an interlinking converter model are included, with parameters as shown in Table 3.2. The control strategy is implemented in a hardware simulation using an FPGA controller. The interlinking converter voltage and current on both sides are sampled through the AN706 board connected to the RTLAB AO board. The adaptive bidirectional droop control is carried out in the FPGA controller, and the control signals for the interlinking converter switches are generated. The control signals are inputted through the RTLAB DI board connected to the AN706 board, forming a closed-loop simulation.

The effectiveness of the proposed adaptive bidirectional droop control strategy is verified through a semi-physical simulation in different strong and weak situations of AC/DC hybrid microgrids. The load capacity of the microgrid varies with the adjustable DG penetration level in AC/DC hybrid microgrids, and load changes in weak microgrids are more likely to cause significant deviations or even exceeding limits in frequency and DC voltage. In such cases, a stronger microgrid is required to provide more power support to a weaker microgrid. To verify the effectiveness of the proposed strategy in the above situations, the power response characteristics of the interlinking converter during load changes with different DG penetration levels

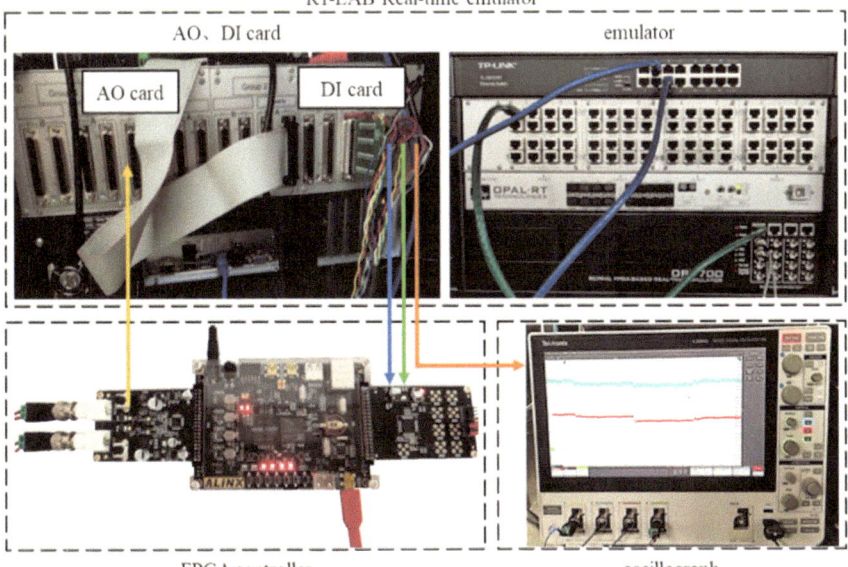

Fig. 3.11 RT-LAB hardware in loop simulation experiment platform

in AC/DC microgrids are simulated, and the proposed method is compared with the fixed coefficient bidirectional droop method.

3.6.1 Scenario 1: Weak AC Microgrid and Strong DC Microgrid

At the initial moment, the AC and DC loads are both 0.3 MW. Only DG1 is put into operation in the AC microgrid, while DG4 and DG5 are put into operation in the DC microgrid. The initial power of each DG has been set to achieve source-load balance within the AC/DC microgrid. At 2 s, the AC load increases to 0.6 MW, and at 6 s, the DC load also increases to 0.6 MW. The experimental results are shown in Fig. 3.12.

As shown in Fig. 3.12, the AC load increases between 2 and 6 s, and only DG1 in the AC microgrid has a weak load capacity. Adjusting it according to the traditional fixed coefficient method would result in a significant drop in AC frequency. However, using the adaptive coefficient control method proposed in this chapter can deliver more supporting power to the AC microgrid. Compared to the traditional method, the power transmission increases by 0.038 MW and the overall deviation decreases by 28.7%. After 6 s, both the AC and DC loads increase to 0.6 MW. Although the power deficit is equal between AC and DC, the AC microgrid is weaker and experiences a greater frequency drop. The method proposed in this chapter can adaptively provide

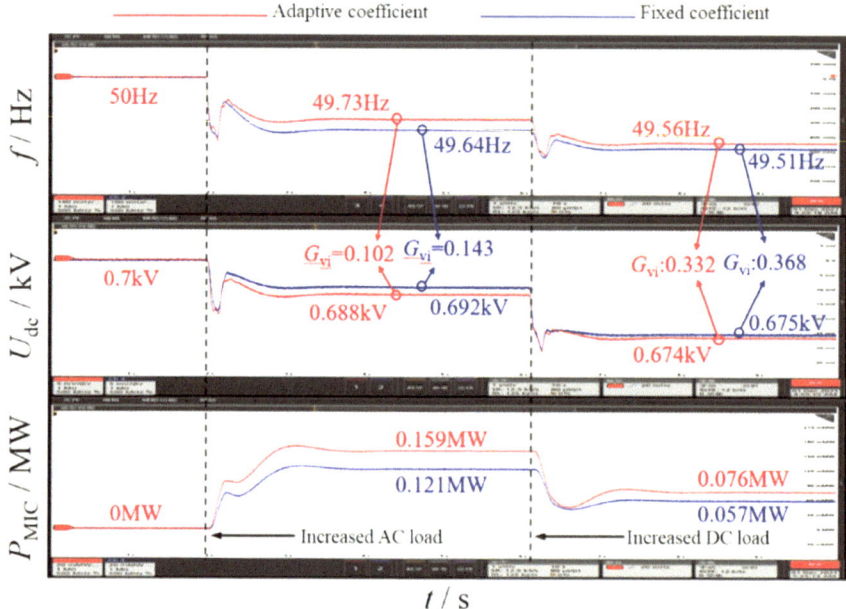

Fig. 3.12 Experimental results of scenario 1

more power support to the AC side. Compared to the traditional method, the power transmission increases by 0.019 MW and the overall deviation decreases by 9.8%.

3.6.2   Scenario 2: Strong AC Microgrid and Weak DC Microgrid

At the initial moment, the AC and DC loads were both 0.3 MW. AC microgrid DG1 and DG2 were put into operation, while only DG4 was put into operation in the DC microgrid. The initial power of each DG has been set to ensure source-load balance within the AC-DC microgrid. At 2 s, the DC load increased to 0.6 MW, and at 6 s, the AC load also increased to 0.6 MW. The experimental results are shown in Fig. 3.13.

As shown in Fig. 3.13, during the 2~6 s period, the DC load increases and only DG4 in the DC microgrid has a weaker load capacity. Adjusting it using the traditional fixed coefficient method would result in a significant voltage drop. However, by using the adaptive coefficient control method proposed in this chapter, more support power can be provided to the DC microgrid. Compared to the traditional method, the power transmission is increased by 0.035 MW and the overall deviation is reduced by 32.6%. After 6 s, both the AC and DC loads increase to 0.6 MW. Although the shortfall in AC and DC power is equal, the DC microgrid is weaker and experiences more voltage drop. The proposed method can adaptively provide more power support to the DC

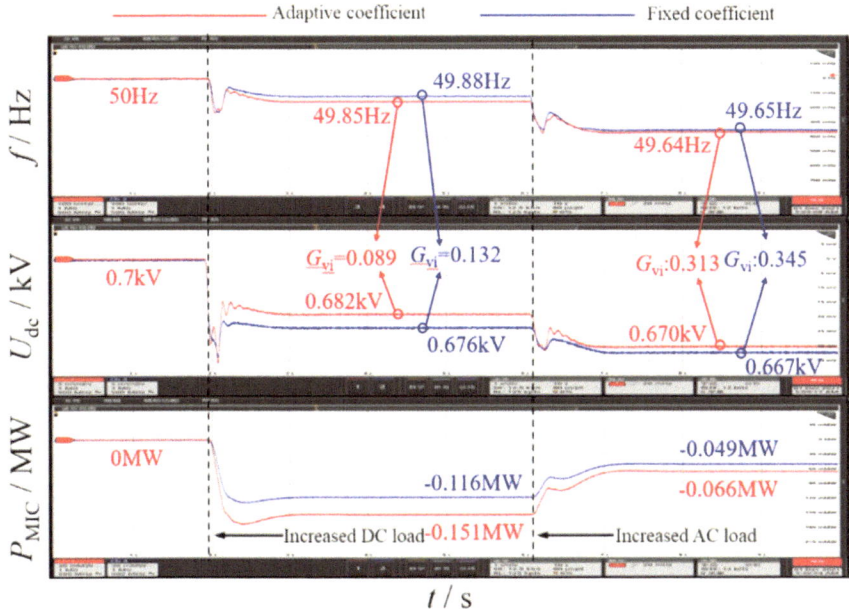

Fig. 3.13 Experimental results of scenario 2

side. Compared to the traditional method, the power transmission is increased by 0.017 MW and the overall deviation is reduced by 9.3%.

The results of the semi-physical simulation verification show that using the adaptive bidirectional droop control strategy can achieve better compensation effects for both AC frequency and DC voltage balance. Combining theoretical analysis and simulation, the results indicate that the proposed strategy can fully utilize the power mutual assistance capability between microgrids and achieve power optimization distribution under various operating states with source and load fluctuations in the AC/DC microgrid.

3.7 Conclusion

The interlinking converter is a core device for flexible interconnection between AC microgrids and DC microgrids, playing a key role in balancing power between AC and DC microgrids and improving frequency and voltage stability. In response to the problem that traditional bidirectional droop control strategies cannot dynamically adjust the priority of AC frequency control and DC voltage control, An adaptive bidirectional droop control strategy for the interlinking converter is proposed in this chapter, which dynamically adjusts the adaptive weighting coefficient based on the deviation of AC frequency and DC voltage to achieve a balanced compensation for AC frequency and DC voltage deviation, and effectively reducing the overall deviation index of the AC/DC microgrid. Through theoretical analysis, simulation, and experimental verification, the following main conclusions are drawn:

(1) When dealing with fluctuations in AC or DC loads, the proposed adaptive bidirectional droop control strategy enables bidirectional power flow between AC and DC side, fully leveraging the power mutual assistance capability between AC and DC microgrids and maintaining frequency and voltage within the optimal range.

(2) The proposed strategy does not require communication, making it convenient for plug-and-play of distributed energy sources and improving the flexibility of hybrid microgrid systems. By setting a deadband as the activation condition, frequent actions of the interlinking converter are avoided, thus the system reliability is enhanced.

(3) Compared to traditional bidirectional power droop control strategies, the proposed method exhibits better balance of AC frequency and DC voltage in hybrid microgrids with varying strengths. It also optimizes the control for the side with larger power imbalance, effectively improving the reliability and stability of the hybrid microgrid system.

References

1. Yang, X., Su, J., Lyu, Z., et al.: Overview on micro-grid technology. Proc. CSEE **34**(01), 57–70 (2014)
2. Huang, W., Wu, P., Tai, N., et al.: Architecture design and control method for flexible connected multiple microgrids based on hybrid unit of common coupling. Proc. CSEE **39**(12), 3499–3514 (2019)
3. Liu, Y., Chen, X., Li, B., et al.: State of art of the key technologies of multiple microgrids system. Power Syst. Technol. **44**(10), 3804–3820 (2020)
4. Sang, B., Zhang, T., Liu, Y., et al.: Energy management system research of multi-microgrid: a review. Proc. CSEE **40**(10), 3077–3093 (2020)
5. Yang, W., Miao, S., Zhang, S., et al.: Distributed autonomous economic control strategy for AC/DC hybrid microgrid cluster. Proc. CSEE **41**(03), 857–868 (2021)
6. Chen, J., Niu, B., Zhang, J., et al.: Hybrid control strategy for the seamless transfer of microgrids. Proc. CSEE **5**(17), 4379–4387 (2015)
7. Zhu, Y., Jia, L., Cai, B., et al.: Overview on topologies and basic control strategies for hybrid AC/DC microgrid. High Volt. Eng. **42**(09), 2756–2767 (2016)
8. Malik, S.M., Ai, X., Sun, Y., et al.: Voltage and frequency control strategies of hybrid AC/DC microgrid: a review. IET Gener. Transm. Distrib. **11**(2), 303–313 (2017)
9. Nejabatkhah, F., Li, Y.: Overview of power management strategies of hybrid AC/DC microgrid. IEEE Trans. Power Electron. **30**(12), 7072–7089 (2015)
10. Lan, Z., Tu, C., Xiao, F., et al.: The power control of power electronic transformer in hybrid AC-DC microgrid. Trans. China Electrotech. Soc. **30**(23), 50–57 (2015)
11. Yang, J., Jin, X., Yang, X., et al.: Overview on power control technologies in hybrid AC-DC microgrid. Power Syst. Technol. **41**(01), 29–39 (2017)
12. Li, X., Li, Z., Guo, L., et al.: Flexible control and stability analysis of AC/DC microgrids clusters. Proc. CSEE **39**(20), 5948–5961 (2019)

13. Tang, L., Zeng, C., Miao, H., et al.: One novel control strategy of the AC/DC bi-directional power converter in micro-grid. Power Syst. Protect. Contr. **41**(14), 13–18 (2013)
14. Zhou, W., Dai, Y., Bi, D., et al.: Coordinative control strategy for hybrid AC-DC microgrid. Electr. Power Autom. Equip. **35**(10), 51–57 (2015)
15. Wang, X., Sun, K., Li, Y.: Review of control strategies for bidirectional interfacing converters in hybrid AC/DC microgrid. J. Power Supp. **14**(02), 70–79 (2016)
16. Liu, J.: Research on Control Strategy of Bidirectional AC/DC Power Converter in AC/DC Hybrid Microgrid. PhD thesis, Taiyuan University of Technology, Taiyuan (2014)
17. Loh, P.C., Li, D., Chai, Y.K., et al.: Autonomous operation of hybrid microgrid with ac and dc subgrids. IEEE Trans. Power Electron. **28**(5), 2214–2223 (2013)
18. Xie, W., Zhu, Y., Du, S., et al.: Power control of interconnected converters in AC/DC hybrid microgrid. Electr. Power Constr. **37**(10), 9–15 (2016)
19. Eghtedarpour, N., Farjah, E.: Power control and management in a hybrid AC/DC microgrid. IEEE Trans. Smart Grid **5**(3), 1494–1505 (2014)
20. Shi, J., Li, Y., Wang, Z., et al.: Flexible power flow control strategy for interlinking converter in AC /DC hybrid microgrid. Electr. Power Autom. Equip. **38**(11), 107–113 (2018)
21. Liu, Z., Miao, S., Fan, Z., et al.: Power control and voltage fluctuation suppression strategy of the bidirectional AC/DC converter in the islanding hybrid microgrid. Proc. CSEE **39**(21), 6225–6238 (2019)
22. Zhu, J., Li, R., Wu, B., et al.: Virtual synchronous generator operation of interlinking converter between AC and DC microgrids. Power Syst. Protect. Contr. **45**(11), 28–34 (2017)
23. Li, F., Qin, W., Ren, C., et al.: Virtual synchronous motor control strategy for interfacing converter in hybrid AC/DC microgrid. Proc. CSEE **39**(13), 3776–3788 (2019)
24. He, L., Li, Y., Shuai, Z., et al.: A flexible power control strategy for hybrid AC/DC zones of shipboard power system with distributed energy storages. IEEE Trans. Industr. Inf. **14**(12), 5496–5508 (2018)
25. Zhou, Q., Shahidehpour, M., Li, Z., et al.: Two-layer control scheme for maintaining the frequency and the optimal economic operation of hybrid AC/DC microgrid. IEEE Trans. Power Syst. **34**(1), 64–75 (2019)
26. Nguyen, T., Yoo, H., Kim, H.: A droop frequency control for maintaining different frequency qualities in a stand-alone multimicrogrid system. IEEE Trans. Sustain. Energy **9**(2), 599–609 (2018)

Chapter 4
Flexible Connected Multiple Port Microgrids

4.1 Introduction

Port microgrid is an organic combination of the distributed generator (DG), energy storage, and load, with two modes of operation: grid-connected and islanded, and is one of the most important ways to effectively use renewable energy [1, 2]. Microgrids are positioned in medium and low-voltage distribution networks and support plug-and-play and seamless switching by efficiently coordinating the operation of microp-ower, energy storage, and load to provide self-control, protection, and management [3, 4]. The microgrid provides a platform for liaison and interaction between the grid, DG, energy storage, and load due to its flexible mode of operation and high dependability of power supply. With the ongoing implementation and promotion of new energy policies and the increasing penetration of DG, particularly renewable energy generation [5–7], it is difficult for a single microgrid to fully consume and effectively utilize a large quantity of dispersed connected DG [8].

As an extension of a single microgrid's structure and function, multiple microgrids (MMGs) can increase the DG consumption and control capacity by clustering the operation of numerous sub-microgrids [9], as well as the microgrid's reliability and economics. Because of this, the mode of microgrid building has evolved from a single small-scale demonstration to the clustering of numerous microgrids in locations with abundant distributed energy supplies [10, 11].

MMGs are combinations of sub-microgrids aggregated to increase the interconnected interactive coordination of sub-microgrids. The EU Framework Plan for Science and Technological Development FP6 pioneered the control architecture of MMGs, which consists of a distribution network management layer, a centralized control layer, and a microgrid control layer [9]. Literature [10] suggested an AC contact line-based interconnection strategy for MMGs to accomplish reactive power synergy during normal system operation and frequency control during islanded operation while enabling the sub-microgrid black start. Direct connection of several

© The Author(s) 2023
W. Huang et al., *Energy Management of Integrated Energy System in Large Ports*,
Springer Series on Naval Architecture, Marine Engineering, Shipbuilding and Shipping
18, https://doi.org/10.1007/978-981-99-8795-5_4

microgrids across stations, voltage levels, and control systems to form complementary interconnections is challenging to do using the AC interconnection approach [11].

MMGs coordinate many microgrids by complementing one another to achieve energy balance, mutual aid, and other operational objectives under various operating scenarios [12, 13]. Current research on MMGs focuses on control, optimization, and management in order to improve the reliability and economics of operation through the design and development of practical control and operation systems. Literature [14] reduces the control hierarchy of MMGs into two layers and employs the cooperative operation of several microgrids to enhance reliability by creating a standby capacity mechanism. Based on the peculiarities of the operation mode of MMGs, the literature [15] suggests an intertemporal strategy and real-time scheduling strategy to achieve the flexible and economical operation of the system. To meet the operational aims and needs of MMGs, the previous research presented coordinated operation solutions from various perspectives. Nevertheless, the sub-microgrids are connected to the grid in an AC synchronous manner and coordinate with one another in a uniform manner, which severely restricts the operational flexibility and economic potential of MMG. Actual operation of AC interconnection MMGs reveals that the sub-microgrids have a significant impact on one another and that the control flexibility is limited. In some cases, the parallel operation of multiple microgrids is unable to perform the originally specified regulation and control functions, and the islanding mode makes it difficult to operate in a stable manner due to microgrid disturbances or abnormalities.

The use of multi-terminal DC to construct flexible interconnection of sub-microgrids is a novel approach to structurally altering MMGs, and there are few pertinent research results and papers. ABB's Hrithik Majumder first proposed connecting various numbers and types of microgrids using voltage-sourced converters [16], but he did not elaborate on the synergy and interaction of MMGs. MD. Jahangir Hossain et al. recommended the use of multi-terminal flexible DC technology to regulate MMG islanding operating currents in order to increase system operating stability [17], but they did not take the flexibility of operating modes into account. The literature [18, 19] proposed using droop-controlled flexible converters to connect different areas of the distribution network to achieve reasonable power distribution and stable system operation; the literature [20] combined flexible DC transmission technology with distribution networks to establish a multi-terminated flexible DC distribution network across station areas; and the related studies provided technical ideas for the application of flexible DC in distribution networks. Currently, the cost of voltage source converters based on mature topology architecture is gradually decreasing, and it is increasingly used in soft switches, active filters, and low-voltage DC distribution networks. The investment cost and technical economy meet the requirements, and the technical economy is still improving [20]. MMGs use voltage source converters to establish interconnection structures, which not only improves the access and consumption capacity of new energy sources, but also reduces the investment cost and improves the technical economy.

To improve the flexibility and coordination of MMGs operation, this chapter first analyzes the typical connection mode of MMGs and then proposes a hybrid unit of common coupling (HUCC) for the microgrid to establish the flexible interconnection structure of MMGs; based on the combination of the switch, bidirectional converter, and power conditioning unit in HUCC, the connection mode of MMGs flexible interconnection system is proposed with a flexible interconnection string. The connection mode and control mode of flexible interconnection system are proposed based on the combination of the switch, bi-directional converter, and power regulation unit in HUCC; three operation modes of grid connection, islanding, and emergency and their switching methods are analyzed; the multi-layer control system of flexible connected MMGs (FCMMGs) is established; and the control methods of central layer, an interface layer, and a micro network layer are presented. The simulation model is built, and the simulation results indicate that FCMMGs and their control scheme have good operating performance and are an important approach to access and consume widely distributed power sources.

4.2 Typical Structure and Characteristics of MMGs

Microgrids offer the benefits of independence and autonomy, coordination and optimization with high efficiency and dependability, but their scale and operational capacity are determined by DG penetration, control methods, and the capacity of the primary power source [3, 4]. In regions with a significant number of DGs, several independently dispersed sub-microgrids serve as the foundation for resolving this technical issue, and the electrical connection between them is enhanced using a rational connection mechanism and organizational structure. MMGs generally use AC interconnection [7, 12] with the following three characteristics: (1) the presence of two or more adjacent sub-micro-networks; (2) the existence of energy and information interactions between sub-micro-networks; and (3) the ability of each sub-micro-network to operate independently or collaboratively to accomplish common goals.

Figure 4.1 depicts the usual configuration of a MMGs system based on AC connection, which consists of four independent sub-microgrids from MG1 to MG4; each sub-microgrid has complete control and protection devices [21] and can access different types of DGs, energy storage, and local loads, etc. MMGs provide the foundation for energy mutualization, frequency and voltage support, and cooperative operation between sub-microgrids.

As depicted in Fig. 4.1 According to the "source-feed" connection of access, the fundamental structure of MMGs can be categorized into one of three types.

(1) Same-source and same-feeder (SSSF) type: several microgrids are connected to the same feeder and interchange power with the same superior power source, such as MG1 and MG2;

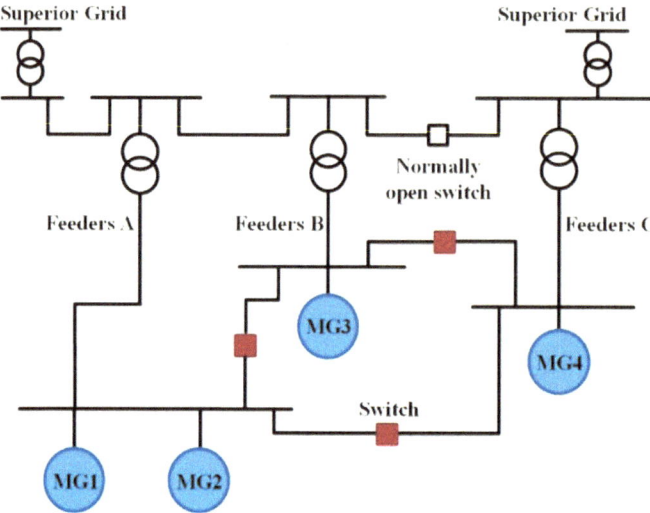

Fig. 4.1   A typical MMG structure and organization

(2)   Same-source and different-feeder (SSDF) type: many microgrids are connected
      to the grid by separate feeders, but are connected to the same power source, and
      the electrical connection between multiple microgrids is near, as in MG1 and
      MG3;
(3)   Different-source and different-feeder (DSDF) type: multiple microgrids are
      connected to the grid via different feeders and connected to different supe-
      rior power sources, and the superior power sources are typically in the unlisted
      state between them, and the electrical relationship between sub-microgrids is
      weak, as in MG1 and MG4.

Currently, MMGs generally establish interconnection relationships through
communication methods. The three types of structures, SSSF, SSDF, and DSDF, form
the basis of interconnecting MMGs, and the three basic structures can be switched
through a point of common coupling (PCC) according to operational needs. As
shown in Fig. 4.2, PCC, as a key node connecting MMGs, can adjust the topology
of MMGs by switching states, controlling the grid connection and disconnection
between microgrids and between MMGs and higher-level power grids. At the same
time, PCC can provide measurement information on electrical quantities and operational status, supporting the coordinated control and operation of a group of microgrids. As the number of interconnected microgrids continues to increase, the combination of basic structures and switching of PCC states enables MMGs to adapt to
complex operating environments.

MMGs feature significant electromagnetic coupling between AC interconnected
sub-microgrids, and there are technical issues with synchronous grid connection,
fault isolation, flexible operation, etc. MMGs achieve flexible connections by decoupling active and reactive components via voltage source converters and configuring a

Fig. 4.2 Structure and
function of microgrid PCC

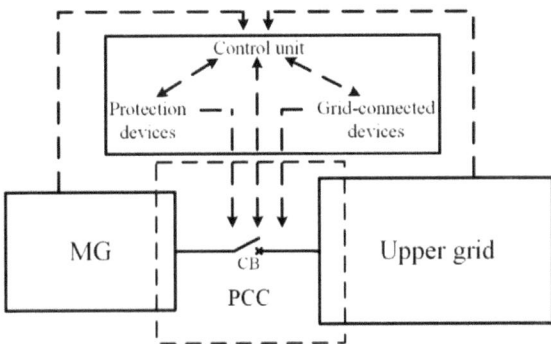

set amount of energy storage to form HUCCs with two types of grid connection interfaces, AC and DC. Compared with AC interconnection, the HUCC of MMGs can, on
the one hand, smooth out disturbances and quickly isolate faults, reduce the impact
of disturbances, and improve the operational stability of the system and individual
microgrid; on the other hand, through the decoupling control of active and reactive
components, it increases the control dimension of the system, has the capability of
asynchronous interconnection, meets the demand of multiple operation modes, and
reduces the impact of disturbances on the system and individual microgrid.

4.3 Flexible Interconnection of MMGs

4.3.1 HUCC Structure and Operation Mode

The flexibility and coordinated operation of MMGs are based on their interface
technology. This chapter proposes a hybrid point of common coupling (PCC) unit
that integrates both AC and DC connection methods. On the one hand, microgrids
are connected to the higher-level power grid through the AC interface, fully utilizing
the support function of the power grid and improving the frequency and voltage
stability of MMGs. On the other hand, flexible interconnection between microgrids
is achieved through the DC interface, fully leveraging the compatibility and control
capabilities of the DC unit, and improving the scalability and transient stability of
multi-microgrid systems.

HUCC is an organizational and control unit for interconnecting MMGs. As shown
in Fig. 4.3, it consists of AC and DC interfaces, power regulation units, control and
protection systems, etc. To ensure the reliability and flexibility of HUCC operation, a
three-bus structure is adopted for its primary wiring, where the common connection
bus serves as a bridge between the microgrid and AC/DC interfaces, responsible for
connecting the AC bus, DC bus, and power regulation unit. The AC and DC buses
provide external interfaces, with the AC interface typically used for connection to
the public grid and the DC interface used for interconnecting microgrids. To flexibly

combine the SSSF, SSDF, and DSDF structures and minimize the isolated fault area, corresponding types of circuit breakers are configured for the AC and DC interfaces, respectively.

As indicated in Fig. 4.3, the DC bus is connected to the common connection bus via the converter, and the voltage source type is preferred for the converter in order to increase the control dimension. HUCC can configure modular multi-level converter (MMC) to increase the efficiency and power quality of multi-microgrid operation due to MMC's small size, low loss, and high output waveform quality. The power regulation unit primarily employs an energy storage system (ESS) to optimize the regulation of sub-microgrid grid characteristics, such as power fluctuations, etc. To

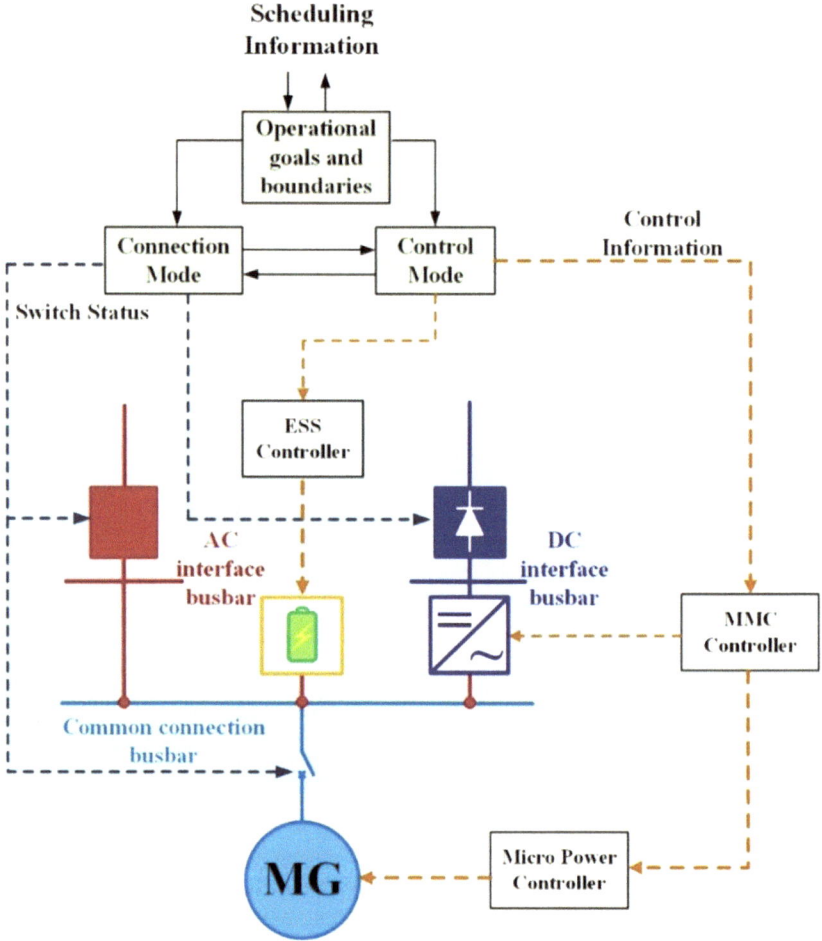

Fig. 4.3 Configuration fundamentals of the HUCC

increase the economics, the energy storage system can be constructed individually or by utilizing existing energy storage equipment in the microgrid.

As a unified interface for the flexible interconnection of MMGs, the HUCC relies on a coordinated combination of switching states, MMC control modes, and quick ESS response to perform its duties. Based on the actual operation of MMGs, the operation mode of the HUCC consists of two layers: connection mode and control mode, and the switching of the HUCC connection mode is based on the combination of switching states, which can be categorized into three types: AC mode, DC mode, and hybrid mode, with the following operating characteristics.

(1) AC mode: AC mode is used to connect sub-microgrids to the upper-level distribution networks, and the steady voltage and frequency of distribution networks is utilized to support MMGs and increase system operating stability.
(2) DC mode: Based on MMC to realize flexible interconnection between mutual mass sub-microgrids, MMC control mode and characteristics can effectively solve the technical problems of AC connection, such as power balance and electromagnetic coupling. This mode enables access to sub-microgrids and improves DG's consumption capacity.
(3) Hybrid mode: Hybrid mode is a combination of AC and DC modes with the benefits of 2 independent modes; also, AC and DC connections in hybrid mode can be backed up by each other, making hybrid mode the most prevalent operating mode of FCMMGs.

Matching the connection mode, HUCC employs the flexible control of MMC and ESS to enhance the operational stability and dependability of the system. Due to the limited capacity of individual MMCs in MMGs, the active power-DC voltage (P-U) droop control is used for the DC interface MMCs of controllable sub-microgrids to jointly participate in the voltage stability control of the interconnection system [24, 25]; for uncontrollable sub-microgrids (microgrids that cannot operate independently and require voltage and frequency support from the grid) then constant AC voltage/ frequency (v/f) control is used [26]. The ESS functions as a regulator for power balancing and fluctuation reduction and collaborates with the MMC to meet the multi-flexible microgrid's control objectives. The HUCC control mode is determined by the operating needs of MMGs and is subdivided as follows: P–U droop mode and v/f mode.

(1) *P–U* droop mode: MMC adopts *P–U* droop control, which can provide voltage support for DC-connected networks and accomplish flexible power dispatching, and ESS as an auxiliary smoothing fluctuation employs P–U mode.
(2) *v/f* mode: MMC utilizes fixed AC voltage and frequency control, ESS functions primarily for peak regulation and valley filling, and sub-microgrid HUCC without an adjustable power source often utilizes this mode.

As the core of flexible interconnection in MMGs, HUCC not only has structural compatibility but also has the ability to coordinate control and optimize operation.

Compared with the PCC interface, HUCC uses the DC interface to isolate the electro-magnetic coupling between different microgrids, supports asynchronous interconnection of multiple microgrids, and can provide power to passive systems. By decoupling the power components in the multi-microgrid system through MMC, HUCC achieves separate control of active and reactive power, improving the controllability and flexibility of the system operation.

4.3.2   Flexible Interconnection Solution for HUCC-Based MMGs

The AC interface is utilized to connect the microgrid to the higher-level grid, while the DC interface is used for the flexible connectivity of many sub-microgrids. The flexible interconnection of MMGs is based on the classic multi-microgrid structure, and the configuration and interconnection regulations of MMGs are established by the three types of fundamental units, SSSF, SSDF, and DSDF. The flexible interconnection rules of MMGs are outlined in the next section.

(1) Several sub-microgrids under the basic structure of SSSF have the same "source-feeder" characteristics, are interconnected by DC interfaces, and can be connected to the higher grid via AC interfaces, allowing for grid-connected or islanded operation;

(2) The connection rules of sub-microgrid under SSDF basic structure are the same as (1);

(3) The sub-microgrid under DSDF basic structure uses a DC interface to form a flexible interconnection on the one hand, and on the other hand, the AC interfaces of its sub-microgrid are all connected to the superior grid due to the access to different power sources.

Figure 4.4 depicts the FCMMGs developed in accordance with the above flexible interconnection rules, based on the usual construction of MMGs depicted in Fig. 4.1. MMGs have a more adaptable mode of operation due to their connectivity layout. During normal operation, FCMMGs can operate in two modes: grid-connected and islanded. In the event of an emergency, HUCC can be utilized to rapidly reconfigure the system in order to restore power to important locations and ensure power quality.

4.3.3   Operation Modes and Mode Switching of FCMMGs

(1)   Operation Modes of FCMMGs

FCMMGs offer versatile operation modes, the most common of which being grid-connected mode, islanding mode, and emergency mode. The HUCC connection and control modes corresponding to each mode are distinct. Take the MMGs depicted

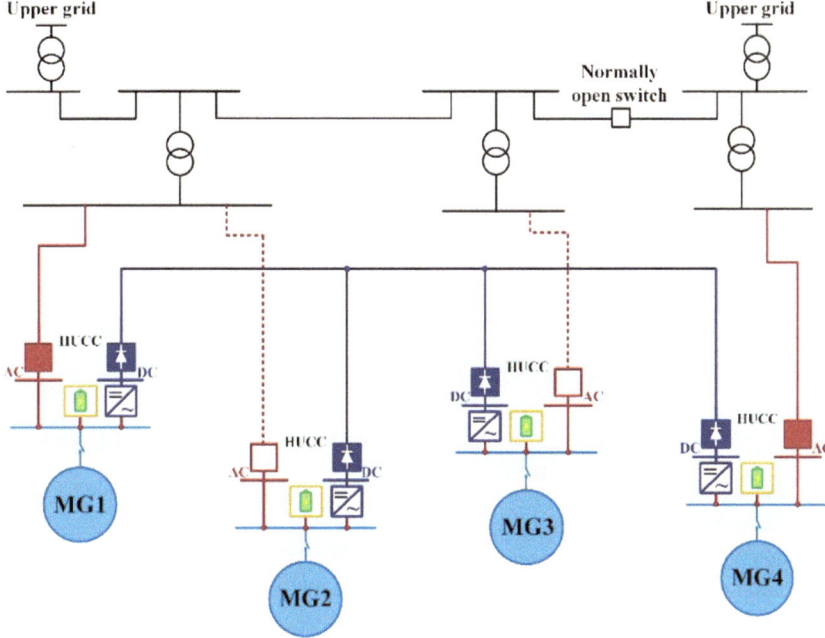

Fig. 4.4 Structure of HUCC-based FCMMGs

in Fig. 4.4 as an example, where MG1, MG2, and MG4 are controllable microgrids and MG3 is a non-controllable microgrid; Table 4.1 illustrates the associated HUCC states and control techniques for various operation modes.

Table 4.1 Operation modes of FCMMGs

Sub-microgrids	HUCC connection and control mode	Grid-connected operation	Islanding operation	Emergency operation
MG1	AC/DC switch	1/1	0/1	0/0
	MMC	P–U droop	P—U droop	/
	ESS	PQ	PQ	PQ
MG2	AC/DC switch	0/1	0/1	0/0
	MMC	P–U droop	P–U droop	/
	ESS	PQ	PQ	PQ
MG3	AC/DC switch	1/1	0/1	/
	MMC	v/f	v/f	/
	ESS	PQ	PQ	/
MG4	AC/DC switch	0/1	0/1	0/0
	MMC	P-U droop	P-U droop	/
	ESS	PQ	PQ	PQ

Note that in Table 4.1, 1 indicates the switch is closed, whereas 0 indicates the switch is open. To maximize DG utilization, FCMMGs typically operate in grid-connected mode, as depicted in Fig. 4.4 and Table 4.1, where the HUCC of MG1 and MG4 is in hybrid connection mode and the HUCC of MG2 and MG3 is in DC connection mode. MG1 and MG4 of DSDF exchange electricity with the superior grid via the AC interface, while relying on the superior grid to provide voltage and frequency support

When there is a defect or anomaly in the higher-level grid, HUCC swiftly modifies the connection mode and MMGs enter islanding operation. During islanding operation, the HUCC of all sub-microgrids adopts DC connection mode and disconnects the AC connection with the superior AC grid, the sub-microgrids operate in cooperation with one another to improve operational stability. The MMGs grid structure and control mode are essentially identical to grid-connected operation, as shown in Table 4.1.

MMGs have emergency operating modes such as black start and fault recovery, in addition to grid-connected and islanding operation modes. The structure of MMGs in emergency mode is highly variable, and each sub-microgrid can operate independently in the islanding state as depicted in Table 4.1, or in local interconnection states such as MG1 and MG2 interconnection, depending on operational requirements. Consequently, the control mode in emergency mode is also highly variable. Each sub-microgrid system is in a decentralized islanding state when the system fails and needs to be "black-started" by the microgrid. After all sub-microgrid DC connections have been completed, some HUCCs convert to hybrid connection and return to grid-connected mode, which can offer voltage and frequency support for the higher-level grid, as the power supply is progressively restored.

(2) **Operation Mode Switching of FCMMGs**

FCMMGs have three typical operation modes that can be changed based on operational objectives and restrictions when various types of system events occur. FCMMGs will switch between grid-connected mode, islanding mode, and emergency mode, as illustrated in Table 4.2. When there is a defect in the superior grid or low power quality, FCMMGs will disconnect from the superior grid and switch from grid-connected operation to islanding operation to ensure the safe and stable functioning of MMGs and good power quality. During the islanding operation, the power output of each sub-microgrid DG and energy storage system will be regulated to maintain the power balance of MMGs. Similarly, when a defect or low power quality in MMGs negatively impacts the higher-level grid, MMGs will convert to islanding operations to protect the higher-level system. When the aforementioned faults and anomalies have been rectified, FCMMGs will reconnect to the higher-level power grid and resume grid-connected operation.

Contingency MMGs extreme fault removal, switch to islanding, then switch to grid connection MMGs severe fault removal—switch to islanding, then switch to grid connection.

Table 4.2 Operation modes switch conditions of FCMMGs

Operation mode	Grid-connected operation	Islanding operation	Emergency
Grid-connected operation	/	Fault or power quality violation in the superior grid MMGs fault or power quality exceeds limit Higher-level grid maintenance, etc.	MMGs severe failure
Islanding operation	Fault or power quality violation in the superior grid MMGs fault or power quality exceeds limit Higher-level grid maintenance, etc.	/	MMGs severe failure
Emergency operation	MMGs severe fault excision, switch to isolated island first, then switch to grid connection	MMGs severe fault removal	/

In addition, when there is a significant fault in MMGs and the system cannot maintain grid-connected or islanded operation, MMGs will be delisted and each sub-microgrid will enter emergency operation mode. After the fault has been cleared, each isolated microgrid will be black-started, gradually restoring flexible interconnection operation and eventually returning to grid-connected or islanded operation mode.

Due to the strong adaptability of the control methods used in MMC and ESS in HUCC, and the use of peer-to-peer control in controllable microgrids, the control methods of HUCC and microgrids can autonomously adapt during the mode switching of FCMMGs. During the switching process, only the suppression of impacts needs to be considered to improve the stability of FCMMGs mode switching. The switch from grid-connected mode to islanded mode includes two types: planned switching and unplanned switching. Planned switching adjusts the control of HUCC and microgrids to exchange power with the higher-level power grid through the AC interface until the power exchange drops to zero, and then disconnects the connection when the voltage crosses zero, smoothly switching modes. Unplanned switching usually refers to abnormalities or faults in the higher-level distribution network or MMGs, where the AC interface circuit breaker isolates the fault. At this time, the MMGs system responds quickly and ensures power balance within the system by adjusting the microgrids and HUCC, improving the stability of MMGs switching from grid-connected to islanded mode. The switch from islanded mode to grid-connected mode is a planned switch. MMGs control the ESS of HUCC, adjust the voltage amplitude, phase angle, and frequency of the AC interface bus, and close the

AC circuit breaker when the grid connection conditions are met, smoothly switching to grid-connected mode.

4.4 HUCC-Based Control Strategies for FCMMGs

4.4.1 FCMMGs Control Architecture

The control approach of MMGs is crucial to their flexible connectivity and flexible functioning. Taking into account various operation scenarios, FCMMGs employ a three-layer design for coordination and control, as depicted in Fig. 4.5. The interface layer control uses the center layer command to determine the switching state, MMC control mode, and ESS control method, where MMC is used to meet the operation target under specific conditions.

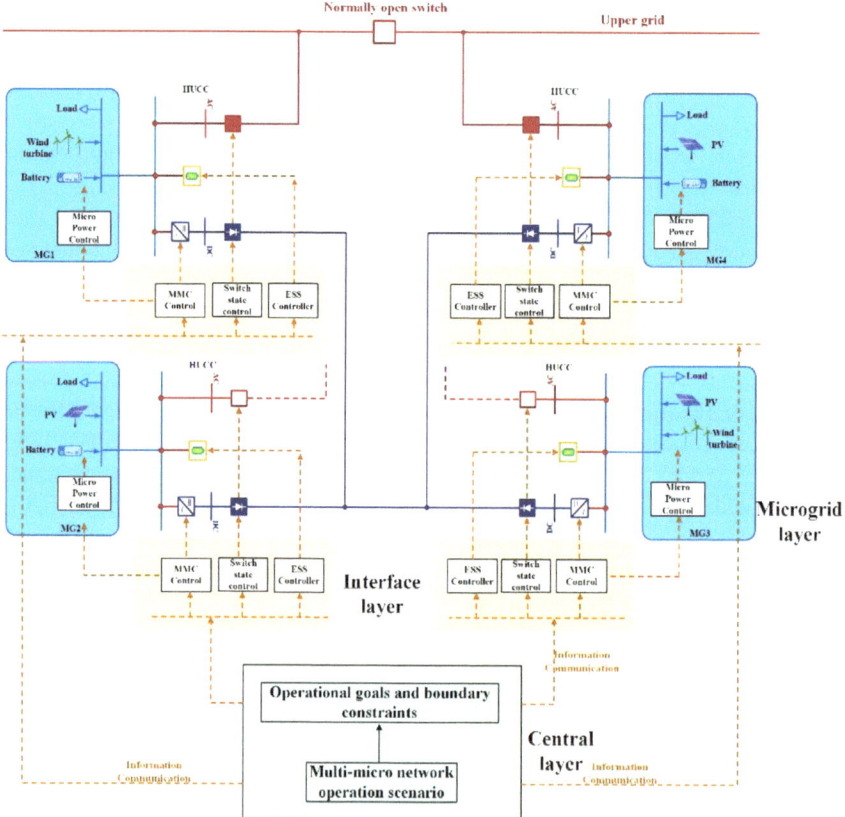

Fig. 4.5 Control architecture of HUCC-based MMGs

Figure 4.5 depicts the control system architecture and control information flow of FCMMGs. The central layer controller generates control targets and constraints based on operational scenarios and transmits control information to the interface layer and the microgrid layer; the interface layer consists of the switch controller, the MMC controller, and the ESS controller; the switch controller controls the HUCC connection mode and switches accordingly; the MMC controller primarily completes the cooperative operation of MMGs; and the ESS controller suppresses power outages.

After the flexible interconnection of MMGs, power exchange between microgrids is carried out in the form of DC, but the overall operation of MMGs needs to cooperate with the scheduling strategy of the higher-level AC distribution network. The distribution network scheduling system collects real-time information and predictive information of micro power sources and loads in FCMMGs through the central layer, mainly achieving short-term power balance and long-term energy management of MMGs. The former is mainly used to control the voltage and frequency within the system and the smooth switching of MMGs operating modes, while the latter mainly performs economic dispatch of microgrid groups to achieve optimized operation. The distribution scheduling system formulates scheduling strategies and issues them to the central layer of MMGs. The central layer combines operating constraints and goals to achieve power balance, stability, and economic operation.

4.4.2 Control Technique for the Central Layer of FCMMGs

(1) Grid-Connected Operation Mode

The goal of FCMMGs is to fully utilize and maximize the use of renewable energy under reliable and stable conditions. When FCMMGs are grid-connected, they use the voltage and frequency support provided by the higher-level distribution network, and optimize the distribution of sub-microgrid currents through HUCC to improve the utilization rate of DG. Therefore, the operating goals of FCMMGs can be summarized as:

1. Maximizing the consumption of renewable energy generation;
2. Flexibly controlling the active and reactive power between multiple microgrids, and optimizing the control of power distribution.

To achieve these operating goals, DG in each sub-microgrid operates at maximum power, the MMC in HUCC allocates excess power between sub-microgrids, and the ESS smoothes power fluctuations. The remaining power or power shortage of FCMMGs is consumed or supplemented by the higher-level distribution network, usually through local consumption of FCMMGs. When regulating the grid-connected operation of FCMMGs, the system should meet operating constraints to ensure stability, safety, and quality, and the grid-connected operation constraints of FCMMGs are shown in Eq. (4.1):

$$\begin{cases} \left| P_T^i \right| \le P_{T\,\max}^i \\ \left| P_C^j \right| \le P_{C\,\max}^j \\ \left| P_E^j \right| \le P_{E\,\max}^j \\ P_T^j + P_C^j - P_L^j + P_E^j + P_G^j = 0 \\ \left| \frac{f_{MG}^j - f_{MG}^{j\,*}}{f_{MG}^{j\,*}} \right| \le \varepsilon_1 \\ \left| \frac{U_{MG}^j - U_{MG}^{j\,*}}{U_{MG}^{j\,*}} \right| \le \varepsilon_2 \\ \left| \frac{U_{DC}^j - U_{DC}^{j\,*}}{U_{DC}^{j\,*}} \right| \le \varepsilon_3 \\ T_{THDMG}^j \le \delta \end{cases} \tag{4.1}$$

where the superscript $i \in [1, N]$, $j \in [1, M]$ denotes the ith AC feeder transformer and the jth microgrid, respectively; N denotes the total number of AC feeder transformers; M denotes the total number of microgrids; P_T^i and $P_{T\,\max}^i$ denotes the power and maximum power of AC feeder transformers; P_T^j is the grid-connected transformer of the jth microgrid, the positive direction of power flow to the microgrid; P_C^j and $P_{C\,\max}^j$ denotes the power and maximum power of MMC, the are the total harmonics limits of the sub-microgrid. The variables P_E^j and $P_{E\,\max}^j$ denotes the the power and maximum power of the ESS, with the positive direction indicating the energy storage discharge direction; P_G^j denotes the total power of DG in the microgrid. P_L^j is the total load of the microgrid; U_{DC}^j and $U_{DC}^{j\,*}$ denote the DC voltage and rated voltage of HUCC. f_{MG}^j and $f_{MG}^{j\,*}$ denote the frequency and rated frequency of the microgrid, respectively; U_{MG}^j and $U_{MG}^{j\,*}$ denote the AC voltage and rated voltage of HUCC, respectively; T_{THDMG}^j denotes the total distortion of the sub-microgrid j; $\varepsilon_1, \varepsilon_2, \varepsilon_3$ and δ represent the limits of frequency deviation, AC voltage deviation, DC voltage deviation, and total harmonic distortion of the sub-microgrid, respectively.

(2) Islanded Operation Mode

When FCMMGs operate in island mode, the higher-level distribution network no longer provides voltage and frequency support for MMGs. Therefore, the operating goals of FCMMGs are:

1. Stable operation of each sub-microgrid, including voltage stability and frequency stability;
2. Smoothing the impact of renewable energy output fluctuations;
3. Minimizing the power outage load in MMGs.

To achieve these operating goals, each sub-microgrid first meets the power demand of local loads and uses energy storage to balance the output fluctuations of renewable energy. Secondly, the central layer coordinates the MMC of the interface layer to dispatch the power between sub-microgrids and achieve complementary balance. The constraints of multi-microgrid system island operation are shown in Eq. (4.2), where the power of energy storage and MMC should not exceed the limit, and the DC voltage deviation and the power quality of each sub-microgrid should be controlled within the specified range.

$$
\begin{cases}
\left| P_C^j \right| \leq P_{C\,\text{max}}^j \\
\left| P_E^j \right| \leq P_{E\,\text{max}}^j \\
P_C^j - P_L^j + P_E^j + P_G^j = 0 \\
\left| \frac{U_{DC}^j - U_{DC}^{j\,*}}{U_{DC}^{j\,*}} \right| \leq \varepsilon_3 \\
T_{\text{THDMG}}^j \leq \delta
\end{cases}
\tag{4.2}
$$

(3) Contingency Operation Mode

In addition to normal operation mode, to improve the reliability and stability of the system power supply, FCMMGs should also have the ability to operate in emergency mode. The emergency operation mode of FCMMGs mainly refers to the process of black starting the system or gradually interconnecting sub-microgrids to restore power supply in the event of a serious fault. The overall goal of FCMMGs' emergency operation is to improve the transient stability of the multi-microgrid system, which can be divided into two aspects:

1. Increase the system inertia of the sub-microgrid that has been restored power supply;
2. Reduce transient impacts during the power restoration process.

To achieve these operating goals, the constraints of emergency operation should be satisfied as shown in Eq. (4.3):

$$
\begin{cases}
\left| P_C^j \right| \leq P_{C\,\text{max}}^j \\
\left| P_E^j \right| \leq P_{E\,\text{max}}^j \\
U_{DC'} < \sigma \\
T_{U_{DC}}^j \leq \Delta t \\
\left| \frac{f_{\text{NMG}}^j - f_{\text{NMG}}^{j\,*}}{f_{\text{NMG}}^{j\,*}} \right| \leq \varepsilon_1 \\
\left| \frac{U_{\text{NMG}}^j - U_{\text{NMG}}^{j\,*}}{U_{\text{NMG}}^{j\,*}} \right| \leq \varepsilon_2
\end{cases}
\tag{4.3}
$$

where the subscript NMG denotes the microgrid during emergency operation; $T_{U_{DC}}^j$ is the DC voltage recovery time; Δt denotes the permitted recovery time; $U_{DC'}$ denotes the DC voltage variation rate; σ denotes the DC voltage variation rate limit; and represents the DC voltage variation rate.

4.4.3 Control Strategies for the Interface Layer and Microgrid Layer of FCMMGs

FCMMGs' grid-connected, islanded, and emergency operation modes are controlled through the interface layer and microgrid layer of HUCC and sub-microgrids based on the goals and constraints formulated by the central layer. As shown in Fig. 4.6, there are interface layer and microgrid layer control strategies for controllable and uncontrollable sub-microgrids. The interface layer receives instructions from the central layer and selects the HUCC connection status and operation mode according to the strategy shown in Table 4.1. The DC interface MMC of the controllable sub-microgrid adopts $P–U$ droop control, while the DC interface MMC of the uncontrollable sub-microgrid adopts v/f control. The ESS in the sub-microgrid HUCC adopts a unified PQ control strategy.

As a key component in the control of FCMMGs, the microgrid layer is the regulatory basis for the interdependent energy production, storage, and mutual assistance of the system. The controllable sub-microgrid adopts a peer-to-peer control strategy, where controllable micro-sources have equal status in control and participate in voltage/frequency regulation. Controllable micro-sources use $P–f$ and $Q–U$ droop control methods, and intermittent micro-sources use PQ control as shown in Fig. 4.6a. The droop control used in the peer-to-peer control mode can automatically participate in power distribution and is easy to implement plug-and-play and mode switching for microgrids. The uncontrollable sub-microgrid is supported by the MMC for voltage and frequency, and intermittent micro-sources in the sub-microgrid use PQ control as shown in Fig. 4.6b.

Compared with grid-connected and islanding modes, emergency mode control is relatively complex and involves structural changes and control strategy adjustments. During emergency operation, the controllable sub-microgrid operates independently, and the MMC in HUCC does not work, while the ESS participates in power adjustment as needed. During black start-up, the controllable micro-sources first start to restore the islanded operation of the controllable sub-microgrid, then start the MMC to raise its DC voltage to the set value, establish connection relationships among multiple sub-microgrids, and increase the output power to the set value, restoring MMGs to islanded operation. Finally, based on the status of the superior distribution network, the system can restore grid-connected operation or provide black start-up power for the upper-level network.

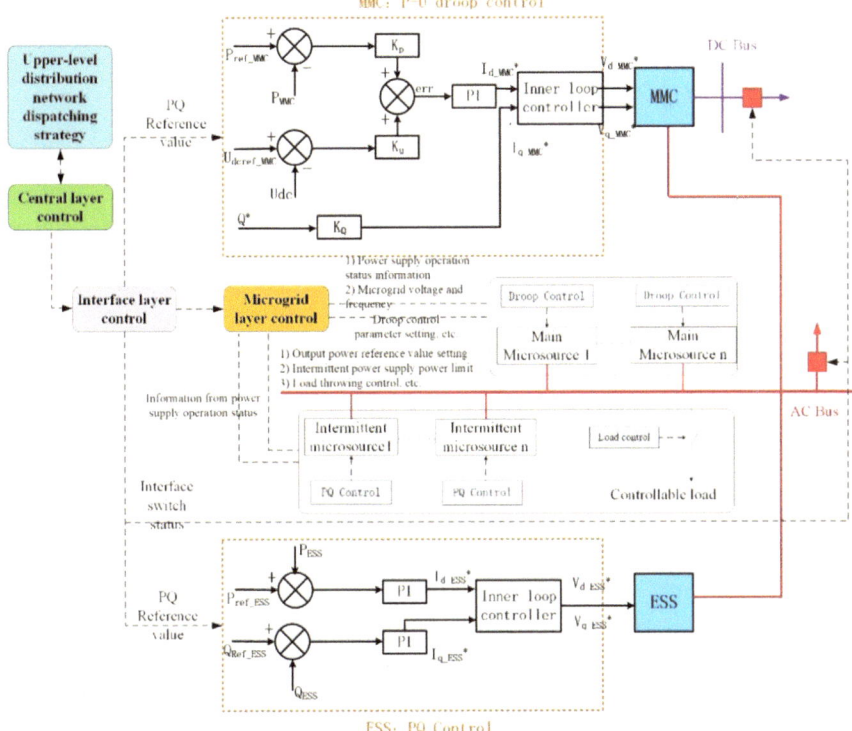

(a) Controllable sub-microgrid (with controllable micro-power)

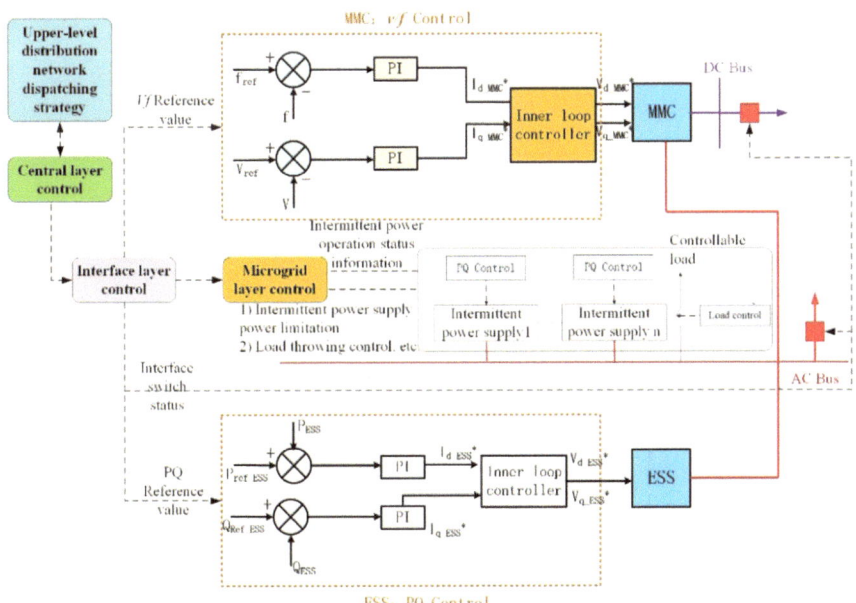

(b) An unmanageable sub-microgrid (with intermittent micro-power only)

Fig. 4.6 FCMMG control scheme for the interface layer and microgrid layer

4.5   Simulations

4.5.1   System Architecture and Settings for MMGs

The FCMMGs simulation system shown in Fig. 4.7 is based on the multi-microgrid demonstration project in Rizhao Port. To fully verify the multi-scenario operation capability of MMGs, the simulation system includes four sub-microgrids, where MG1, MG2, and MG4 are controllable sub-microgrids, and MG3 is an uncontrollable sub-microgrid. The HUCC, AC interface switch, DC interface switch, ESS, and MMC corresponding to MG1 to MG4 are HUCC1 to HUCC4, ACB1 to ACB4, DCB1 to DCB4, ESS1 to ESS4, and MMC1 to MMC4, respectively.

MMGs form a flexible interconnection structure through HUCC. The AC voltage at the grid connection point for MG1 and MG3 is 10 kV, while for MG2 and MG4, it is 10.5 kV. The frequency of the sub-microgrid is 50 Hz, and the MMC in HUCC uses 21 levels with a DC voltage of $\pm$ 2.5 kV and a capacity of 1 MW. The DG type, power, and load parameters for each sub-microgrid are shown in Table 4.3.

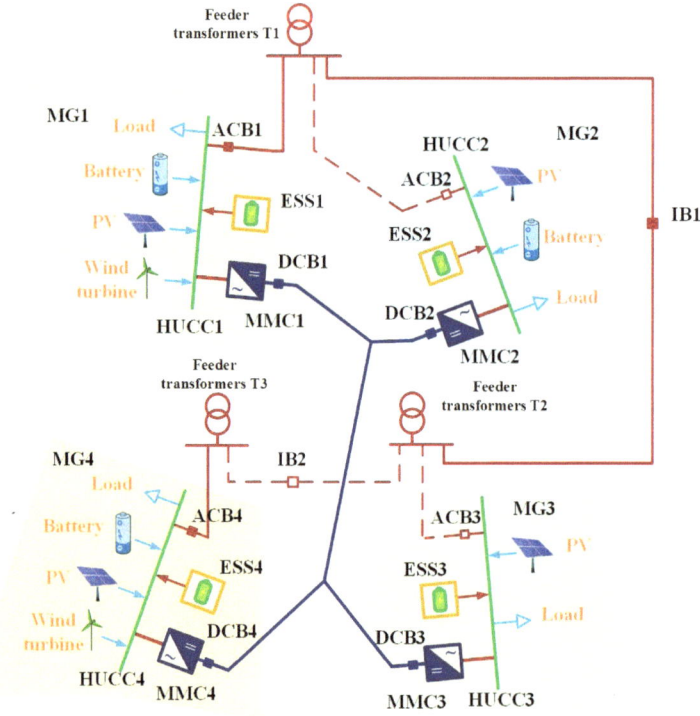

Fig. 4.7 FCMMG simulation model

Table 4.3 Configuration parameters of the simulation model

Sub microgrid	Micro power (MW)			Load	
	PV	Wind power	Battery	Load capacity (MW)	Power factor
MG1	0.2	0.1	0.1	0.90	0.9
MG2	0.4	0.3	0.2	0.40	0.9
MG3	0.3	/	/	0.55	0.9
MG4	0.3	0.2	0.2	0.40	0.9

MG1, MG2, and MG4 have the ability to operate in both grid-connected and islanded modes, using peer-to-peer control. Controllable micro-sources operate under droop control, while intermittent micro-sources (wind and solar) operate under PQ control. MG3 does not have the ability to operate independently, and its photovoltaic system uses PQ control. The MMCs in HUCC1, HUCC2, and HUCC4 operate under P-U droop control, while HUCC3 uses v/f control. The ESS in HUCC uses PQ control to mainly suppress power fluctuations.

Based on PSCAD/EMTDC, a simulation model of FCMMGs is established for steady-state, transient, and emergency operation analysis. According to the IEEE 519-1992 and IEC 61000-2-2 standards, the operating boundary constraint parameters of MMGs are 2%, 10%, 10%, 5%, and 0.5 s for $\varepsilon_1, \varepsilon_2, \varepsilon_3$ and δ respectively.

4.5.2 Simulation Results

(1) Steady-State Operation Simulation of FCMMGs System

As shown in Fig. 4.7, during the parallel operation of FCMMGs, MG1 and MG4 are connected to the upstream distribution network through AC interfaces, while MG2 and MG3 are flexibly interconnected through DC interfaces with MG1 and MG4. The control methods of DG, energy storage and MMC in each sub-microgrid are shown in Table 4.1. The energy storage systems in MG1, MG2 and MG4 can flexibly operate to absorb (compensate for) the surplus (shortage) power of the FCMMGs system and maintain the power balance of the system. Distributed new energy generation (wind and solar) in MG2 and MG4 operates at maximum power, and in addition to meeting their own load demand, can also transmit surplus power to the connected sub-microgrid and upstream distribution network through HUCC. To improve the operational efficiency, each sub-microgrid balances reactive power locally, so the reactive component of MMC can be set to 0, and the active component is matched according to the power difference.

As shown in Fig. 4.8, the simulation results of the parallel operation of FCMMGs show that the system operates stably at the rated working state from 0.6 to 1 s, and the DC voltage is stable and the output power is constant, with the power exchange between FCMMGs and the upstream distribution network (AC connection) being basically 0. During the stable operation, the maximum deviation of the DC voltage is 3.1%; MG1 and MG4 are supported by the upstream distribution network for voltage and frequency, with small deviations during operation; The frequency and voltage deviations of MG2 and MG3 are relatively large, with maximum values of 0.2% and 0.3%, respectively.

During the islanded operation of FCMMGs, MG1 and MG4 are disconnected from the AC connection and flexibly interconnected through DC interfaces. As shown in Fig. 4.9, the simulation results of islanded operation show that MG1 to MG4 operate stably, with stable DC voltage and effective support formed between sub-microgrids, and flexible power mutual assistance. Due to the lack of support from the upstream distribution network, the maximum deviation of the DC voltage is 3.9%, and the maximum deviation of frequency and voltage between sub-microgrids becomes 0.5% and 0.7%, respectively.

(2) **Transient Operation Simulation of FCMMGs System**

To verify the transient operation performance of the flexible interconnection structure and control method of MMGs, the simulation system shown in Fig. 4.7 is used to simulate two scenarios to verify the transient operation capability of the system during power adjustment, and to compare and analyze the simulation results with those of MMGs system using AC interconnection.

Scenario 1: Parallel operation mode, MG1 to MG4 operate in parallel

During the parallel operation of FCMMGs, the system structure is complete, and MG1 to MG4 are all in parallel operation. At 0.7 s, the new energy output of MG2 decreases by 0.2 MW, and its DC output power is adjusted from 0.5 to 0.3 MW, resulting in a power shortage in the system. Due to the $P–U$ droop control of the MMC in FCMMGs, the DC voltage of the system will experience a brief drop process, and at the same time, the MMC of each sub-microgrid will automatically adjust the DC output power until the power on the DC interconnection line reaches a new equilibrium point. As shown in Fig. 4.10, the voltage and power regulation and response curves of FCMMGs during the system adjustment process show that during the system adjustment process, the DC voltage quickly reaches a new stable working point after a brief decrease and remains stable, and the DC output power of each sub-microgrid is also quickly adjusted to a new stable working point according to their respective droop control curve parameters, while the voltage and frequency of each sub-microgrid remain basically stable. The entire transient adjustment process of FCMMGs is relatively short, with a stable time of DC voltage and DC output power of 0.12 s and an power variation overshoot of 25%.

To verify the transient operational performance of FCMMGs and for the sake of generality, a simulation case of interconnected AC MMGs was designed for comparison. The AC MMGs have the same parameters as the FCMMGs, but their interconnected structure and controllable elements are different. The simulation was conducted for 0.7 s under the scenario where the output power of MG4 renewable energy source increased by 0.2 MW. Interconnected AC MMGs refer to a cluster of

Fig. 4.8 Simulation results
of grid-connected operation

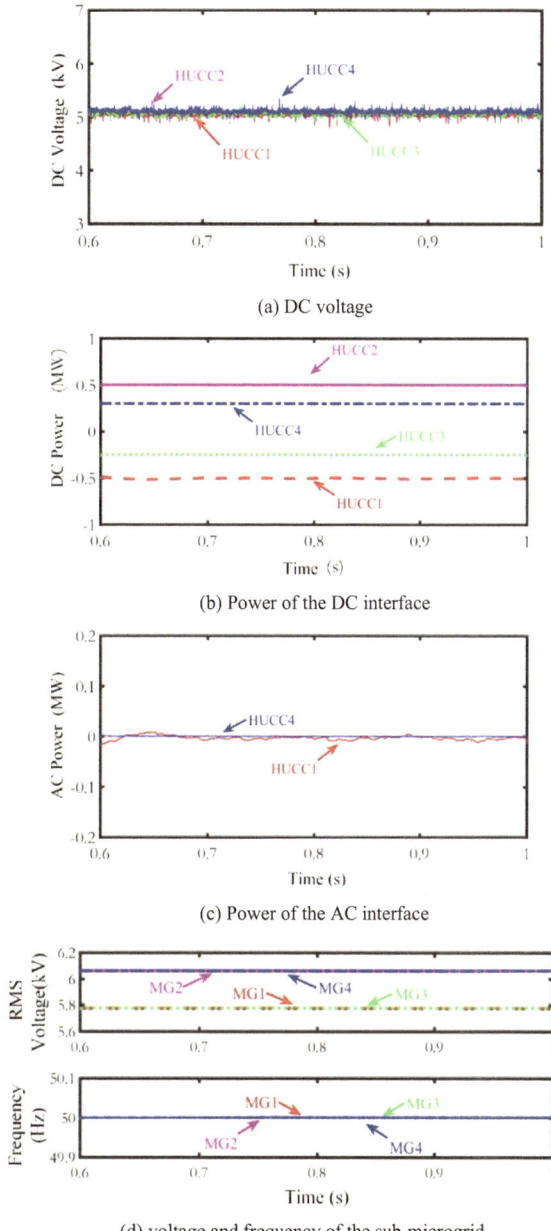

(a) DC voltage

(b) Power of the DC interface

(c) Power of the AC interface

(d) voltage and frequency of the sub-microgrid

Fig. 4.9 Simulation results
during islanded operation

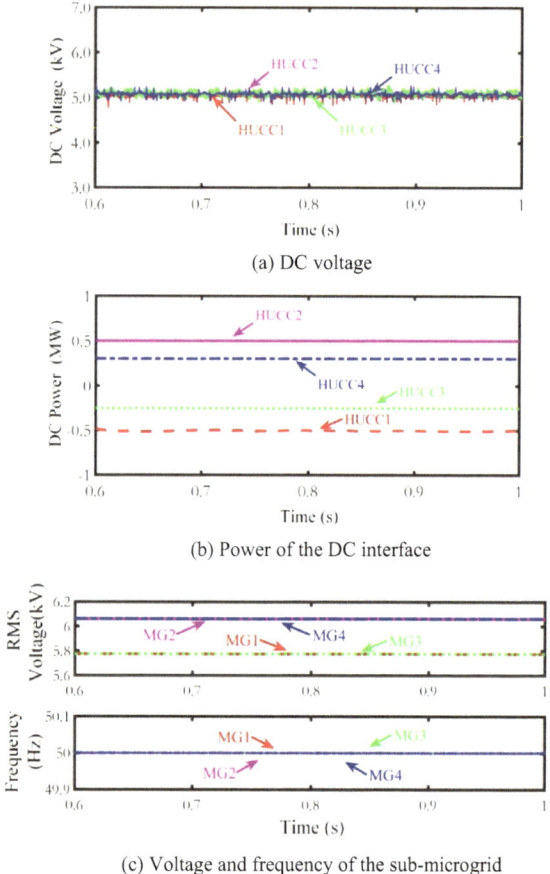

(a) DC voltage

(b) Power of the DC interface

(c) Voltage and frequency of the sub-microgrid

microgrids that are interconnected with each other to form a community system, as
described in literature [10]. Based on the design methods of interconnected AC microgrids in literatures [8] and [11], the DCBs 1-4 of the system shown in Fig. 4.7 were
disconnected and the ACBs 1-4, IB1, and IB2 were closed to form an interconnected
AC microgrid.

Figure 4.11 shows the power adjustment response curves of the FCMMGs and
the interconnected AC MMGs. In the adjustment process of the FCMMGs, due to
the excess DC power in the system, the DC voltage gradually increased and rapidly
reached a new steady-state operating point, which remained stable. The DC output
power of the sub-microgrid also quickly reached a new steady-state operating point
through autonomous adjustment, as shown in Fig. 4.11a. The entire transient adjustment process of the FCMMGs was relatively short, with a stabilization time of 0.09 s
for the DC voltage and DC output power, and an overshoot of 23% in power change.
As shown in Fig. 4.11b, the transient adjustment process of the interconnected AC
microgrid was relatively long, with a stabilization time of about 0.66 s, an overshoot

Fig. 4.10 Voltage and
power curves of FCMMGs
during MG2 power adjusting

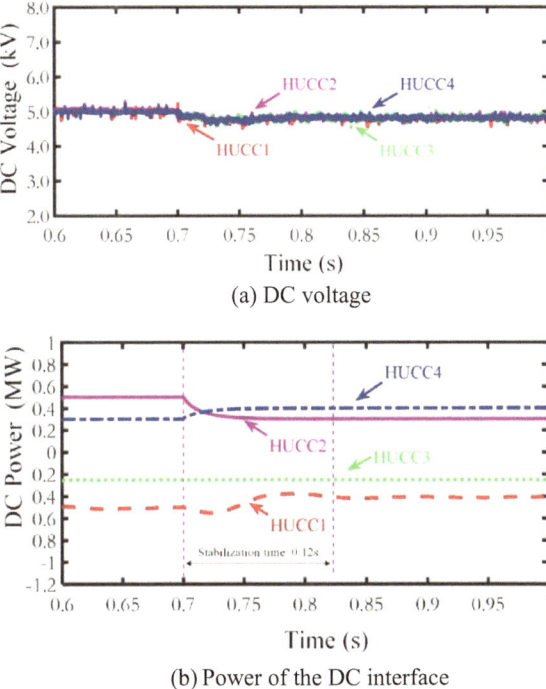

(a) DC voltage

(b) Power of the DC interface

of 30% in power change, and oscillations during the adjustment process. Due to the
increase in DG output power, the voltage at the internal node of MG4 increased,
and the active power of the load increased. After stabilization, the power of the AC
interconnection line was 0.497 MW. The comparison shows that the flexible interconnection method has better transient characteristics for power adjustment in scenarios
with fluctuating renewable energy generation under grid-connected mode.

Based on the above simulation and analysis, the change in renewable energy output
of the sub-microgrid leads to a change in DC output power. The DC voltage and DC
output power of each sub-microgrid will be autonomously adjusted according to their
respective MMC droop control curve parameters. Compared with the interconnected
AC MMGs, the FCMMGs have faster adjustment speed, shorter transient process,
and higher stability.

Scenario 2: Islanding mode, MG4 out of operation

Considering the change in operating mode to islanding and the structural adjustment
of the MMGs (MG4 out of operation), the transient operational performance of
the flexible interconnection and interconnected AC MMGs was compared through
simulation. In the islanding operation of the FCMMGs, when MG2 increased its
power output by 0.1 MW at 0.7 s, the DC voltage, HUCC interface power, and AC
voltage and frequency change curves of the sub-microgrid are shown in Fig. 4.12. It
can be seen that the system operates stably during power adjustment, with values of

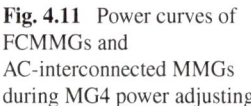

Fig. 4.11 Power curves of FCMMGs and AC-interconnected MMGs during MG4 power adjusting

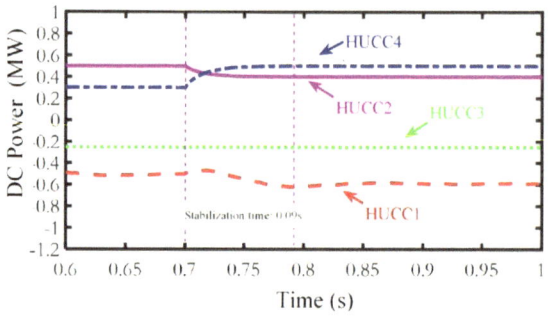

(a) DC power interface of FCMMGs

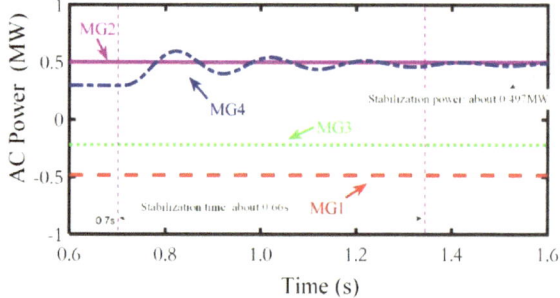

(b) Power of the tie-line for AC-interconnected MMGs

ε_1, ε_2 and ε_3 are 0.07%, 1.5%, and 5.8% respectively, and the transient adjustment time is about 0.1 s.

Figure 4.13 shows the voltage, frequency, and interconnection line power of the interconnected AC MMGs during power adjustment in islanding mode with MG4 out of operation. During the power adjustment process, the voltage and frequency of the interconnected AC MMGs continue to fluctuate, and when stabilized, the frequency deviation is 0.12% and the voltage deviation is 5.5%, both of which are larger than those of the FCMMGs.

Compared with the interconnected AC MMGs, the flexible interconnection MMGs can balance the system power and suppress disturbances through the control of MMC, energy storage, and sub-microgrid during transient operation. The operation control of FCMGs is more flexible and stable in grid-connected mode, islanding mode, and network structure changes compared to the control of AC interconnection MMGs.

(3) Emergency Operation Simulation of FCMMGs System

To verify the emergency operation ability of FCMMGs, a severe fault was simulated by disconnecting the HUCC interconnection and DC switches of MG1 to MG4. MG1, MG2, and MG4 were placed in isolated island operation mode, while MG3 was out of service. After the fault was cleared, MG1, MG2, and MG4 resumed DC

Fig. 4.12 Power adjusting curves during gird-connected operation of FCMMGs

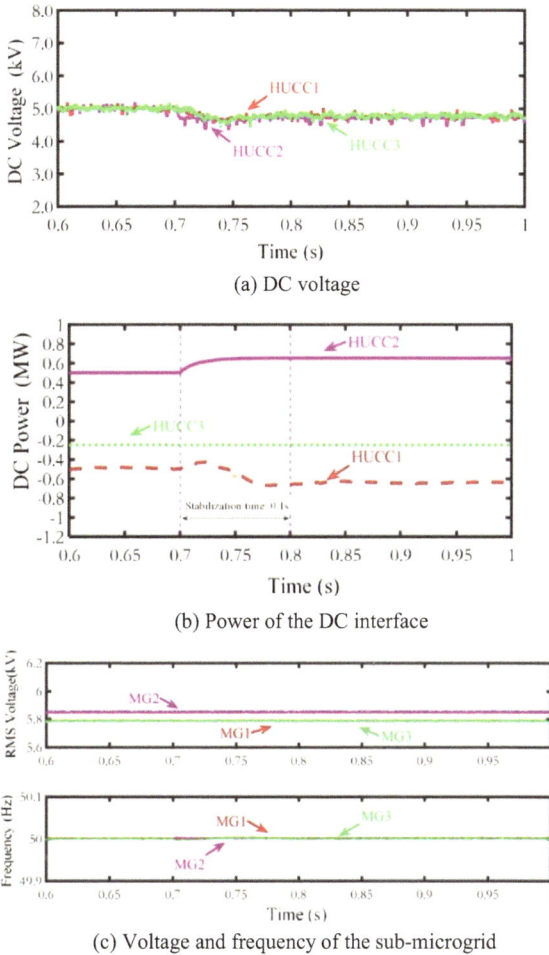

(a) DC voltage

(b) Power of the DC interface

(c) Voltage and frequency of the sub-microgrid

connection, and MG3 was reconnected to the grid and restored power supply after 0.7 s, as shown in Fig. 4.14.

As shown in Fig. 4.14, when MG3 is not connected to the grid, FCMMGs operate at a new equilibrium point with stable DC voltage and output power after MG1, MG2, and MG4 are connected in DC. When MG3 is reconnected to the grid after 0.7 s, the DC voltage and power of the system will undergo transient adjustment due to the P-U droop control of the MMCs in MG1, MG2, and MG4. As shown in Fig. 4.14b, after MG3 is connected to the multi-microgrid, the DC voltage of the system drops slightly, and the DC output power of each microgrid increases, and MG3 gradually restores power supply.

During the emergency operation of FCMMGs, the transient process is stable, and after autonomous adjustment, it can quickly stabilize to a new equilibrium point. The indicators of FCMMGs' emergency operation are shown in Table 4.4, and the

Fig. 4.13 Power adjusting
Curves during
gird-connected operation of
AC-connected MMGs

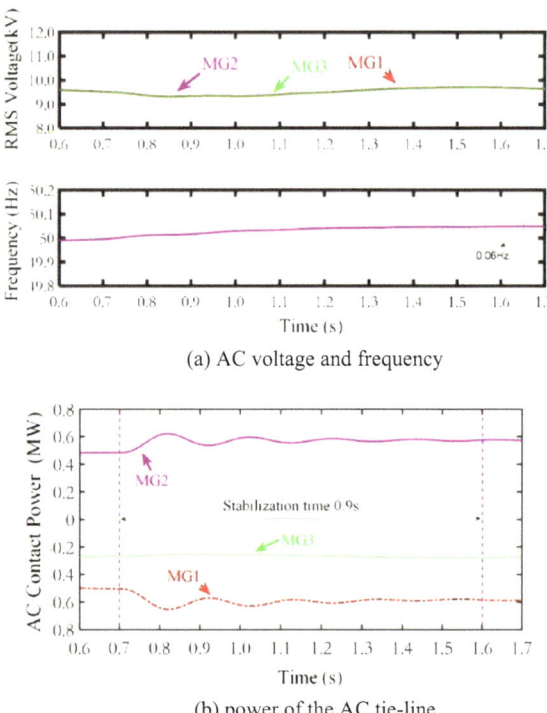

(a) AC voltage and frequency

(b) power of the AC tie-line

frequency deviation, AC voltage deviation, DC voltage adjustment time, and total voltage harmonic are all better than the requirements.

Fig. 4.14 Simulation results during emergency operation

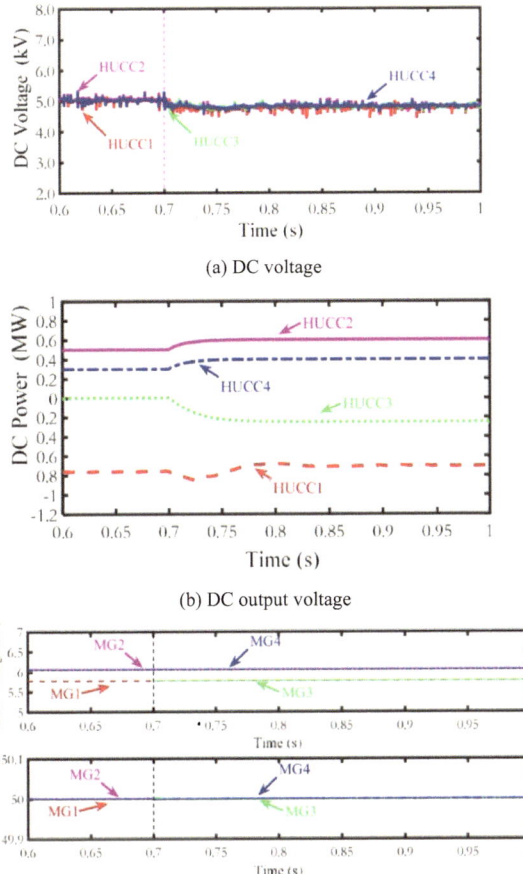

(a) DC voltage

(b) DC output voltage

(c) Voltage and frequency of the sub-microgrid

Table 4.4 Operation indexes during emergency operation

Sub microgrid	Frequency deviation maximum (%)	AC voltage deviation maximum (%)	DC voltage adjustment time (s)	Total harmonics of voltage (%)
MG1	0.9	1.1	0.15 s	1.2
MG2	1.2	0.9		1.0
MG3	1.0	1.0		1.4
MG4	1.0	1.2		0.9

4.6   Conclusion

This chapter systematically studied the basic structure, unified interface, flexible interconnection scheme, and control system and methods of multi-microgrids. A hybrid public connection unit structure with integrated AC and DC interfaces was designed, and a flexible interconnection multi-microgrid system scheme was proposed, which expanded the diversity and flexibility of multi-microgrid interconnection. A multi-layer control system for flexible interconnection multi-microgrids was constructed, and control methods for the central layer, interface layer, and microgrid layer were proposed. The control methods fully considered the operating characteristics of each sub-microgrid and distributed power source, achieved the coordinated control of the hybrid public connection unit and sub-microgrid, and completed the operational objectives and constraints formulated by the central layer.

Based on the hybrid public connection unit, the flexible interconnection multi-microgrid integrates AC and DC connection modes. On the one hand, it fully utilizes the frequency and voltage support function of the AC interface to improve the operational stability of the multi-microgrid system. On the other hand, by connecting each sub-microgrid through the DC interface, it reduces the coupling effect between sub-microgrids, improves the scalability and stability of the multi-microgrid system, and fully utilizes the compatibility and control ability of the DC unit to expand the control dimension, thereby improving the flexibility and coordination of multi-microgrid operation.

The simulation results of the flexible interconnection multi-microgrid in port show that the proposed flexible interconnection scheme and control methods meet the operational requirements in different scenarios and have good operational performance. The hybrid public connection unit is suitable for connecting multi-microgrid groups containing high-penetration distributed power sources and requiring high operational flexibility, and the flexible interconnection multi-microgrid formed by it is a feasible way to scale up the consumption of distributed new energy. With the continuous development of the application technology of voltage source converters in distribution networks, the application of flexible interconnection multi-microgrids not only improves the stability, flexibility, and reliability of the system but also optimizes the technical and economic performance of the system. It is a supplement and extension to the application research of existing multi-terminal DC technology in distribution networks.

References

1. Huang, W., Tai, N., Yang, X.: Inverse-time low-impedance protection scheme for microgrids. Proc. CSEE **34**(1), 105–114 (2014)
2. Nisar, A., Thomas, M.S.: Comprehensive control for microgrid autonomous operation with demand response. IEEE Trans. Smart Grid **8**(5), 2081–2089 (2017)

3. Huang, W., Tai, N., Fan, C., et al.: Study on structure characteristics and designing of microgrid. Power Syst. Protect. Contr. **40**(18), 149–155 (2012)
4. Guo Li, Fu.X., Li, X., et al.: Coordinated control of battery storage and diesel generators in isolated AC microgrid systems. Proc. CSEE **32**(25), 70–78 (2012)
5. Jianhua, B., Songxu, X., Jun, L., et al.: Roadmap of realizing the high penetration renewable energy in China. Proc. CSEE **35**(14), 3699–3705 (2015)
6. Yu, Y., Luan, W.: Smart grid review. Proc. CSEE **29**(34), 1–8 (2009)
7. Zhou, N., Jin, M., Wang, Q., et al.: Hierarchical coordination control strategy for multi-microgrid system with series and parallel structure. Autom. Electr. Power Syst. **37**(12), 13–18 (2013)
8. Zhao, M., Chen, Y., Shen, C., et al.: Characteristic analysis of multi-microgrids and a pilot project design. Power Syst. Technol. **39**(6), 1469–1476 (2015)
9. European Commission: Advanced Architecture and Control Concepts for More Microgrids. Nuno Melo, Microgrids Project Consortium (2009)
10. Hatziargyriou, N.: Microgrids: Architectures and Control. John Wiley and Sons Ltd., United Kingdom, pp. 165–203 (2014)
11. Pei, W., Du, Y., Li, H., et al.: Novel solution and key technology of interconnection and interaction for large scale microgrid cluster integration. High Volt. Eng. **41**(10), 3193–3203 (2015)
12. Wang, S., Wu, Z., Yuan, S., et al.: Method of multi-objective optimal dispatching for regional multi-microgrid system. Proc. CSU-EPSA **29**(5), 14–20 (2017)
13. Xie, M., Ji, X., Ke, S., et al.: Autonomous optimized economic dispatch of active distribution power system with multi-microgrids based on analytical target cascading theory. Proc. CSEE **37**(17), 4911–4921 (2017)
14. Lyu, T., Ai, Q., Sun, S., et al.: Behavioural analysis and optimal operation of active distribution system with multi-microgrids. Proc. CSEE **36**(1), 122–132 (2016)
15. Jiang, R., Qiu, X., Li, D., et al.: Economic operation of smart distribution network containing multimicrogrids and energy storage system. Power Syst. Technol. **37**(12), 3596–3602 (2013)
16. Majumder, R.: Bag G (2014) Parallel operation of converter interfaced multiple microgrids. Int. J. Electr. Power Energy Syst. **55**, 486–496 (2014)
17. Hossain, M.J., Mahmud, M.A., Milano, F., et al.: Design of robust distributed control for interconnected microgrids. IEEE Trans. Smart Grid **7**(6), 2724–2735 (2016)
18. Peyghami, S., Mokhtari, H., Loh, P.C., et al.: Distributed primary and secondary power sharing in a droop-controlled LVDC microgrid with merged AC and DC characteristics. IEEE Trans. Smart Grid **9**(3), 2284–2294 (2018)
19. Lu, X., Guerrero, J.M., Sun, K., et al.: An improved droop control method for DC microgrids based on low bandwidth communication with DC bus voltage restoration and enhanced current sharing accuracy. IEEE Trans. Power Electron. **29**(4), 1800–1812 (2014)
20. Wu, J., Wu, D., Zhu, J., et al.: Grounding method design of multi–terminal flexible DC distribution. Proc. CSEE **37**(9), 2551–2561 (2017)
21. Yuen, C., Oudalov, A., Timbus, A.: The provision of frequency control reserves from multiple microgrids. IEEE Trans. Industr. Electron. **58**(1), 173–183 (2011)
22. Wu, P., Huang, W., Tai, N., et al.: A novel design of architecture and control for multiple microgrids with hybrid AC/DC connection. Appl. Energy **210**, 1002–1016 (2018)
23. Ge, L., Lu, W., Yuan, X., et al.: Back-to-back VSC-HVDC based loop-closed optimal operation for active distribution network. Autom. Electr. Power Syst. **41**(6), 135–141 (2017)
24. Pei, W., Deng, W., Zhang, X., et al.: Potential of using multiterminal LVDC to improve plug-in electric vehicle integration in an existing distribution network. IEEE Trans. Industr. Electron. **62**(5), 3101–3111 (2015)
25. Wu, J., Wu, D., Zhu, J., et al.: Grounding method design of multi-terminal flexible DC distribution. Proc. CSEE **37**(9), 2551–2561 (2017)
26. Cai, X., Zhao, C., Pan, H., et al.: Control and protection strategies for MMC-HVDC supplying passive networks. Proc. CSEE **34**(3), 405–414 (2014)

Chapter 5
Smooth Control Strategy for Port-Ship Islanding/Grid-Connected Mode Switching

5.1 Introduction

As an important distribution center for ship berthing, the smooth access of ships to shore power (cold ironing) affects the stable operation of the port microgrid. Ship shore power technology uses shore power instead of ship auxiliary machinery to provide all the electricity needed during the ship's stay in the port, thereby controlling air pollution in the port area [1, 2]. The ship shore power integrated power supply system operates in an asynchronous interconnection manner with multiple terminal power sources. Therefore, the grid connection process between the ship's power grid and shore power could bring significant power fluctuations, which may affect the normal operation of electrical equipment on both sides of the ship and even damage the equipment, leading to grid connection failure, posing great challenges to the safe, reliable, and economical operation of the port microgrid [3–6].

The flexible interconnection of the port microgrid and ship microgrid, using flexible multi-state switches (FMS), provides a new approach for safe, reliable, and fast grid connection and disconnection of ships. On one hand, the port microgrid itself, using flexible control technology, can optimize the distribution and balance of energy to a certain extent, enabling bidirectional power flow with the ship microgrid [7]. On the other hand, the FMS, as a fully controllable device with a back-to-back structure, can achieve real-time regulation of the voltage phase angle for grid connection and disconnection, enhancing the operability of ship access. Additionally, the application of FMS improves the charging and discharging capabilities and power balance of the port-ship interconnected microgrid, further enhancing the coordination and interaction between ships and the port microgrid, enabling reliable access for different types of ships, especially electric ships. This technology has already been applied in coastal ports in the East China Sea.

© The Author(s) 2023
W. Huang et al., *Energy Management of Integrated Energy System in Large Ports*,
Springer Series on Naval Architecture, Marine Engineering, Shipbuilding and Shipping
18, https://doi.org/10.1007/978-981-99-8795-5_5

The coordinated control of FMS is crucial for improving the operational performance of the port microgrid. During transient fluctuations or shocks [8], it is necessary to explore the coordinated control between FMS and the port-ship interconnected microgrid to ensure smooth switching and continuous stable operation of ship access to shore power. FMS is an important link for ship access to shore power and power exchange [9–13], and the transient process is influenced by both. When the ship microgrid and the port microgrid are interconnected, the operating mode of FMS needs to be urgently switched. Considering the transient impact of the connection time on the system, FMS should switch from P-Q control to U_{ac}-f control in the shortest possible time, providing stable and continuous voltage and frequency support for the ship microgrid to achieve load transfer [14].

The detection and communication delays are involved in the emergency mode switching process, which affects the frequency, voltage stability, and FMS control switching [15–17]. During the delay, FMS remains in the P-Q mode. If there is an imbalance between the power supply and load, it can cause significant changes in frequency and voltage [15]. Due to the impact of the switching delay, the reference phase of FMS control switching may experience sudden changes, causing significant frequency shocks. Additionally, the output signal of FMS control undergoes sudden changes before and after the delay, causing significant voltage shocks. The aforementioned switching fluctuations and shocks can lead to the failure of ship access to shore power and, in severe cases, even system instability. Therefore, smooth control of the mode switching process becomes a key technical challenge for the flexible interconnection of the port microgrid and ship microgrid.

Traditional islanding/grid-connected mode switching has been extensively studied [18–21]. The most commonly used switching method is parallel computation of output signals and direct switching using P-Q control and U_{ac}-f control methods [22]. However, switching delays can cause abrupt changes in the output signal. Estimating and setting the initial values of control after switching in advance [23] ignores the impact of load fluctuations on the initial values. As the penetration rate of distributed generation (DG) increases, the deviation of the initial values will be larger. Therefore, the above method is more suitable for scenarios where the source-load power is relatively stable. Phase-locked loop control [24], initial phase angle [25], and pre-synchronization control [26] can reduce the impact caused by FMS mode switching, but due to the limitations of controller response speed, there are still deviations in actual scenarios. Existing inverter control methods mostly solve the problem of output signal mutation, but it is difficult to simultaneously consider the voltage and frequency fluctuations and impacts caused by source-load imbalance during islanding/grid-connected mode switching.

In order to address the issues of switching delay, initial phase angle, and the impact caused by control mode switching in FMS, this chapter proposes a smooth control strategy for port-ship islanding/grid-connected mode switching. The frequency and voltage variations in the microgrid during the switching delay are analyzed, and an adaptive droop coefficient control method is proposed to prevent frequency and voltage violations. A smooth switching method, which includes coordinated phase compensation, coordinated control switching, and pre-synchronization, is proposed to mitigate the instantaneous impact during the switching process. A port-ship islanding/grid-connected simulation model is established based on PSCAD/EMTDC to validate the effectiveness of the proposed strategy.

5.2   Flexible Interconnected Port-Ship Microgrid and Operation Mode Based on FMS

5.2.1   Flexible Interconnected Port-Ship Microgrid Based on FMS

The flexible interconnected microgrid based on FMS combines the control advantages of FMS and microgrid, greatly improving the cluster access of DG, the scale consumption capacity, and the steady-state operation efficiency of the new distribution network. As shown in Fig. 5.1, the flexible interconnected microgrid based on FMS consists of back-to-back converters, with MMC1 and MMC2 representing the two-side converters of FMS, C0 representing the common DC capacitor, AC1 and AC2 representing the power supply of the two ends of FMS, K representing the circuit breaker switch, solid lines representing power flow, and dotted lines representing information communication. As shown in Fig. 5.1a, the normal flexible "soft connection" between the distribution network feeders is realized to achieve microgrid interconnection and grid connection. The intelligent distribution system sets the power reference value according to the feeder current distribution and power quality, and the system operates normally. K1 and K2 are closed, and power exchange between the microgrids on both sides of FMS can be realized. As shown in Fig. 5.1b, if the upper-level power grid on one side of FMS is abnormal or power failure occurs due to a fault, the intelligent distribution system sends a switching command to FMS. Taking MMC1 side as an example, K1 is opened, and FMS switches to $U_{ac}\text{-}f$ control mode to provide voltage and frequency support to the microgrid at PCC1, realizing uninterrupted power supply and safe and stable operation of distributed power sources.

5.2.2   Operation Modes of the Flexible Interconnected Port-Ship Microgrid

The two operating modes of the classified microgrid are shown in Table 5.1. In the power regulation mode, FMS adopts the $P\text{-}Q$ control mode, and outputs active power and reactive power according to the reference command of the intelligent distribution system. The ESS can use droop control to adjust the microgrid frequency and voltage in real-time. In the emergency control mode, FMS adopts the $U_{ac}\text{-}f$ control mode

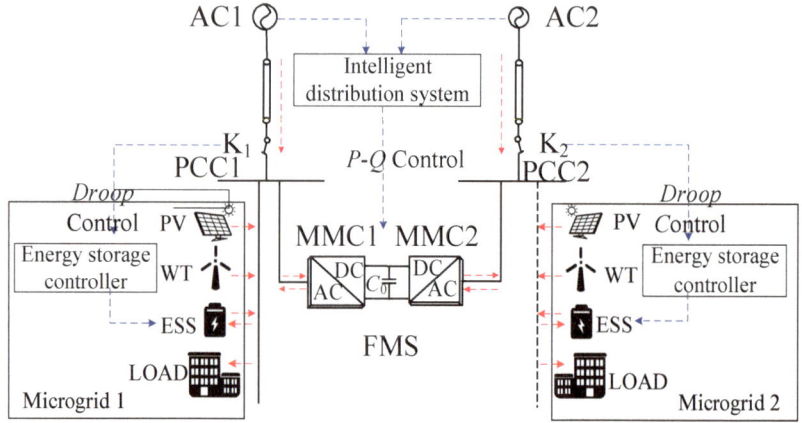

(a) Microgrid interconnection and Interconnection and grid-connection

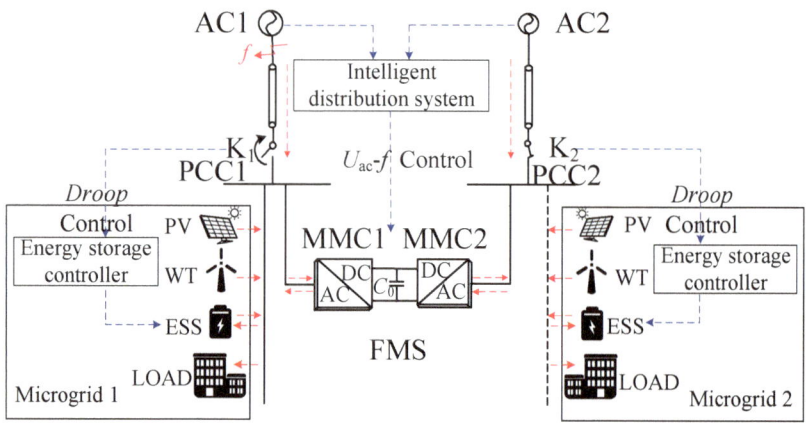

(b) Abnormal or fault power loss

Fig. 5.1 Interconnected microgrids via FMS

to keep the microgrid voltage and frequency unchanged. The ESS still uses droop control to mitigate the impact during the switching process.

5.2.3 *Emergency Switching of Flexible Interconnected Ship-Port Microgrid*

As shown in Fig. 5.2, when a fault occurs, the microgrid mode is switched from power regulation mode to emergency control mode. However, switching delay occurs due to

Table 5.1 Operating modes of interconnected microgrids via FMS

Method	Control object	Control mode	Scenarios	Control purpose
Power regulation	FMS	*P-Q*	Interconnection and Interconnection and grid-connection	Control the output power of FMS
	ESS	*Droop*		Real-time control of microgrid voltage frequency
Emergency control	FMS	U_{ac}-f	Abnormal or fault power loss	Keep the microgrid voltage/frequency constant
	ESS	*Droop*		Smooth the impact of the switching process

factors such as detection, research and judgment delay and communication delay. In this delay stage, the voltage of PCC node loses power grid voltage clamping. When the microgrid source and load imbalance is large, it is easy to cause voltage and frequency exceeding limits, threatening the safe and stable operation of the microgrid. This process is a transition state. The ESS still uses droop control to effectively mitigate the voltage and frequency fluctuations within the FMS switching delay. When the fault is recovered, the microgrid mode is switched from the emergency control mode to the power regulation mode. Due to frequency fluctuation during the switching process, there is a phase difference between the power supply phase angle of the grid and the microgrid PCC phase angle, causing a momentary impact during grid connection.

In addition to the voltage and frequency fluctuations within the switching delay, there will also be significant impacts during the emergency mode switching. Firstly, because the U_{ac}-f control mode cannot lock the phase of the grid power source, an initial reference phase needs to be set. Due to the switching delay and changes in network topology, the reference phase setting may experience a sudden change during the switching process, causing significant frequency impact. Secondly, the sudden change in control mode will cause transient adjustment of the converter control switching process, further causing a sudden change in the PWM output signal during the switching moment, causing significant voltage impact. The switching impact described above will cause significant voltage and frequency fluctuations at

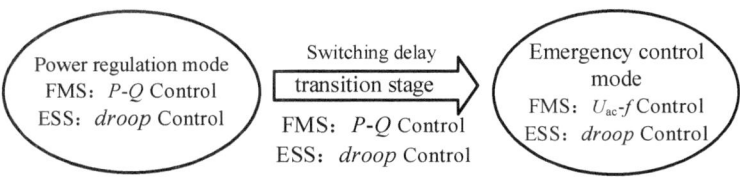

Fig. 5.2 Emergency switching process of interconnected ship-port microgrids

the PCC of the microgrid connected to FMS. How to effectively mitigate the above fluctuations and impacts is the key to achieving smooth switching.

5.3   Emergency Mode Switching Control Strategy for Interconnected Port-Ship Microgrid

5.3.1   Mode Emergency Switching

A control strategy for smoothing mode switching was proposed to minimize fluctuations and mitigate the impact of emergency mode transitions. Figure 5.3 illustrates the stage of emergency mode switching. During normal operation, the flexible interconnected microgrid operates in power regulation mode. MMC1 in the FMS utilizes the P-Q control mode to facilitate accurate power exchange among microgrids, while MMC2 adopts the U_{dc}-Q control mode to maintain voltage stability of the DC bus and ensure power balance throughout the entire system. The ESS is controlled using droop control. In the event of a fault, the FMS receives an instruction for emergency mode switching from the intelligent power distribution system, which is subject to a time delay (t_2-t_3). The FMS changes its mode switching control mode at the moment of switching (t_3). As depicted in Fig. 5.4, the FMS transitions from the P-Q control mode to the U_{ac}-f control mode, while the ESS collaboratively handles voltage and frequency fluctuations in the microgrid. Once the fault is resolved (t_4), the FMS receives a pre-synchronization instruction, which primarily considers phase angle changes and utilizes pre-synchronization to approximate the phase angle of the grid. After achieving the grid-connection condition, the FMS switches the control mode at t_5.

5.3.1.1   Mode Control Strategy for Emergency Switching

(1)   Adaptive *droop* control

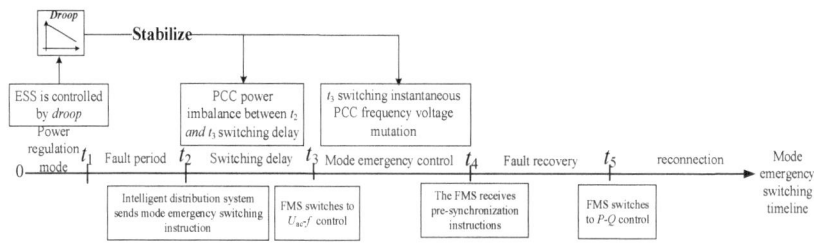

Fig. 5.3   Process of emergency switching of the interconnected port microgrids via FMS

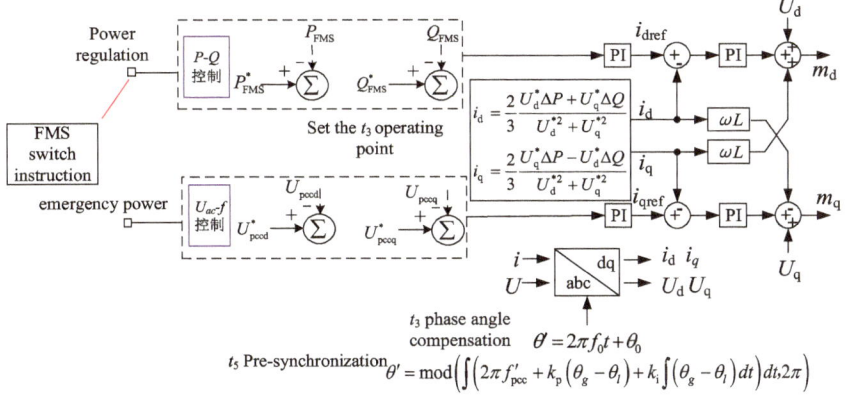

Fig. 5.4 Operation mode switching of FMS

The ESS employs *droop* control to mitigate frequency and voltage fluctuations in the power grid, as described by control equation (5.1):

$$\begin{cases} P_{ESS} = \frac{f_0 - f_{pcc}}{m} + P_0 \\ Q_{ESS} = \frac{U_0 - U_{pcc}}{n} + Q_0 \end{cases} \tag{5.1}$$

where P_{ESS} and Q_{ESS} are ESS output active power and reactive power respectively, P_0 and Q_0 are ESS active power and reactive power ratings respectively, U_{pcc} and f_{pcc} are voltage and frequency of PCC respectively, U_0 and f_0 are rated voltage and frequency respectively, m and n are droop coefficient.

As shown in Fig. 5.5, during normal system operation, K_1 is closed, and the system operates in power regulation mode, where the grid power stabilizes the voltage and frequency of the system. FMS and DG operate in a *P-Q* control mode, effectively acting as a current source characteristic power supply. The ESS utilizes *droop* control to regulate the voltage and frequency of the system, effectively acting as a controlled voltage source characteristic power supply. The AC load is equivalently represented as a constant impedance load. Based on the power flow relationship at the point of common coupling (PCC), we can derive the following equation:

$$\begin{cases} P_{load} = P_{FMS} + P_{DG} + P_{ESS} + P_s = \frac{U_{pcc}^2}{R} \\ Q_{load} = Q_{FMS} + Q_{DG} + Q_{ESS} + Q_s = \left(\frac{1}{2\pi f_{pcc} L} - 2\pi f_{pcc} C \right) U_{pcc}^2 \end{cases} \tag{5.2}$$

where P_{FMS} and Q_{FMS} respectively represent the active and reactive power output of FMS, P_{load} and Q_{load} respectively represent active and reactive load, P_{DG} and Q_{DG} respectively represent the active and reactive power output of DG, P_s and Q_s represent the active and reactive power of the grid. R, L and C are load equivalent inductance, resistance and capacitance, respectively.

As shown in Fig. 5.6, within the switching delay, K_1 is opened, and the system enters a transitional state. Under P-Q control, DG and FMS still act as current source characteristic power supplies. The load power varies with PCC voltage and system frequency. Similarly, we can obtain the following relationship:

$$\begin{cases} P'_{load} = P_{FMS} + P_{DG} + P'_{ESS} = \dfrac{U'^2_{pcc}}{R} \\ Q'_{load} = Q_{FMS} + Q_{DG} + Q'_{ESS} = \left(\dfrac{1}{2\pi f'_{pcc}L} - 2\pi f'_{pcc}C \right) U'^2_{pcc} \end{cases} \tag{5.3}$$

where, P'_{load} and Q'_{load} are active load and reactive load after fault, P'_{ESS} and Q'_{ESS} are active load and reactive load after fault, f'_{pcc} and U'_{pcc} are PCC frequency and voltage after fault respectively.

In combination with (5.2) and (5.3), we can obtain:

$$\begin{cases} f'_{pcc} = m\left(\dfrac{U^2_{pcc} - U'^2_{pcc}}{R} - P_s \right) + f_{pcc} \\ \dfrac{U'_{pcc} - U_{pcc}}{R} + Q_s = \dfrac{1}{2\pi L}\left(\dfrac{U^2_{pcc}}{f_{pcc}} - \dfrac{U'^2_{pcc}}{f'_{pcc}+c} \right) - 2\pi C\left(U^2_{pcc}f_{pcc} - U'^2_{pcc}f'_{pcc} \right) \end{cases} \tag{5.4}$$

From Eq. (5.4), we can determine the voltage and frequency fluctuations within the switching delay of the microgrid and their relationship to the power loss when the feeder line loses grid power after a fault. Grid-connected inverters should possess

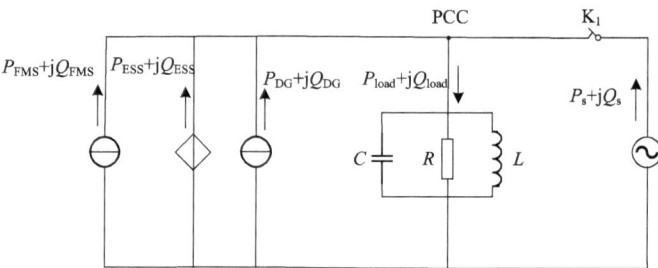

Fig. 5.5 Equivalent circuit of microgrid under power regulation mode

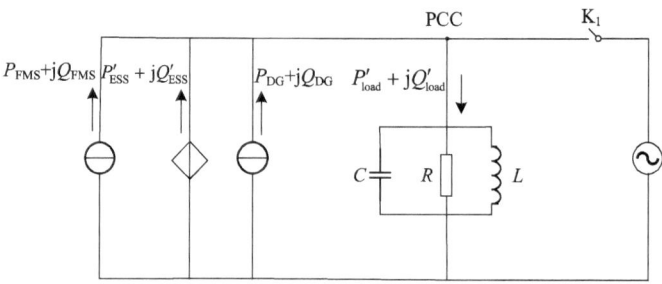

Fig. 5.6 Equivalent circuit of port microgrid under transition state

anti-islanding functionality, which means that the voltage and frequency variations during the switching delay must stay outside the triggering range of the anti-islanding device. Under the assumption that the communication system is not faulty, this delay is generally kept below 2 s. Setting voltage and frequency operating range parameters can be done based on other standards to ensure that distributed energy sources remain connected during the microgrid's emergency mode switching process:

$$
\begin{cases}
aU_{pcc} < U'_{pcc} < bU_{pcc} \\
f_{pcc} + c < f'_{pcc} < f_{pcc} + d
\end{cases}
\tag{5.5}
$$

where a, b, c, d are constants, $a < 1, b > 1, c < 0, d > 0$.

By solving Eqs. (5.4) and (5.5) simultaneously, we obtain the range of droop coefficients that ensures the anti-islanding device does not malfunction:

$$
\begin{cases}
m \in \left[\dfrac{\lambda c R}{(1-b^2)U_{pcc}^2 - P_s R}, \dfrac{dR}{(1-b^2)U_{pcc}^2 - P_s R} \right] \\
n \in \left[\dfrac{(a-1)U_{pcc}}{\frac{1}{2\pi L}\left(\frac{U_{pcc}^2}{f_{pcc}} - \frac{b^2 U_{pcc}^2}{f_{pcc}+c} \right) - 2\pi C\left[U_{pcc}^2 f_{pcc} - b^2 U_{pcc}^2 (f_{pcc}+d) \right] - Q_S} \cdots, \right. \\
\left. \dfrac{(b-1)U_{pcc}}{\frac{1}{2\pi L}\left(\frac{U_{pcc}^2}{f_{pcc}} - \frac{a^2 U_{pcc}^2}{f_{pcc}+d} \right) - 2\pi C\left[U_{pcc}^2 f_{pcc} - a^2 U_{pcc}^2 (f_{pcc}+c) \right] - Q_S} \right]
\end{cases}
\tag{5.6}
$$

Microgrid voltage frequency is rated value during normal operation, $U_{pcc} = U_0$, $f_{pcc} = f_0$.

When a fault occurs, the grid power abruptly drops to zero. In order to quickly compensate for the power imbalance and accelerate the response speed, a smaller value for the droop coefficient should be chosen. This corresponds to selecting the minimum value boundary in Eq. (5.6). However, considering that the boundary values of the droop coefficient may affect the reliability of the switching process, a margin factor λ is introduced to enhance its reliability, namely:

$$
\begin{cases}
m = \dfrac{\lambda c R}{(1-b^2)U_{pcc}^2 - P_s R} \\
n = \dfrac{\lambda(a-1)U_{pcc}}{\frac{1}{2\pi L}\left(\frac{U_{pcc}^2}{f_{pcc}} - \frac{b^2 U_{pcc}^2}{f_{pcc}+c} \right) - 2\pi C\left[U_{pcc}^2 f_{pcc} - b^2 U_{pcc}^2 (f_{pcc}+d) \right] - Q_S}
\end{cases}
\tag{5.7}
$$

(2) Collaborative smooth switching method

1. Collaborative phase angle compensation

At time t_3, FMS's MMC2 continues to utilize U_{dc}-Q control, while MMC1 switches to U_{ac}-f control. Since both FMS modeling and control are implemented in the dq coordinate system, and the abc/dq and dq/abc transformations require a reference phase, the absence of grid power after the PCC switch opens prevents obtaining synchronized phase information through a PLL. Therefore, it is necessary to manually provide the reference phase for the controller. During grid-connected operation, the phase of MMC1 is dependent on the grid phase. Considering the phase angle variation during the switching delay, a direct switch may cause significant impacts. Hence, compensation for the phase angle difference is required. In other words:

$$\theta' = 2\pi f_0 t + \theta_0 \tag{5.8}$$

where θ' sets phase angle for U_{ac}-f control; θ_0 is the phase angle at the control switching moment.

The ESS collaborates with FMS for phase angle compensation. When a deviation occurs between the manually set reference phase angle and the actual phase angle, it results in a sudden change in microgrid frequency. The ESS quickly responds by generating or absorbing active power to restore the frequency, thereby reducing momentary impacts. This is expressed in Eq. (5.9).

$$\begin{cases} \frac{d\Delta\theta}{dt} = 2\pi \left(f'_{pcc} - f_{pcc} \right) = 2\pi \Delta f_{pcc} \\ \Delta f_{pcc} = -m\Delta P_{ESS} \end{cases} \tag{5.9}$$

2. Collaborative control switching

Prior to the mode switching, the current loop's PI controller has reached a steady state, and the output signal is at a steady value. However, after the switching, the current PI controller needs to gradually adjust from a zero state to reach a steady state. This adjustment process takes a certain amount of time, resulting in a control switching impact. To accelerate the FMS control switching and adjustment speed, the steady-state operating point (5.10) is calculated using the steady-state inverse model. Based on the imbalance between source and load in the disrupted feeder line, the required power ΔP and ΔQ for FMS are determined. Combining this with the voltage reference value, the operating point is computed to provide command values for the current inner loop, thereby reducing the adjustment time.

$$\begin{cases} i_d = \frac{2}{3} \frac{U_d^* \Delta P + U_q^* \Delta Q}{U_d^{*2} + U_q^{*2}} \\ i_q = \frac{2}{3} \frac{U_q^* \Delta P - U_d^* \Delta Q}{U_d^{*2} + U_q^{*2}} \end{cases} \tag{5.10}$$

The control switching between the ESS and FMS results in a switching impact ΔU due to the sudden change in the control output signal. This can cause a voltage variation ΔU_{pcc} in the microgrid. To mitigate this, the ESS quickly responds by generating or absorbing reactive power ΔQ_{pcc} to restore the voltage, thereby reducing momentary impacts. This relationship is illustrated in Eq. (5.11).

$$\begin{cases} \Delta U_{pcc} = \Delta U + (i_2 - i_1)\left[R_T + j\left(L_T + \frac{L_0}{2} \right) \right] \\ \Delta U_{pcc} = -n\Delta Q_{ESS} \end{cases} \tag{5.11}$$

where i_1 is the output current of the converter before switching, i_2 is the output current of the converter after switching, L_0 is the inductance of the bridge arm, and R_T and L_T are the equivalent inductance and resistance of the connecting line and transformer respectively.

3. Pre-synchronization

When the grid needs to be restored after a disruption, there is a phase difference between the grid power phase angle and the faulted feeder line phase angle due to frequency fluctuations during the control process. To avoid the instantaneous impact caused by the phase difference during grid connection, a phase pre-synchronization strategy is employed to gradually align the feeder line phase angle with the grid phase angle. Once the phase angle difference falls within the allowable range for grid connection, the system resumes grid operation. The phase angle control equation is as follows:

$$\theta' = \mathrm{mod}\left(\int \left(2\pi f'_{pcc} + k_p\left(\theta_g - \theta_l\right) + k_i \int \left(\theta_g - \theta_l\right)dt\right)dt, 2\pi\right) \tag{5.12}$$

where θ_g is the phase angle of power supply in the grid, θ_l is the phase angle of planned grid connection, and mod is the complementary function.

5.3.2 Flow of Smooth Control for Emergency Switching of Operation Modes

The flowchart in Fig. 5.7 illustrates the process of smooth control for mode emergency switching. In the absence of faults, the real-time monitoring of grid power transmission is conducted based on the current operational state. The droop coefficient is adaptively adjusted according to Eq. (5.7). When a fault occurs, FMS receives a mode emergency switching signal from the intelligent power distribution system and switches to U_{ac}-f control mode. ESS collaborates to mitigate the switching impact. After the fault is resolved, FMS receives a pre-synchronization signal to adjust the phase angle. Once the grid connection conditions are met, the switches are closed, and FMS switches to P-Q control mode.

5.4 Simulations

The operational scenario of the flexible inter-connected microgrid, as shown in Fig. 5.1, was constructed in PSCAD/EMTDC. The simulation was set up as follows: the system operates normally from 0 to 1.8 s, at 1.8 s, a fault occurs in the 10 kV feeder line, and at 2 s, the grid PCC switch opens, triggering the control mode switching by FMS after the switching delay. At 5 s, the system resumes grid operation. The

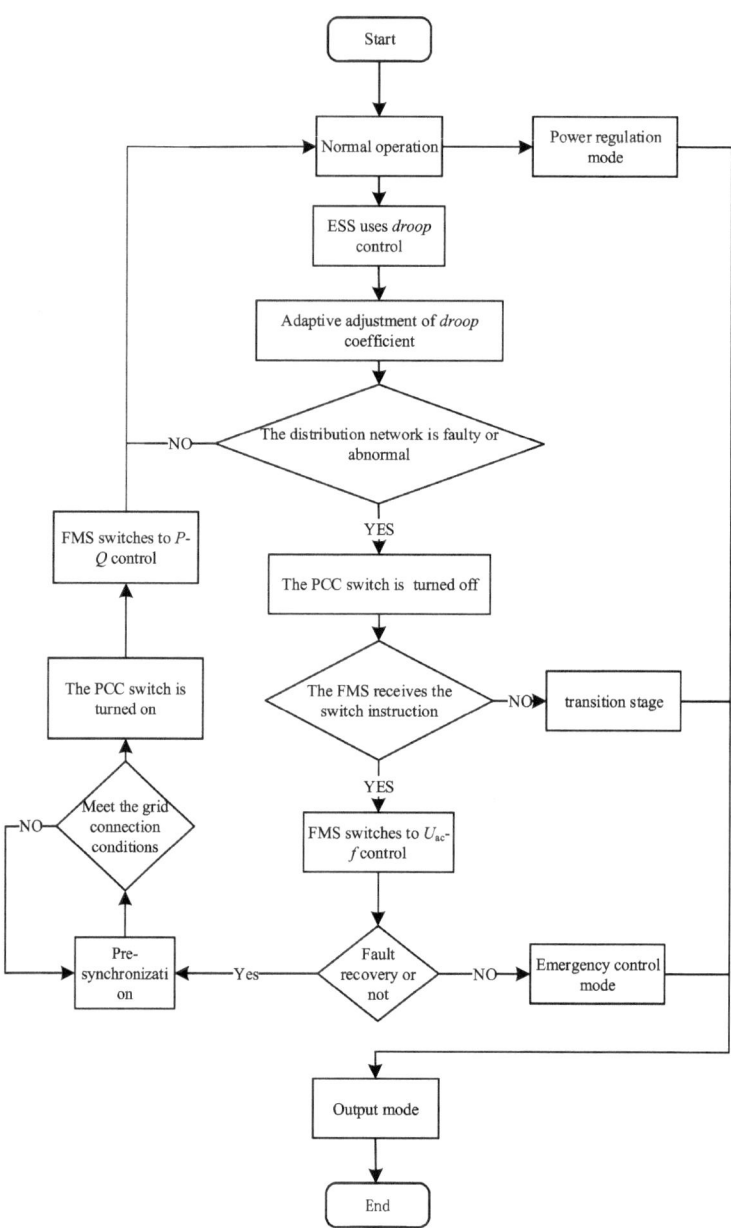

Fig. 5.7 Smooth control for emergency switching of operation modes

Table 5.2 Parameters of examples and test

System parameter	Value
PCC nominal voltage U_0 (kV)	10
Equivalent inductance of AC side L_T (p.u.)	0.018
Equivalent resistance of AC side R_T (p.u.)	0.006
Bridge arm inductance L_0 (p.u.)	0.0025
FMS nominal Capacity S (MVA)	2
DC side voltage U_{dc} (kV)	20
PCC nominal frequency f_0 (Hz)	50

source-load ratio is defined as the ratio of the active power between the DG and the load. Taking the parameters of a university energy network as an example, the specific parameters are listed in Table 5.2.

According to Fig. 5.7, considering a safety margin factor $\lambda = 1.1$ and referring to IEEE Std. 2000.929 [27] Sect. 5.3, which specifies the operating frequency range of the distributed power system as 59.3–60.5 Hz, the following parameters are designed: c = −0.7 Hz and d = 0.5 Hz. Referring to the voltage disturbance requirements in Sect. 5.1.1 of IEEE Std. 2000.929, the PCC operating voltage can range from 88 to 137%. Based on this, the parameters a = 0.88 and b = 1.37 are selected. The active power droop coefficient for ESS is calculated as 0.82 Hz/MW, and the reactive power droop coefficient is 6.3 kV/MVar. In the power regulation mode, the FMS output power is set to 0.05 MW for active power and 0 MVar for reactive power. Different scenarios with varying switching delays and source-load ratios are shown in Table 5.3. Simulation results are obtained to compare the switching effects between the proposed mode switching smoothing control strategy and the hard switching of the microgrid mode without smoothing control.

(1) Scenario 1

As shown in Fig. 5.8, at 1.8 s, a fault occurs, resulting in a rapid voltage drop in the microgrid. The FMS power transmission becomes unstable. At 2 s, the PCC switch opens, and the microgrid enters a transitional state. Within a delay of 0.2 s, due to the source being smaller than the load, the voltage and frequency of the microgrid continue to decrease, while the FMS power transmission remains at its initial value. At 2.2 s, the microgrid enters an emergency power supply mode, with the voltage and frequency reaching their lowest points. The FMS provides voltage and frequency support to the microgrid, gradually restoring the voltage and frequency. The FMS

Table 5.3 Simulation scenarios

Scenarios	Switching delay (s)	Source-load ratio	DG power (MW)	Load power (MW)
1	0.2	0.5	0.3	0.6
2	0.2	2	0.3	0.15
3	0.5	2	0.3	0.15

output power compensates for the difference between the source and the load, which is 0.3 MW. By employing the strategy proposed in this chapter, during the transitional state of the microgrid, the ESS compensates for the shortfall in active power, resulting in a rapid recovery of the dropped voltage and a significant reduction in frequency fluctuations. The voltage and frequency experience only slight impacts during the switching moment, far smaller than the impacts caused by hard switching of the microgrid mode.

(2) Scenario 2

As shown in Fig. 5.9, by varying the source-to-load ratio, the microgrid enters a transitional state where, within a delay of 0.2 s, the source is greater than the load, resulting in a continuous increase in the voltage and frequency of the microgrid. At 2.2 s, the microgrid enters an emergency power supply mode, with the FMS providing voltage and frequency support to the microgrid. The voltage and frequency gradually recover, and the FMS absorbs power equal to the difference between the source and the load, which is 0.15 MW. By employing the strategy proposed in this chapter, during the transitional state of the microgrid, the ESS absorbs excess active power, resulting in a rapid recovery of the increased voltage and a significant reduction in frequency fluctuations. The voltage and frequency experience only slight impacts during the switching moment, far smaller than the impacts caused by hard switching of the microgrid mode.

(3) Scenario 3

As shown in Fig. 5.10, by varying the magnitude of the switching delay, the microgrid enters a transitional state where, within a delay of 0.5 s, the source is greater than the load, resulting in a continuous increase in the voltage and frequency of the microgrid. The longer the delay, the more severe the voltage and frequency instability. At 2.5 s, the microgrid enters an emergency power supply mode, with the FMS providing voltage and frequency support to the microgrid. The voltage and frequency gradually recover, and the FMS absorbs power equal to the difference between the source and the load, which is 0.15 MW. By employing the strategy proposed in this chapter, during the transitional state of the microgrid, the ESS absorbs excess active power, resulting in a rapid recovery of the increased voltage and a significant reduction in frequency fluctuations. The voltage and frequency experience only slight impacts during the switching moment, far smaller than the impacts caused by hard switching of the microgrid mode.

To quantify the comparative effects between the proposed strategy and hard switching of the microgrid mode, the maximum voltage and frequency deviations of the microgrid are considered, along with the system recovery time as the performance metric. The system recovery time is defined as the maximum value between the voltage recovery time and the frequency recovery time. The former is calculated based on the time it takes for the voltage to recover to within 90–110% of the rated voltage, while the latter is calculated based on the time it takes for the frequency to recover to within 99–101% of the rated frequency. Refer to Table 5.4 for detailed results.

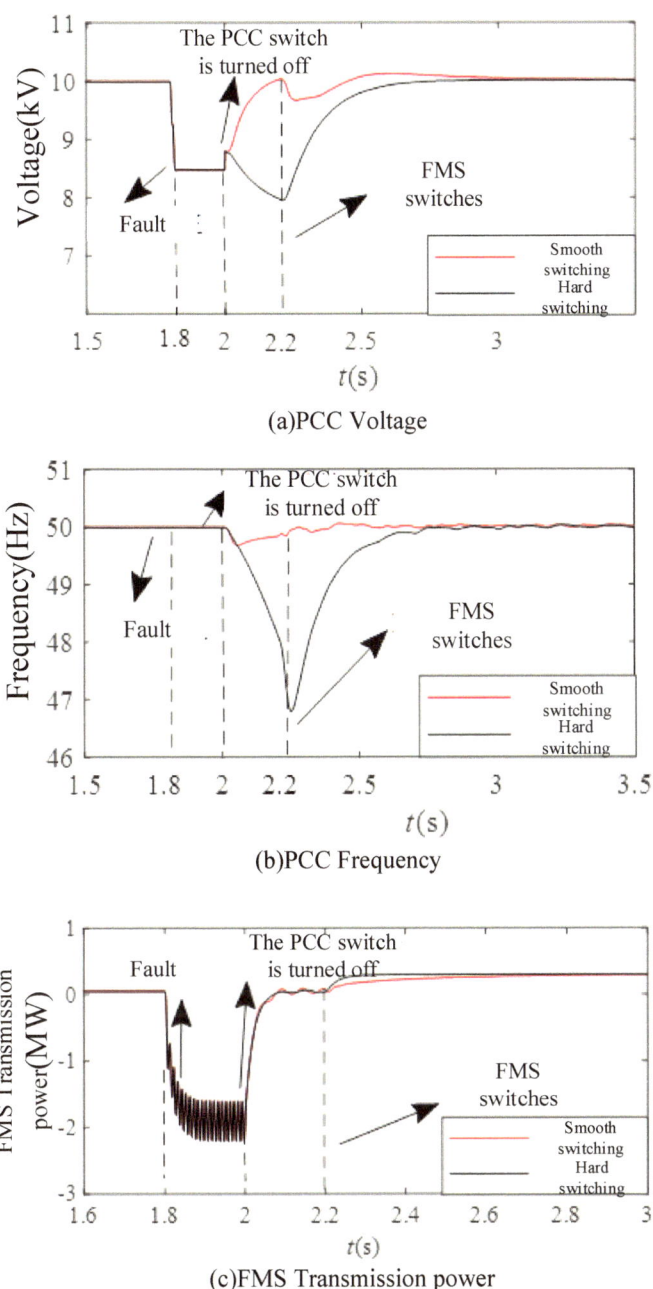

(a)PCC Voltage

(b)PCC Frequency

(c)FMS Transmission power

Fig. 5.8 Switching process of scenario 1

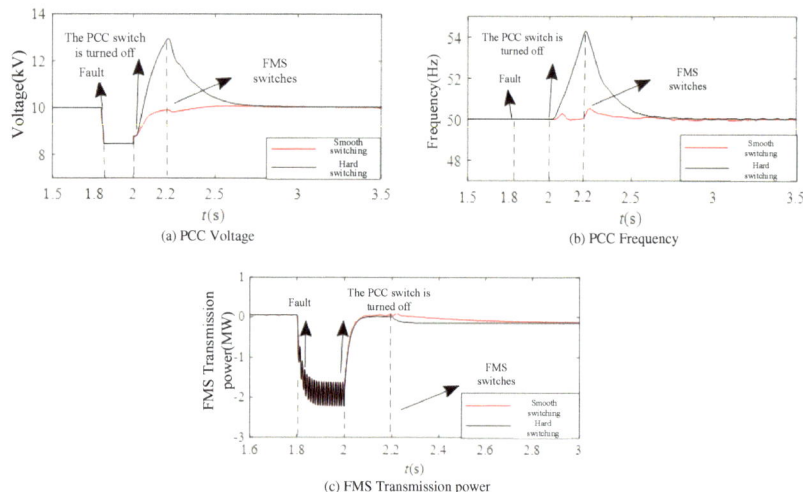

Fig. 5.9 Switching process of scenario 2

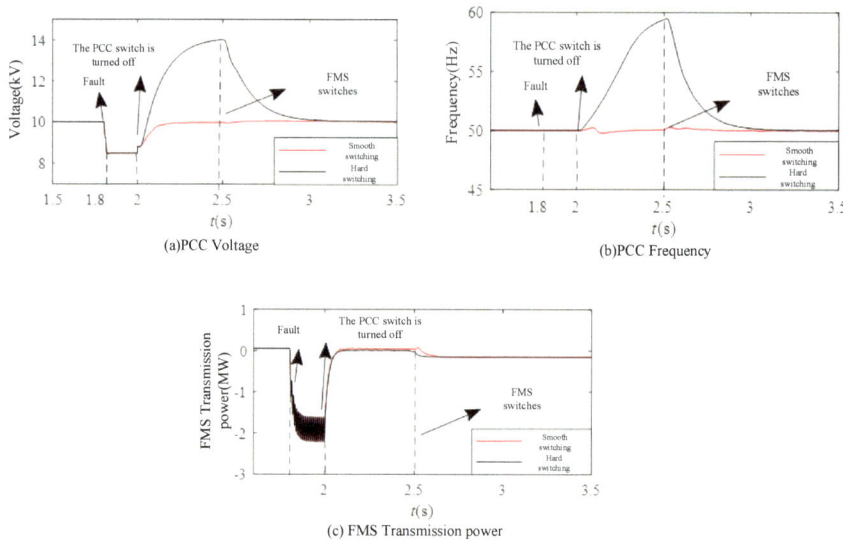

Fig. 5.10 Switching process of scenario 3

Comparing the three scenarios mentioned above, the microgrid mode switching process is influenced by the switching delay and the source-load ratio. Longer switching delays and higher source-load ratios result in larger voltage and frequency deviations, as well as longer recovery times. Hard switching of the mode cannot meet the requirements for safe and stable operation of the microgrid. However, by

Table 5.4 Comparison of strategies

Scenarios	Strategies	Maximum voltage offset (kV)	Maximum frequency offset (Hz)	System recovery time (s)
1	Hard switching	2.04	3.21	0.42
	Smooth switching	0.32	0.33	0.06
2	Hard switching	2.92	4.27	0.43
	Smooth switching	0.04	0.22	0.05
3	Hard switching	4.01	9.49	0.81
	Smooth switching	0.05	0.23	0.08

adopting the proposed smooth switching strategy, significant reductions in frequency and voltage fluctuations during the mode transition process can be achieved. Moreover, the recovery time for both frequency and voltage is substantially shortened. This enables smooth mode switching of the microgrid in different scenarios, providing support for the safe and stable operation of distributed generation (DG) in large-scale integration and cluster consumption.

To further validate the correctness of the theoretical analysis and simulation results, a flexible interconnected microgrid model was constructed on the experimental platform of the RT-LAB hardware-in-the-loop testbed, as shown in Fig. 5.11. The experimental platform consisted of a computer, RT-LAB OP7000 hardware simulation platform, Tektronix MSO44 oscilloscope, an FPGA development board (XILINX FPGA ZYNQ7020) as the controller, and an AN706 module with a maximum sampling frequency of 200 K as the controller sampling module. The experimental parameters are listed in Table 5.2. Scenario 1 was selected for the experiment, where a fault occurred at 1.8 s, the PCC switch opened at 2 s, and FMS mode switching took place at 2.2 s with a 0.2 s delay. The experimental results are shown in Fig. 5.12.

As shown in Fig. 5.12, through the comparison of microgrid frequency, voltage, and FMS transmission power during smooth switching and hard switching, the proposed strategy effectively raises the microgrid voltage and maintains the microgrid frequency within the 2–2.2 s switching delay. In contrast, hard switching exacerbates the loss of control over voltage and frequency within the switching delay. At 2.2 s, hard switching results in the lowest voltage and frequency, while under the proposed strategy, the microgrid voltage quickly recovers, and the frequency remains within the operating range. After 2.2 s, the FMS transmission power smoothly increases, compensating for the imbalanced source-load situation, and the switching current increases. The switching process and simulation results based on the RT-LAB platform for Scenario 1 are consistent, further validating the effectiveness of the proposed strategy.

RT-LAB

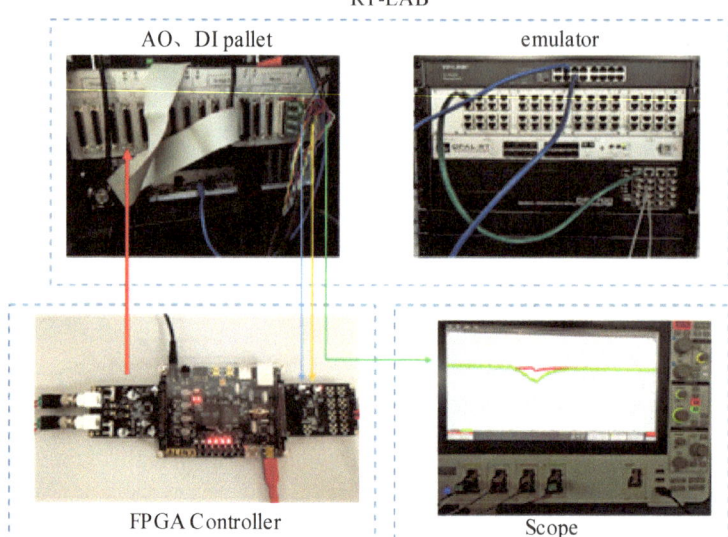

Fig. 5.11 RT-LAB test platform

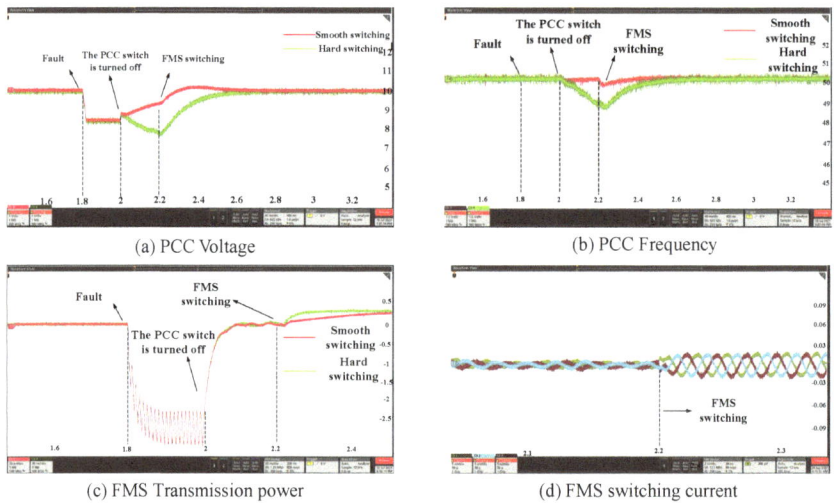

(a) PCC Voltage

(b) PCC Frequency

(c) FMS Transmission power

(d) FMS switching current

Fig. 5.12 Switching process of scenario 1 on RTLAB platform

5.5 Conclusion

Flexible interconnected microgrids are a key approach to improving the operational efficiency and distributed energy integration capability of the port microgrid. An emergency smooth control strategy for flexible interconnected microgrid modes is proposed in this chapter, which includes a microgrid energy storage adaptive droop coefficient control method and a coordinated phase angle compensation-coordinated control switch-pre-synchronization smooth transition method. These methods address the significant voltage and frequency fluctuations and disturbances in [microgrids caused by switching delays, phase shifts, and abrupt changes in control modes. The main conclusions are as follows:

(1) The microgrid energy storage adaptive droop control can quickly suppress imbalanced power within the switching delay, preventing voltage and frequency violations. Coordinated phase angle compensation is used to prevent frequency shocks caused by instantaneous phase angle variations during switching. Coordinated control switching is employed to mitigate voltage shocks caused by sudden changes in output signals, thereby reducing the regulation time.

(2) The process of microgrid mode switching is influenced by the switching delay and the ratio of source to load. The longer the switching delay and the larger the source-to-load ratio, the greater the voltage and frequency deviations. Compared to the hard switching process of microgrid modes, the proposed strategy significantly reduces the frequency and voltage deviations during the switching process, thereby shortening the recovery time.

(3) The flexible interconnected microgrid structure and operational modes enable the coordination and complementarity of the FMS (Flexible Manufacturing System) and microgrid control capabilities. The adoption of the proposed strategy allows for a smooth transition between microgrid modes, which is advantageous for the safe and stable operation of Distributed Generation (DG) with large-scale integration and cluster-based consumption. It holds significant importance in enhancing the continuous and stable operational capabilities of the power distribution grid based on flexible interconnected technologies.

References

1. Paul, D., Peterson, K., Chavdarian, P.: Designing cold ironing power systems: electrical safety during ship berthing. IEEE Ind. Appl. Mag. **20**(3), 24–32 (2014)
2. Wu, Z., Ye, X., Xing, M.: Discussion on the key technology of ship shore power. Electrotech. Appl. **32**(6), 22–26 (2013)
3. Ji, H., Wang, C., Li, P., et al.: Robust operation of soft open points in active distribution networks with high penetration of photovoltaic integration. IEEE Trans. Sustain. Energy **10**(1), 280–289 (2019)

4. Pamshetti, V.B., Singh, S., Thakur, A.K., et al.: Multistage coordination Volt/VAR control with CVR in active distribution network in presence of inverter-based DG units and soft open points. IEEE Trans. Ind. Appl. **57**(3), 2035–2047 (2021)
5. Wang, C., Song, G., Li, P., et al.: Optimal siting and sizing of soft open points in active electrical distribution networks. Appl. Energy **18**(9), 301–309 (2017)
6. Dong, X., Liu, Z., Li, P., et al.: Intelligent distribution network control technology based on multi-terminal flexible distribution switch. Proc. CSEE **8**(31), 86–92 (2018)
7. Lu, Z., Pei, X., Wang, C., et al.: A voltage over limit suppression method based on soft normally-open point in distribution network. Autom. Electr. Power Syst. **43**(12), 214–224 (2019)
8. Wang, C., Sun, C., Li, P., et al.: SNOP-based operation optimization and analysis of distribution networks. Autom. Electr. Power Syst. **39**(9), 82–87 (2015)
9. Huang, W., Wu, P., Tai, N., et al.: Architecture design and control method for flexible connected multiple microgrids based on hybrid unit of common coupling. Proc. CSEE **39**(12), 3499–3513 (2019)
10. Hossain, M.J., Mahmud, M.A., Milano, F., et al.: Design of robust distributed control for interconnected microgrids. IEEE Trans. Smart Grid **7**(6), 2724–2735 (2016)
11. Zhang, S., Pei, W., Yang, Y., et al.: Optimization and analysis of multi-microgrids integration and aggregation operation based on flexible DC interconnection. Trans. China Electrotech. Soc. **34**(5), 1025–1037 (2019)
12. Li, X., Guo, L., Li, Y., et al.: Flexible interlinking and coordinated power control of multiple DC microgrids clusters. IEEE Trans. Sustain. Energy **9**(2), 904–915 (2018)
13. Huo, Q., Li, M., Su, M., et al.: Flexible multi-state switch application scenario analysis. Autom. Electr. Power Syst. **45**(8), 13–21 (2021)
14. Zhong, C., Wei, L., Yan, G., et al.: Seamless transfer strategy of operation mode for microgrid based on collaborative control of voltage and current. Autom. Electr. Power Syst. **43**(5), 129–135 (2019)
15. Jia, L., Zhu, Y., Du, S., et al.: Analysis of the transition between multiple operational modes for hybrid AC/DC microgrids. CSEE J. Power Energy Syst. **4**(1), 49–57 (2018)
16. Li, Y.: Nondetection zone analytics for unintentional islanding in a distribu-tion grid integrated with distributed energy resources. IEEE Trans. Sustain. Energy **10**(1), 214–225 (2019)
17. Zhong, C., Wei, L., Yan, G., et al.: A seamless transfer control strategy of distributed generation with considering unintentional islands. Trans. China Electrotech. Soc. **32**(9), 130–139 (2017) (in Chinese)
18. Shi, Y., Lai, J., Su, J., et al.: Control strategy of seamless transfer for microgrid operation mode. Autom. Electr. Power Syst. **40**(8), 85–91 (2016)
19. Yoon, S., Oh, H., Choi, C.: Controller design and implementation of indirect current control based utility-interactive inverter system. IEEE Trans. Power Electron. **28**(1), 26–30 (2013)
20. Ochs, D.S., Mirafzal, B., Sotoodeh, P.: A method of seamless transitions between grid-tied and stand-alone odes of operation for utility-interactive three-phase inverters. IEEE Trans. Ind. Appl. **50**(3), 1934–1941 (2014)
21. Chen, J., Niu, B.Z., et al.: Hybrid control strategy for the seamless transfer of microgrids. Proc. CSEE **35**(17), 4379–4387 (2015)
22. Chen, J., Chen, X., Feng, Z., et al.: A control strategy of seamless transfer between grid-connected and islanding operation for microgrid. Proc. CSEE **34**(19), 3089–3097 (2014)
23. Guo, L., Wu, C., Wang, Y., et al.: A new method based on feed forward control for seamless switching of dual-mode inverter. Actal Energiae Solaris Sinica **36**(7), 1596–1601 (2015)
24. Aithal, A., Li, G., Wu, J., et al.: Performance of an electrical distribution network with soft open point during a grid side AC fault. Appl. Energy **22**(7), 262–272 (2018)
25. Wu, T., Zheng, Y., Wu, H., et al.: Power transfer and multi-control mode of a distribution network based on a flexible interconnected device. IEEE Access **7**, 148326–148335 (2019)
26. Ma, Z., Chen, J., Fang, Z., et al.: A seamless transfer control strategy of SNOP for the critical load safety under network faults. In: 2020 IEEE 9th International Power Electronics and Motion Control Conference, Nanjing, pp. 2348–2351 (2020)

27. IEEE Standards Coordinating Committee 21 (2000) IEEE Std. 2000. 929, IEEE Recommended Practice for Utility Interface of Photovoltaic (PV) Systems. The Institute of Electrical and Electronics Engineers, Inc., New York (2000)

Chapter 6
Voltage Optimization Method for Port Power Supply Networks

6.1 Introduction

Increasing penetration of renewable energy and electrification of logistics transport in port can reduce carbon emissions in integrated port power supply networks (IPPSNs) but exacerbating overvoltage issues [1–3]. A sharp fluctuation of feeder power and severe voltage violation frequently happens due to the constantly changing distributed power supply and the random access of electric vehicle charging [4]. The overvoltage problems of IPPSNs are more serious when extreme weather events or faults happen [5, 6]. To enhance the voltage stability of IPPSNs and ensure power supply to critical loads, it is of great significance to optimize the voltage of IPPSNs.

Port distribution networks could regulate conventional primary devices such as capacitor banks (CBs), series voltage regulators (SVRs), and on-load tap changers (OLTCs) to reduce voltage violation. It is difficult for conventional regulation methods to cope with rapid power flow changes in IPPSNs due to the slow response and discrete output [7, 8]. Most researches focus on managing the distributed generations or electric vehicles to make full use of their high controllability. Bai [9] and John [10] introduced a voltage regulation method to keep the nodal voltages within the limit with the controlled DG power curtailment. Ma [11] developed a decentralized voltage regulation method by adjusting the active and reactive power of each photovoltaic plant. Gusrialdi [12] and Hou [13] proposed scheduling algorithm to coordinate the charging behavior of electric vehicles and improve the utilization of charging resources on highways. However, the above studies are limited in that they do not modify the power flow distribution throughout the IPPSNs.

Soft open points (SOPs) are power electronic devices that interconnect two feeders as a replacement for normally-open points and can realize the flexible connection between feeders. Recently, SOPs are widely used in active distribution networks to optimize the power flow [14]. The voltage profiles of IPPSNs can be significantly improved using the voltage control strategy of SOPs. Ji [15] and I. Sarantakos [16] have studied the capacity support and optimization method of SOPs to improve

© The Author(s) 2023
W. Huang et al., *Energy Management of Integrated Energy System in Large Ports*,
Springer Series on Naval Architecture, Marine Engineering, Shipbuilding and Shipping
18, https://doi.org/10.1007/978-981-99-8795-5_6

the robustness and reliability of the distribution network. Literature [17] proposes a two-layer control strategy for SOPs in low-voltage distribution networks considering both the system layer and the equipment layer. According to the time scale, voltage optimization approaches for SOPs involvement in distribution networks are separated into long-time optimization methods and real-time optimization methods. The long-time optimization approach usually utilizes global system information for modeling and calculation. Literature [18] applies semidefinite programming (SDP) relaxation to the optimization model to realize rapid three-phase imbalance mitigation in distribution systems. The long-time optimization result is close to the theoretical global optimum but the optimization interval is too long to accommodate the frequent voltage fluctuations. Real-time optimization methods can regulate the power transmission of SOPs within a short time scale. Literature [19] proposes a coordinated scheduling method including fast-timescale scheduling which optimally coordinates the active and reactive power of SOPs. The real-time control for SOPs is usually implemented based on local information such as the bus voltage at each port of SOPs. Literature [20] and [21] use droop control methods to realize real-time local control of SOPs.

Model predictive control (MPC) based on rolling optimization can actively adjust the control strategy when voltage fluctuations occur to maintain the optimization effect [22]. Literature [23] proposes an MPC-based voltage control strategy for DGs that regularly tunes the parameter of local Q–V control curves to achieve better response to frequent voltage fluctuations. Literature [24] proposes a double-time-scale voltage control scheme using MPC to regulate the voltage profile across a network. MPC can effectively cope with the voltage violation problems in IPPSNs and improve real-time control performance.

Combining port grid topology with long-term measurement data to obtain port grid voltage and line loss sensitivity, a model for predicting distribution network voltage and line loss is developed. A real-time optimal control model is proposed, which aims to minimize the voltage deviation as well as the power losses. The active and reactive power transmission of the SOPs is optimized by the MPC approach and the predictive model is corrected through timely feedback.

The main contribution of this chapter is to propose an MPC-based real-time voltage optimization method for SOPs that does not rely on the global information of IPPSNs to reduce the impact of the distributed generation output uncertainty and load fluctuations on the node voltage. SOPs are optimally controlled considering both the voltage level and line loss, thus increasing operational efficiency during the normal situation and ensuring voltage levels under abnormal conditions.

6.2 Power Supply Networks with SOPs

6.2.1 IPPSNs Structure

IPPSNs provide electricity to the port logistics system and surrounding customers with a radial structure. In IPPSNs, more distributed generators (DGs) and charge stations for electric vehicles are installed to meet the demands of the economy and reduce carbon emissions. Meanwhile, increased energy storage and power electronics components have been introduced to the power supply networks in order to enable accurate voltage control. The grid topology of IPPSNs is usually operated in a radial configuration which is similar to that of the traditional distribution networks. Certain loads such as toll gates and charging infrastructures, are highly crucial and require superior power quality. The DGs are sometimes located far away from the heavy load, making it difficult for them to achieve a partial power balance between generation and loads. This would cause more severe overvoltage issues in certain regions. Furthermore, the power quality of these substantial loads cannot be guaranteed under certain extreme situations, posing a threat to IPPSNs.

The voltage violation at the end of the feeder tends to be severe considering the distance to the source node of IPPSNs. By integrating SOPs, the distribution network can transfer power between different feeders and realize fast and continuous power flow control to mitigate voltage deviation and reduce power losses of whole networks. The SOPs could be installed close to the important loads or DGs to achieve a better optimization effect. It can effectively balance out the power variations caused by renewable energy and lessen the effect on the distribution network's voltage. A typical structure of IPPSNs with SOPs is shown in Fig. 6.1.

IPPSNs can flexibly control various equipment to realize synergistic interaction between distributed power and dynamic loads. However, simply employing the global optimization strategy cannot be sufficient to prevent voltage violation due to the high uncertainty of distributed generators and electric vehicle charging. The localized real-time control method with SOPs is necessary to enhance the voltage opt imitation effect and ensure power quality.

6.2.2 Structure and Operating Principle of the SOPs

SOPs can regulate the power flow of connected feeders within a microsecond time scale. The typical structure of SOPs is shown in Fig. 6.2.

By utilizing SOPs, IPPSNs are able to create a flexible connection between feeders with SOPs while the active and reactive power flow is fully controlled with fast adjustment capability. The transmission power of SOPs may be determined by the control center or computed utilizing local control techniques.

Considering the operating principle of SOPs, the power transmission constraint for each port of SOPs can be expressed as:

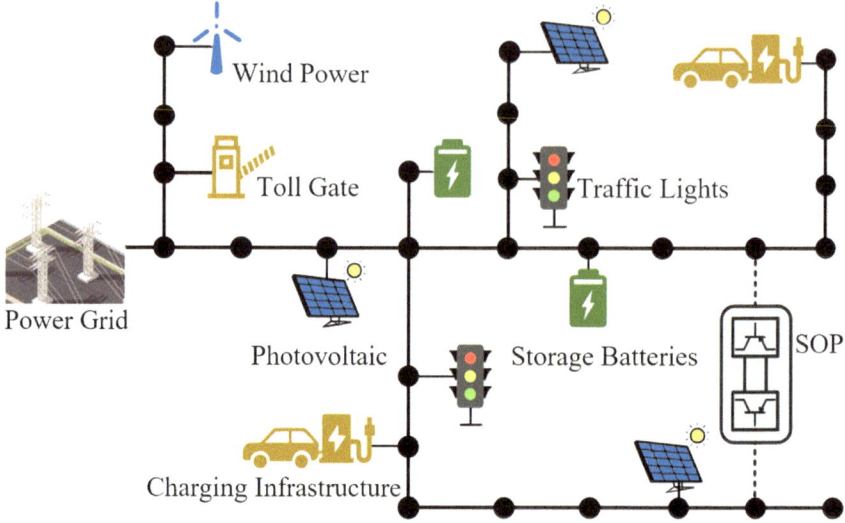

Fig. 6.1 Typical structure of IPPSN with SOPs

Fig. 6.2 Typical structure and application of SOPs

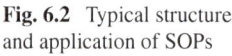
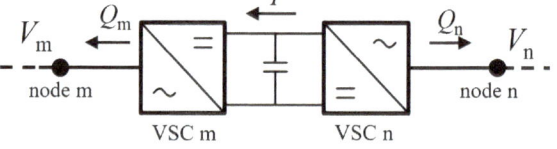

$$\begin{cases} \sqrt{P_m^2 + Q_m^2} \leq S_{\text{sop}}^2 \\ \sqrt{P_n^2 + Q_n^2} \leq S_{\text{sop}}^2 \end{cases} \tag{6.1}$$

The active and reactive power transmissions of SOPs can be controlled separately. To accomplish timely optimization for IPPSNs, a localized real-time control technique is proposed in this chapter. The performance of real-time control of SOPs is constrained by capacity limitations. By simultaneously adjusting the active and reactive power transmission of SOPs, the potential benefits of SOPs are further explored.

The switching loss of SOPs cannot be ignored though the loss is small relative to the transmission power. Note that the SOPs are connected between node m and node n. For convenience, the active power transmission of SOPs could be marked as $P = P_m = -P_n$ ignoring the transmit losses. The switching loss is related to the transmission capacity and can be written as Eq. (6.2).

$$P_{\text{sloss}} = k_{\text{sloss}} \left(\sqrt{P^2 + Q_m^2} + \sqrt{P^2 + Q_n^2} \right) \tag{6.2}$$

6.3 The MPC-Based Voltage Optimization Method

6.3.1 Voltage and Power Losses Model

A sensitivity matrix is introduced to develop a prediction model for the distribution network's voltage and power losses. A linear prediction model is developed for voltage and line loss changes arising from node power injection alterations. When the injected power of the distribution network line changes slightly, it will only affect the voltage distribution on this branch ignoring the higher-order error term [25, 26].

The power losses of the distribution network are shown in Eq. (6.3).

$$P_{\text{loss}} = \sum_{j \in L} R_j \left(\frac{\left(P_j^2 + Q_j^2\right)}{U_j^2} \right) \tag{6.3}$$

Using (6.3) and first-order Taylor expansion, we have the power loss variation shown as (6.4).

$$\Delta P_{\text{loss}} = \sum_{j \in L_i^{path}} \frac{R_j \left(2\Delta P \cdot P_j + \Delta P^2 + 2\Delta Q \cdot Q_j + \Delta Q^2\right)}{U_j^2} \tag{6.4}$$

The sensitivity of line losses to active power and reactive power $\frac{\partial P_{\text{loss}}}{\partial P}$ and $\frac{\partial P_{\text{loss}}}{\partial Q}$ can be written as (6.5) and (6.6).

$$\frac{\partial P_{\text{loss}}}{\partial P} = 2\Delta P \sum_{j \in L_i^{path}} \frac{R_j}{U_j^2} + 2 \sum_{j \in L_i^{path}} \frac{P_j \cdot R_j}{U_j^2} \tag{6.5}$$

$$\frac{\partial P_{\text{loss}}}{\partial Q} = 2\Delta Q \sum_{j \in L_i^{path}} \frac{R_j}{U_j^2} + 2 \sum_{j \in L_i^{path}} \frac{Q_j \cdot R_j}{U_j^2} \tag{6.6}$$

Assuming that the node voltage is the rated voltage V_0, , the node voltage variation can be written as (6.7).

$$\Delta V_i = \frac{\Delta P \sum_{j \in L_i^p} R_j + \Delta Q \sum_{j \in L_i^p} X_j}{V_0} \tag{6.7}$$

The node voltage's sensitivity to injected power changes can be determined using only the feeder voltage and the distribution network impedance. Sensitivity of node voltage to active power and reactive power can be written as (6.8) and (6.9).

$$\frac{\partial V_i}{\partial P} = \sum_{j \in L_i^p} \frac{R_j}{V_0} \tag{6.8}$$

$$\frac{\partial V_i}{\partial Q} = \sum_{j \in L_i^p} \frac{X_j}{V_0} \tag{6.9}$$

The sensitivity matrix is dependent on the operational state of the distribution network, and changes are not discernible in the presence of local disturbances. The voltage and line loss prediction models are established as the foundation of real-time model prediction control, and the sensitivity matrix is calculated using the electrical quantities of the distribution network at a specific time section and does not require frequent updates.

The control variables of the voltage and loss prediction model at time t_0 are denoted as (6.10). The prediction model of voltage and line loss can be written as (6.11) and (6.12).

$$\Delta u(t_0) = [\Delta P(t_0), \Delta Q_n(t_0), \Delta Q_m(t_0)] \tag{6.10}$$

$$V(t_0 + t|t_0) = V(t_0 + t - 1|t_0) + \frac{\partial V}{\partial u}\Delta u(t_0 + t - 1) \tag{6.11}$$

$$P_{loss}(t_0 + t|t_0) = P_{loss}(t_0 + t - 1|t_0) + \frac{\partial P_{loss}}{\partial u}\Delta u(t_0 + t - 1) \tag{6.12}$$

6.3.2 IPPSNs Voltage Optimization Model

The proposed real-time voltage optimization method in this chapter is a local control method that only uses the information from the connected node of the SOPs. The real-time voltage control method proposed in this chapter is a local control method that is based on the operating state of the SOPs for rolling horizon optimization (Fig. 6.3).

To assure the correctness of the prediction model, the optimization process predicts the changes in node voltage and network losses. The active and reactive power transmitted by the SOPs are controlled simultaneously and the prediction model is corrected via feedback correction.

The voltage control method proposed in this chapter is a local control method, which is based on the real-time voltage of the nodes at both ends of the SOPs and the working state of the SOPs for rolling optimization. The real-time optimization objective function is shown as follows.

$$\min : f = \sum_{t=1}^{N_p}\left[\Delta C_{\text{ope,i}} + \alpha \Delta C_{\text{loss,t}}\right] \tag{6.13}$$

$$\Delta C_{\text{ope},t} = P_p \Delta P_t^2 + P_q \Delta Q_{m,t}^2 + P_q \Delta Q_{n,t}^2 \tag{6.14}$$

$$\Delta C_{\text{loss},t} = P_l\left(\frac{\partial P_{\text{loss}}}{\partial P_m}\Delta P_t - \frac{\partial P_{\text{loss}}}{\partial P_n}\Delta P_t + \frac{\partial P_{\text{loss}}}{\partial Q_m}\Delta Q_{m,t} + \frac{\partial P_{\text{loss}}}{\partial Q_n}\Delta Q_{n,t}\right) \tag{6.15}$$

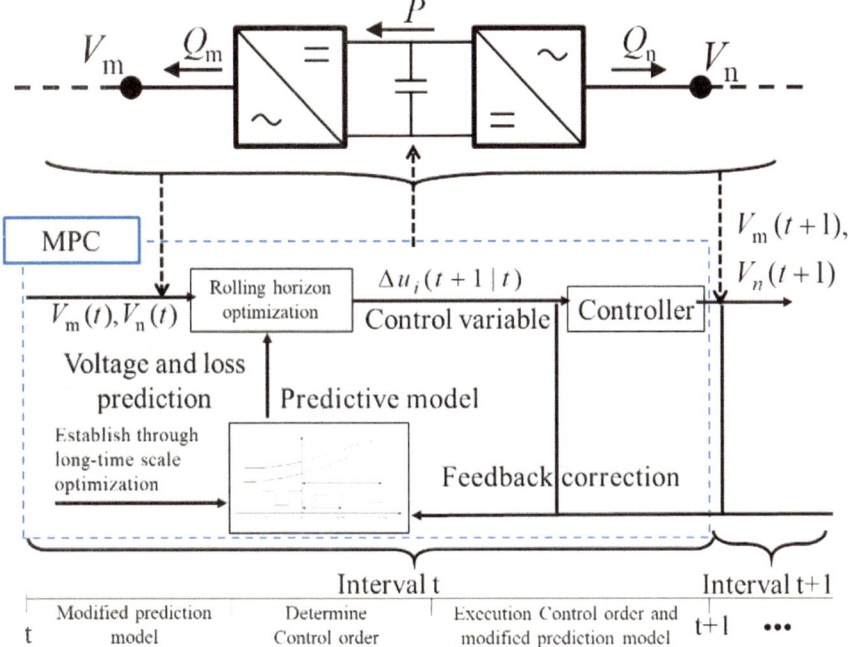

Fig. 6.3 Real-time voltage optimization method in IPPSNs

The line loss coefficient will be reduced when the voltage deviation V_{dev} is larger by controlling the value of the line loss coefficient α. By adjusting the line loss coefficient value in the objective function, the optimal function is modified to ensure that the node voltages at both ends of the SOPs operate within the ideal range.

$$\alpha = e^{-k_v V_{dev}} \tag{6.16}$$

Using voltage restrictions as Eq. (6.17), the node voltage at both ends of the SOPs is constrained:

$$V^{min}(k+i) \le V(k+i|k) \le V^{max}(k+i) \tag{6.17}$$

In order to guarantee that the optimization model has a solution, a voltage asymptotic constraint is implemented when the voltage fluctuation is too large to be adjusted to the target range in a single interval.

$$\begin{cases} V^{min}(k+i) = \left(1 - \frac{N-i}{\rho N} V^{min}\right) \\ V^{max}(k+i) = \left(1 + \frac{N-i}{\rho N} V^{max}\right) \end{cases} \tag{6.18}$$

Considering the operating limit, the power transmission constraint for SOPs of each time step can be written as:

$$\Delta P_i \leq \Delta P_{\max} \tag{6.19}$$

$$\Delta Q_i \leq \Delta Q_{\max} \tag{6.20}$$

To boost the predictive accuracy of the grid voltage and line loss prediction model, the real-time control must adjust the prediction model for the current round based on the previous round's prediction control deviation. This is due to the ever-changing operational status of the distribution network.

If the operation state of the distribution network is stable, the deviation caused by the change in the working state of other nodes of the distribution network is disregarded. And the network loss sensitivity is corrected according to the change in SOPs' working state, as shown in Eqs. (6.21) and (6.22).

$$\frac{\partial \Delta P_{\text{loss}}}{\partial \Delta P}(t_0 + 1) = \frac{\partial \Delta P_{\text{loss}}}{\partial \Delta P}(t_0) + \frac{\Delta P(t_0) \cdot R_b}{U_0^2} \tag{6.21}$$

$$\frac{\partial \Delta P_{\text{loss}}}{\partial \Delta Q}(t_0 + 1) = \frac{\partial \Delta P_{\text{loss}}}{\partial \Delta Q}(t_0) + \frac{\Delta Q(t_0) \cdot R_b}{U_0^2} \tag{6.22}$$

In the actual control process, a feedback correction is introduced in the voltage prediction control while the actual voltage of the distribution network is used as the initial value for a new round of rolling optimal scheduling, constituting a closed-loop voltage control, and the prediction model is corrected according to the control deviation of the previous round to improve the prediction accuracy of the voltage prediction model, as shown in Eqs. (6.23) and (6.24).

$$\Delta V_{\text{err}}(t_0 + 1) = V(t_0 + 1) - V(t_0 + 1|t_0) \tag{6.23}$$

$$V'(t_0 + 1 + t|t_0 + 1) = V(t_0 + t|t_0 + 1) + \frac{\partial V}{\partial u}\Delta u(t_0 + t) + \rho_v V_{\text{err}}(t_0 + 1) \tag{6.24}$$

At this point, the proposed optimization method can be solved quickly by commercial software.

6.4 Implementation of MPC-Based Voltage Optimization Method with SOPs

To cope with the constantly changing distributed power supply and the random access of electric vehicle charging, a real-time voltage optimization control method employing MPC is proposed in this chapter. The flow chart of the MPC-based real-time voltage optimization method is shown in Fig. 6.4.

The voltage and power losses sensitivity matrix are calculated and the prediction model is established using long-time information of IPPSNs. The initial value of the optimization model is set as the actual active and reactive power transmission of

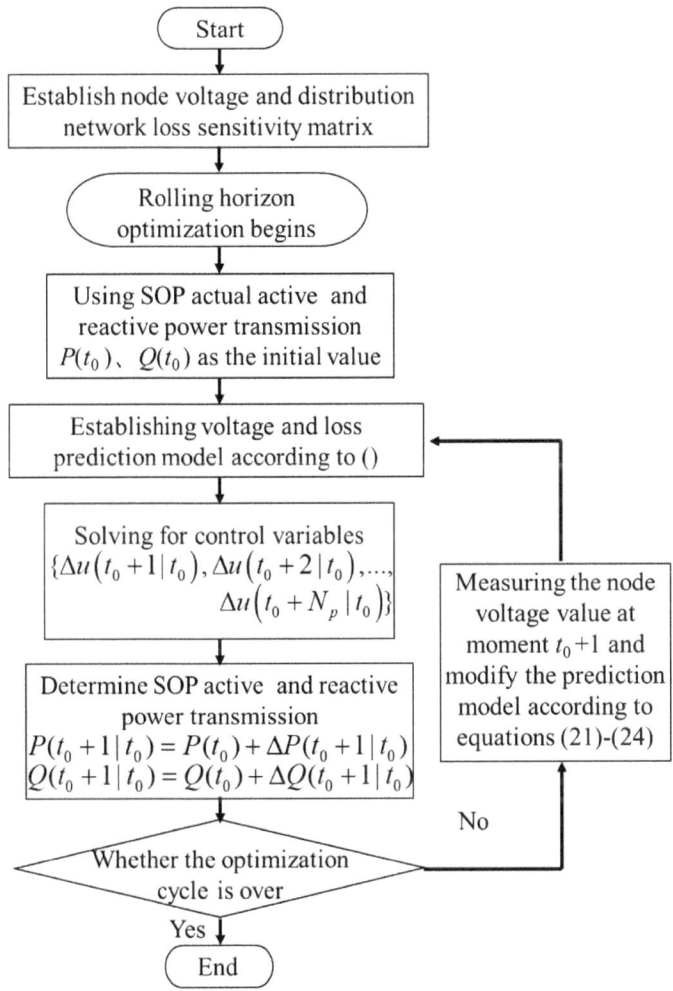

Fig. 6.4 The flowchart of MPC-based voltage optimization method

the SOPs. The objective function aims to minimize the operating costs of SOPs and power losses of IPPSNs. In the phase of real-time voltage optimization, the active and reactive power transmission of the SOPs is continuously modified to meet the voltage limit. A sequence of control variables in the next Np intervals is solved and the first command of the control variable sequence is issued as the active and reactive power transmission command of the SOPs.

The prediction model is corrected during the optimization process. The SOPs could measure the voltage from the connected node and the prediction error is considered as the correction amount. Thus, the MPC-based voltage optimization method can perform better on voltage improvement and power loss reduction.

The distribution network optimization level is evaluated by the network voltage deviation after completing the voltage optimization as shown in Eq. (6.25).

$$V_{dev} = \sum_{t=1}^{N} |V_i - 1| : \left(V_t \geq \overline{V}_{\mathrm{thr}} \| V_t \leq \underline{V}_{\mathrm{thr}} \right) \tag{6.25}$$

6.5 Case Studies

In this section, the case study of the proposed voltage optimization strategy is verified on a port power network in Zhejiang, China. The structure of the case is shown in Fig. 6.5. The simulation is performed in MATLAB software and the optimization solution is performed using YALMIP [27]. The numerical experiments were carried out on a computer with an Intel Core i5-13600 K running at 3.50 GHz and 32 GB of RAM.

The port power supply network system is a typical IPPSN that has a large installed DG capacity and frequent low-voltage phenomena at the end nodes. The IPPSN system includes a substation and 34 branches, of which the rated voltage level is 10 kV. The system contains 5 PV plants with 350 kVA capacity and 4 charging points with 200 kVA installed. An SOP with 500 kVA capacity is installed between nodes 14 and 35 to replace the existing switch. Several toll stations and other electrical loads are located separately in the network.

The loss factor of SOPs is set to 0.02 [28]. The ideal voltage operating range is [0.98 p.u., 1.02 p.u.] and the safe voltage operating range is [0.90 p.u., 1.10 p.u.] [29]. Set each time interval t to 30 s and the total number of time intervals in the solving horizon N_p to 3 rounds. The sampling time interval of voltage measurement is set as 30 s.

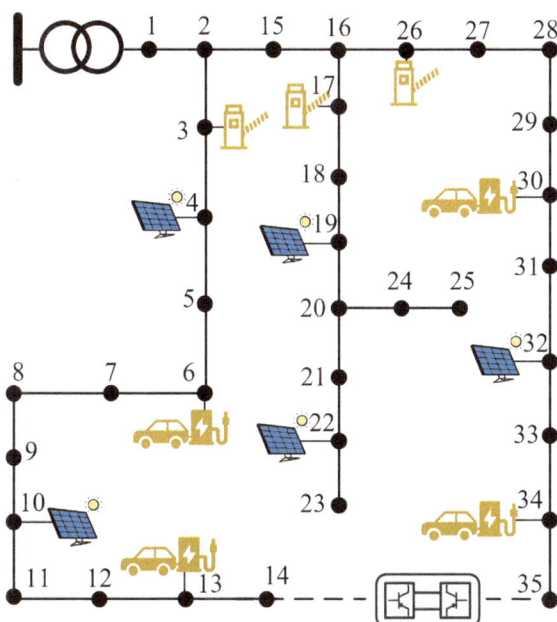

Fig. 6.5 Structure of the modified IPPSNs system

6.5.1   Analysis of All-Day Operation Scenario

Considering the fluctuating distributed power supply and random access of electric vehicles, a case of a typical day of IPPSN is studied in this section. Figure 6.6 shows the daily DG and load operation curves.

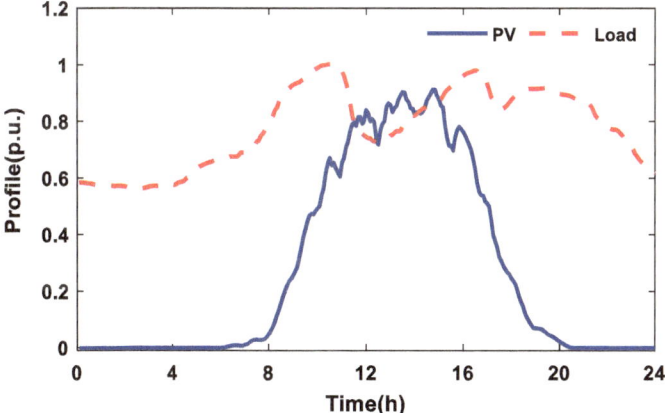

Fig. 6.6 Operation curves of DG and load

Figures 6.7 and 6.8 depict the voltage profiles of IPPSN before and after optimization, respectively. High-penetration PV units lead to frequent voltage fluctuations and voltage deviations. When the load reaches its peak, there is a considerable drop in network voltage. SOP could supply reactive power to support voltage and decrease the active power transmission when the node voltage is less than the lower limit. Noting that the PV power output peak coincides with the load power drop at 12:00 a.m. and causes voltage rise in some areas, the optimization strategy could reduce the reactive power injected into the bus. The nodes connected to SOP could basically remain at ideal voltage during the day while the proposed voltage optimization method is applied to the IPPSN.

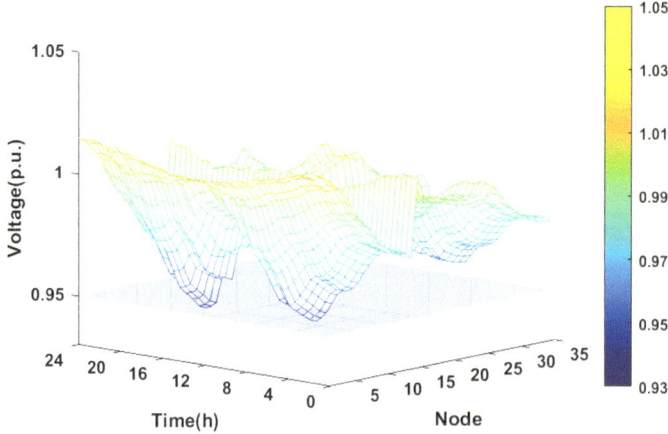

Fig. 6.7 Voltage profiles of all nodes before optimization

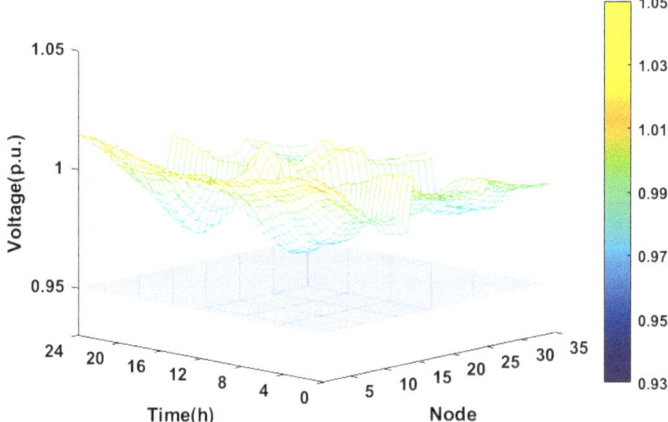

Fig. 6.8 Voltage profiles of all nodes after optimization

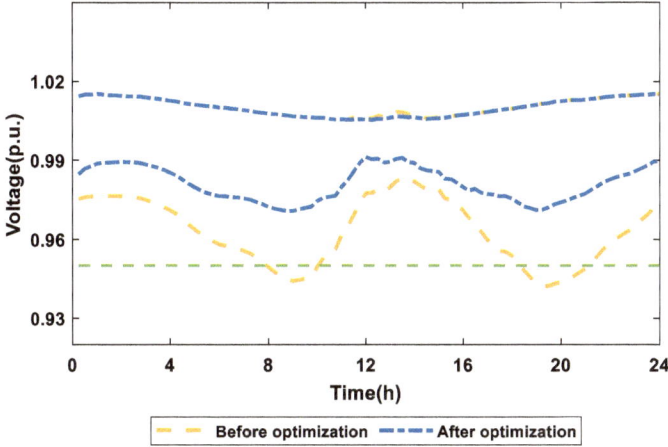

Fig. 6.9 Maximum and minimum system voltages

Table 6.1 Optimization results before and after optimization

Optimization method	Network voltage deviation (p.u)	Network power losses/(kW h)	Minimum voltage of system (p.u.)	Maximum voltage of system (p.u.)
Before optimization	12.30	1230.0	0.9421	1.0151
After optimization	7.511	1179.6	0.9665	1.0151

Figure 6.9 shows the maximum and minimum system voltages before and after optimization. By voltage optimization, the proposed MPC-based real-time SOP control method has significant improvement effects on the voltage profile (Table 6.1).

When the distributed power supply and electric vehicle charging fluctuates in the distribution network, the real-time optimization control method changes the transmitted active power and reactive power simultaneously through SOP to ensure the voltage of the distribution network operates within the acceptable range. The voltage deviation of the IPPSN is reduced by 38.92% and the network power losses are reduced by 4.09%.

6.5.2 Analysis of Comparison Study Under Abnormal Scenarios

In order to evaluate the effectiveness of the suggested strategy, the optimization of SOP is realized by 4 different methods in this section.

Method 1: IPPSN contains no SOP.

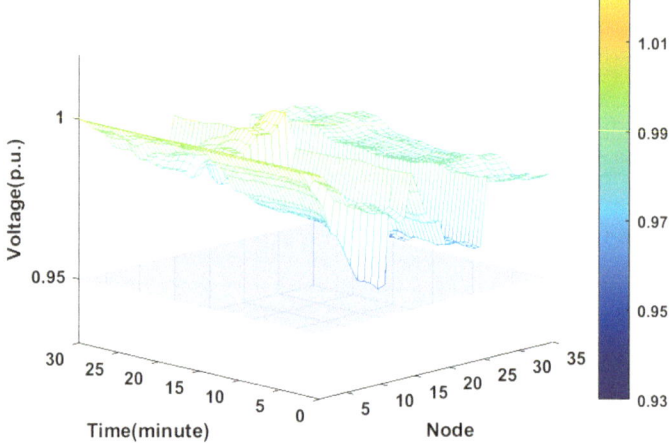

Fig. 6.10 Voltage profiles of all nodes with Method 4

Method 2: A long-time scale optimization method with no real-time control strategy is conducted on SOP. M1's purpose is to reduce the voltage deviation and line loss of the IPPSN while SOP operates in a centralized control mode. Every 30 min, the power transmission command of SOP is optimized.

Method 3: Utilizing long-term scale optimization and droop control to modify SOP's reactive power transfer. Method 2 is a modified version of Method 1 that applies real-time droop control to the reactive power of the SOP as a supplement. Every 30 min, the active power transmission command of SOP is optimized, whereas the reactive power transmission command of SOP is updated every 30 s.

Method 4: The MPC-based optimization method proposed in this chapter. Method 4 is a modified control method based on Method 1 and the active and reactive power transmission command of SOP is optimized every 30 s.

In this case, the PV unit connected to bus 10 appears to be unscheduled offline and re-connects to the system after a few minutes (Figs. 6.10, 6.11 and 6.12).

When the distributed power supply and electric vehicle charging fluctuates in IPPSN, the real-time optimization control technique simultaneously modifies the transmitted active power and reactive power via SOP to maintain the distribution network's voltage within an acceptable range. Table 6.2 presents the optimization results of 4 different methods with unplanned PV off-grid.

The optimization results show that Method 2 using centralized optimization can reduce the power losses and voltage deviations of the distribution network during normal operation, but cannot reduce the voltage deviation when a fault occurs. Method 3 applies droop control to SOP for real-time control and the node voltage deviation is larger and out of the ideal range. The optimization results of Method 2 and Method 3 show that Method 3 has a certain ability to maintain the node voltage during line faults, but cannot guarantee that the voltage is maintained near the ideal operating range. Method 4 introduces real-time optimization based on MPC and

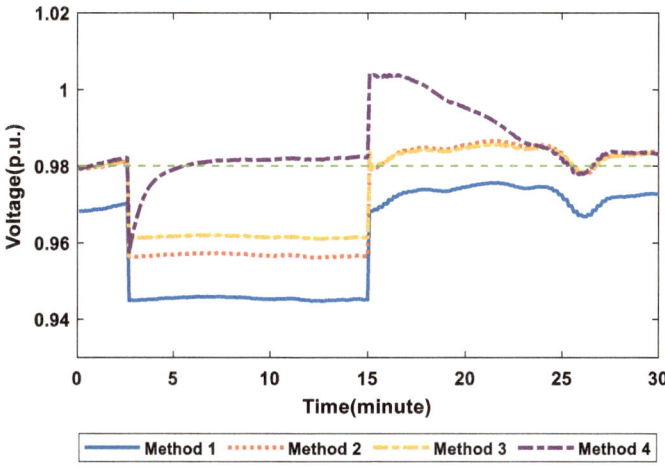

Fig. 6.11 Voltage curves of node 14 during the optimization

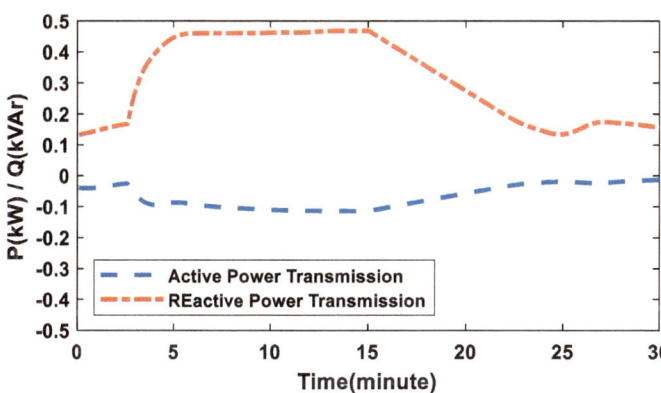

Fig. 6.12 Transmission power curves of SOP of Method 4

Table 6.2 Optimization Results of Different Methods with unplanned PV off-grid

Optimization method	Network power losses (kW h)	SOP power losses (kW h)	Network voltage deviation (p.u.)
Method 1	23.48	0	7.8218
Method 2	20.48	1.46	6.0310
Method 3	20.27	0.65	5.9267
Method 4	16.17	1.15	4.6772

Table 6.3 Optimization results of different methods

Optimization method	Network power losses (kW h)	SOP power losses (kW h)	Network voltage deviation (p.u.)
Method 1	790.39	0	1.6069
Method 2	743.45	2.44	1.3903
Method 3	732.99	19.51	1.1522
Method 4	718.08	23.55	0.6021

enables a rapid response when a fault occurs and supports the node voltage to return to normal levels. The node 14 voltage optimized by Method 4 rises rapidly after the PV is off-grid and is closer to the ideal voltage range [0.98 p.u., 1.02 p.u.] during the off-grid process. The voltage deviation under the control of Method 3 is 5.9267 p.u. while the voltage deviation of Method 4 is 4.6772 p.u.

After the PV is reconnected to the grid, the node voltages at both ends of the SOP optimized by Method 4 basically operate within the ideal voltage range, and the node voltages quickly recover to the optimal level with the optimization objective of minimizing network losses.

To verify the effect to the randomness of PV units' power output, a case with PV power fluctuation at node 4 and node 13 due to cloud cover within 60 min is studied. The optimization results of 4 different methods are listed in Table 6.3.

Comparing the optimization results of Method 1 and Method 2, it can be seen that after the centralized optimization control, the active power loss and voltage deviation of the IPPSN system can be effectively reduced. The comparison between the optimization results of Method 2 and Method 3 shows that the voltage deviation and the power losses of the IPPSN system are reduced while the droop control is introduced. Method 4 could further reduce the voltage deviation of the distribution network by 34.7% compared to Method 3, but the power loss of IPPSN increases slightly.

6.5.3 Analysis of Multiple SOPs Installed

A case study is conducted on an IPPSN with two SOPs to further explore the scalability and performance of the proposed MPC-based voltage optimization approach. The IPPSN structure is similar to Fig. 6.5 while an additional SOP with 500 kVA capacity is installed between bus 25 and bus 31. The daily DG and load operation curves are given in Fig. 6.13.

The optimization results of the IPPSN system with multiple SOPs are given in Table 6.4. This case demonstrates the scalability and effectiveness of the proposed voltage optimization method with multiple SOPs. When the distributed power supply and electric vehicle charging fluctuates in the distribution network, the real-time optimization control with two SOPs could significantly improve the efficiency and

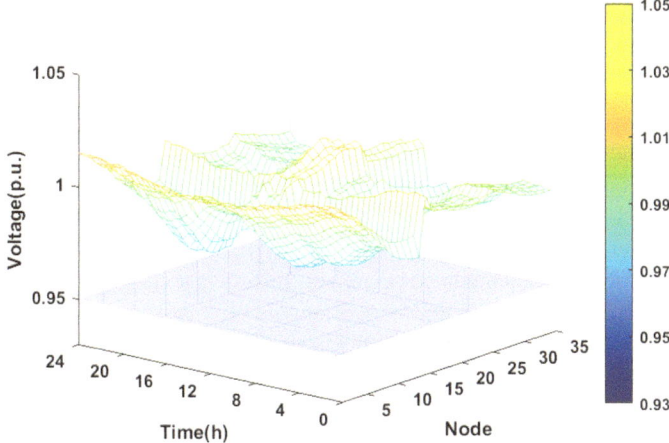

Fig. 6.13 Voltage profiles of all nodes after optimization with multiple SOPs

voltage stability of the IPPSN. The voltage deviation of the IPPSN is reduced by 60.79% and the network power losses are reduced by 5.39% compared to no optimization. The network losses are reduced by 35.79% and the network power losses are reduced by 1.35% compared to one SOP installed in STPSN.

The reasons for efficiency increase mainly include: (1) Multiple SOPs having much more active power transmit and reactive power supply ability when needed. (2) The proposed voltage optimization method could work without cooperation between different SOPs. However, the bound voltage improvement of installing two SOPs to IPPSN is not apparent compared to a single SOP installed. The possible reasons are: (1) The SOPs are not installed in the same area thus there is no additional power supplement for the focus area. (2) The interaction between different SOPs has an effect on the optimization.

Table 6.4 Optimization results before and after optimization with multiple SOPs installed

Optimization method	Network voltage deviation (p.u.)	Network power losses (kW h)	Minimum voltage of system (p.u.)	Maximum voltage of system (p.u.)
Before optimization	12.30	1230.0	0.9421	1.0151
One SOP installed	7.511	1179.6	0.9665	1.0151
Two SOPs installed	4.823	1163.7	0.9665	1.0151

6.6 Conclusion

An MPC-based voltage optimization approach for port power supply networks with SOPs is proposed to address the problem of voltage overrun. It only requires the collection of electrical information from the nodes connected to the SOPs throughout the optimization process. The proposed method could actively change the control strategy when voltage variations occur, effectively lowering the amplitude of voltage fluctuations. By simultaneously regulating the active and reactive power transmitted by the SOPs, the proposed method maximizes the SOP's ability meanwhile ensuring the system voltage runs within the desired range. Different case results demonstrate that the proposed method efficiently eliminates voltage deviations and reduces power losses when PV output power fluctuates or unplanned off-grid operation occurs. It can significantly minimize the voltage deviations of distribution network nodes compared with conventional optimization techniques and regulate the DG volatility in IPPSNs. By simultaneously modifying the active and reactive power transmission of SOPs, the potential advantages of SOPs are investigated further.

References

1. Ma, J., Xu, H., Cheng, P., et al.: A review of the development of resilient highway energy system coping with climate. In: 2021 IEEE 2nd China International Youth Conference on Electrical Engineering, pp. 1–8 (2021)
2. Lyu, S., Chen, S., Zang, H.: Power and traffic nexus: from perspective of power transmission network and electrified highway network. IEEE Trans. Transp. Electr. **7**(2), 566–577 (2021)
3. Lei, C., Lu, L., Ouyang, Y.: System of systems model for planning electric vehicle charging infrastructure in intercity transportation networks under emission consideration. IEEE Trans. Intell. Transp. Syst. **23**(7), 8103–8113 (2022)
4. Poudel, S., Dubey, A.: Critical load restoration using distributed energy resources for resilient power distribution system. IEEE Trans. Power Syst. **34**(1), 52–63 (2019)
5. Guo, C., Ye, C., Ding, Y., et al.: A multi-state model for transmission system resilience enhancement against short-circuit faults caused by extreme weather events. IEEE Trans. Power Delivery **36**(4), 2374–2385 (2021)
6. Yuan, W., Wang, J., Qiu, F., et al.: Robust optimization-based resilient distribution network planning against natural disasters. IEEE Trans. Smart Grid **7**(6), 2817–2826 (2016)
7. Tang, Z., Hill, D.J., Liu, T.: Distributed coordinated reactive power control for voltage regulation in distribution networks. IEEE Trans. Smart Grid **12**(1), 312–323 (2021)
8. Ji, H., Wang, C., Li, P., et al.: A centralized-based method to determine the local voltage control strategies of distributed generator operation in active distribution networks. Appl. Energy **228**, 2024–2036 (2018)
9. Ba, L., Jiang, T., Li, F., et al.: Distributed energy storage planning in soft open point based active distribution networks incorporating network reconfiguration and DG reactive power capability. Appl. Energy **210**, 1082–1091 (2018)
10. John, B., Ghosh, A., Goyal, M., et al.: A DC power exchange highway based power flow management for interconnected microgrid clusters. IEEE Syst. J. **13**(3), 3347–3357 (2019)
11. Ma, W., Wang, W., Chen, Z., et al.: Voltage regulation methods for active distribution networks considering the reactive power optimization of substations. Appl. Energy **284**, 116347 (2021)

12. Gusrialdi, A., Qu, Z., Simaan, M.A.: Distributed scheduling and cooperative control for charging of electric vehicles at highway service stations. IEEE Trans. Intell. Transp. Syst. **18**(10), 2713–2727 (2017)
13. Hou, L., Yan, J., Wang, C., et al.: A simultaneous multi-round auction design for scheduling multiple charges of battery electric vehicles on highways. IEEE Trans. Intell. Transp. Syst. **23**(7), 8024–8036 (2022)
14. Jiang, X., Zhou, Y., Ming, W., et al.: An overview of soft open points in electricity distribution networks. IEEE Trans. Smart Grid **13**(3), 1899–1910 (2022)
15. Ji, H., Wang, C., Li, P., et al.: Robust operation of soft open points in active distribution networks with high penetration of photovoltaic integration. IEEE Trans. Sustain. Energy **10**(1), 280–289 (2019)
16. Sarantakos, I., Zografou-Barredo, N.M., Hu, D., et al.: A reliability-based method to quantify the capacity value of soft open points in distribution networks. IEEE Trans. Power Syst. **36**(6), 5032–5043 (2021)
17. Wang, X., Guo, Q., Tu, C., et al.: A two-layer control strategy for soft open points considering the economical operation area of transformers in active distribution networks. IEEE Trans. Sustain. Energy **13**(4), 2184–2195 (2022)
18. Li, P., Ji, H., Wang, C., et al.: Optimal operation of soft open points in active distribution networks under three-phase unbalanced conditions. IEEE Trans. Smart Grid **10**(1), 380–391 (2019)
19. Yang, X., Xu, C., Zhang, Y., et al.: Real-time coordinated scheduling for ADNs with soft open points and charging stations. IEEE Trans. Power Syst. **36**(6), 5486–5499 (2021)
20. Xu, C., Yuan, X., Xu, Y., et al.: Research on feeder power balancing technology based on SNOP droop control. In: 2019 IEEE 10th International Symposium on Power Electronics for Distributed Generation Systems, Xi'an, pp. 192–196 (2019)
21. Cai, Y., Qu, Z., Yang, H., Zhao, R., et al.: Research on an improved droop control strategy for soft open point. In: 2018 21st International Conference on Electrical Machines and Systems, Jeju, pp. 2000–2005 (2018)
22. Valverde, G., Van Cutsem, T.: Model predictive control of voltages in active distribution networks. IEEE Trans. Smart Grid **4**(4), 2152–2161 (2013)
23. Li, P., Jie, J., Ji, H., et al.: MPC-based local voltage control strategy of DGs in active distribution networks. IEEE Trans. Sustain. Energy **11**(4), 2911–2921 (2020)
24. Guo, Y., Wu, Q., Gao, H., et al.: Double-time-scale coordinated voltage control in active distribution networks based on MPC. IEEE Trans. Sustain. Energy **11**(1), 294–303 (2020)
25. Wang, S., Liu, Q., Ji, X.: A fast sensitivity method for determining line loss and node voltages in active distribution network. IEEE Trans. Power Syst. **33**(1), 1148–1150 (2018)
26. Kumar, P., Gupta, N., Niazi, K.R., et al.: A circuit theory-based loss allocation method for active distribution systems. IEEE Trans. Smart Grid **10**(1), 1005–1012 (2019)
27. Lofberg, J.: YALMIP: A toolbox for modeling and optimization in MATLAB. In: 2004 IEEE International Conference on Robotics and Automation, Taipei, pp. 284–289 (2004)
28. Cao, W., Wu, J., Jenkins, N., et al.: Operating principle of soft open points for electrical distribution network operation. Appl. Energy **164**, 245–257 (2016)
29. Li, P., Ji, H., Yu, H., et al.: Combined decentralized and local voltage control strategy of soft open points in active distribution networks. Appl. Energy **241**, 613–624 (2019)

Chapter 7
Hierarchical Optimization Scheduling Method for Large-Scale Reefer Loads in Ports

7.1 Introduction

With the rapid development of global maritime transportation, ports have become energy and emission-intensive, and under the drive of the "dual carbon" strategy, the electrification level of ports is constantly improving [1]. Refrigerated containers (referred to as "reefers") convert electrical energy into cooling energy through their own refrigeration compressors, maintaining the temperature of the reefers within the allowed range. Large ports can store several thousand or even tens of thousands of reefers at the same time. Due to the impact of the global COVID-19 pandemic, the capacity of cold chain transportation has been reduced, and the density of reefers in port yards has continued to increase [2]. Reefers have become one of the largest energy consumers in ports [3, 4]. According to statistics, reefers at the Port of Los Angeles account for 20% of the total energy demand [5]. In ports where cold chain transportation is the main focus, the proportion of energy used by reefers is even higher. For example, at the Port of Valencia in Spain, the energy demand of reefers accounts for as much as 45%, exceeding logistics equipment to become the largest load in the port [6].

The large number of reefers and high electricity consumption of reefers will increase the system power during peak periods, leading to conflicts between power supply and demand and equipment overload. In addition, disorderly electricity use will also cause large-scale fluctuations in power supply lines, leading to increased losses and reduced efficiency. On the other hand, reefers have certain thermal storage capacity. Reefers with good insulation can maintain an internal temperature rise of about $0.11\,°C/h$ in the non-refrigeration state, which is a controllable flexible load [7]. Therefore, optimizing the electricity consumption of reefers is becoming increasingly important for green ports. It can significantly reduce the peak-to-valley difference of the load while reducing operating costs, thereby improving the economic and safety performance of the port energy system. However, the large scale of the reefer load, high dimensionalities of individual participation in global optimization, and

153

W. Huang et al., *Energy Management of Integrated Energy System in Large Ports*, Springer Series on Naval Architecture, Marine Engineering, Shipbuilding and Shipping 18, https://doi.org/10.1007/978-981-99-8795-5_7

low computational efficiency make it difficult to meet the real-time requirements of reefer group optimization scheduling. In addition, unlike temperature-constrained loads such as air conditioning [8] and electric heat pumps [9], different reefers have different temperature limits, which further increases the complexity of reefer group optimization.

Considering the controllability and thermal storage characteristics of reefers, previous studies have used reefers as flexible response resources to optimize electricity use and peak shaving. For example, literatures [10, 11] established a demand response model for multiple loads including refrigeration systems in ports and incorporated reefers into the port power system scheduling to reduce operating costs. Literature [12] integrated energy storage systems and different flexible loads including reefers in ports to achieve effective management of port energy. These studies modeled all reefers in the port as a whole, without considering the optimization and adjustment flexibility brought by individual differences in reefers, and it is also difficult to ensure that the temperature of each reefer does not exceed the limit. Currently, there are few studies on optimization scheduling considering the different characteristics and constraints of reefers. Literature [13] established an energy consumption model for a single reefer and reduced the peak load of the port by direct load control, but the centralized optimization based on a single reefer not only depends on the central node of centralized scheduling but also faces the problem of high computational dimensions and difficult rapid solutions.

The distribution of reefer loads in ports is relatively concentrated and can be interconnected through a small-scale local communication network to achieve the coordination and autonomy of reefer groups, which is suitable for distributed scheduling methods [14, 15]. Literature [16] proposed a distributed optimization scheduling model and solution algorithm based on multi-agent systems for port loads, where each agent makes autonomous decisions by solving local optimization problems. By dividing a large-scale optimization problem into multiple small-scale local optimization problems for solving, the computational dimensions are reduced, but the improvement in computational speed is limited, which cannot meet the requirements of accurate temperature control of reefers while ensuring overall optimization effects.

In this chapter, a consistent hierarchical optimization scheduling method is proposed to solve the problem of power optimization for large scale reefer groups at green port. Considering the heat exchange process of reefer, the thermoelectric coupling model of reefer is established. The reefers are divided into different clusters according to their electrical characteristics. Combined with dynamic electricity prices, the pre-scheduling strategy of the power consumption of the clusters is optimized with the lowest electricity cost as the goal. The refrigeration efficiency factor was proposed as the consistent state variable, and the dynamic power distribution model of the multi-intelligent body consistent master–slave power of the reefer was established to realize the fast response of the reefer in the cluster, so that the actual power demand of the reefer and the pre-scheduling strategy could be consistent as far as possible and meet their strict temperature constraints. Taking Rizhao Port as an example, the optimization effect and computational efficiency of the proposed method are verified.

7.2   Operation Characteristics and Modeling of Reefers

The front end of the refrigerated container is equipped with a built-in refrigeration compressor, and the temperature inside the container is controlled by power supply through the cold chain plug. As shown in Fig. 7.1a, when the refrigerated container is in the refrigeration state, the refrigeration compressor works, and the cold air blows from the ventilation duct inside the container, flows through the ventilation guide rail at the bottom of the container and surrounds the goods, and then returns to the refrigeration device through the cold air inlet on the front wall, forming a loop. In order to improve the cooling effect, the walls, top and bottom of the container are all lined with insulation materials. The operating characteristics of a single refrigerated container without ordered power supply are shown in Fig. 7.1b, where t_{on} is the time when the refrigerated container is in the refrigeration state, and t_{off} is the time when the refrigerated container is not in the refrigeration state. The operation of the refrigerated container is usually divided into two types: variable frequency operation and fixed frequency operation. After the fixed frequency refrigerated container starts refrigeration, the compressor speed remains unchanged. Under certain working conditions, its refrigeration capacity is fixed. However, as the external environmental conditions change, the working conditions and load inside the container will change, and the fixed frequency operation mode will cause large temperature fluctuations in the container and lower compressor unit efficiency. The variable frequency refrigerated container can change the power output by the frequency converter to achieve continuous and quantitative control of the compressor speed. In recent years, the market share of variable frequency refrigerated containers has been increasing, so in this chapter, it is assumed that the refrigerated container operates in a variable frequency manner.

The load modeling of the reefer is the basis for the study of its demand response mechanism. In the process of thermal-electric coupling modeling of the reefer, an exponential model is used to describe the dynamic temperature characteristics of the reefer [7]. Considering factors such as environmental temperature, solar radiation intensity, and type of goods inside the box, a thermodynamic dynamic model of the

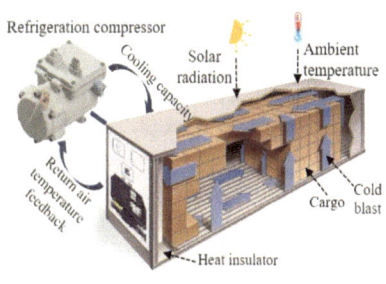

(a) Reefer construction with cold cycle   (b) Operation characteristics

Fig. 7.1 Operation mechanism of a reefer

reefer can be established based on the principle of heat balance, as shown in (7.1).

$$T(t + \Delta t) = T(t) + [T_{amb}(t) - T(t)] \cdot \left(1 - e^{-\frac{\sigma \cdot A \cdot k_t}{10^3 \cdot m \cdot c} \cdot \Delta t}\right) - \frac{P_R(t) \cdot \Delta t}{m \cdot c} \tag{7.1}$$

where Δt is the unit scheduling period(s). T represents the internal temperature of the reefer (°C). T_{amb} is the temperature of the external environment (°C). σ is the correction factor considering the introduction of solar radiation. A is the outer area of the reefer (m^2). k_t is heat transfer coefficient (W/m^2 K). m and c are the mass (kg) and specific heat capacity (kJ/kg K) of the goods in the container respectively. P_R is the refrigerating capacity (kW), and when the refrigeration compressor stops working, $P_R = 0$.

The actual power consumption of the reefer is related to the refrigerating capacity, and the power at all times should not exceed the limit. Generally speaking, the higher the internal temperature of the reefer, the higher the available refrigeration capacity, the higher the maximum power consumption. The power consumption of the reefer satisfies the relation described in (7.2).

$$\begin{cases} StR(t) \cdot P_{eR}^{\min} \leq P_{eR}(t) \leq StR(t) \cdot P_{eR}^{\max} \\ P_R(t) = ERR \cdot P_{eR}(t) \end{cases} \tag{7.2}$$

where $StR(t)$ is the operating state of the reefer, (when reefer operates $StR(t) = 1$, otherwise $StR(t) = 0$). P_{eR} is the electrical power of the reefer. $P_{eR}^{\max}$ and $P_{eR}^{\min}$ are respectively the upper and lower limits of the power consumption of the reefer. ERR is the refrigeration energy efficiency ratio. ERR of the reefer is different at different temperature setting points.

Since the goods in the reefer have certain requirements on the storage temperature, in order not to damage the goods, it is necessary to keep the temperature in the box within a certain range, as shown in (7.3).

$$T^{\min} \leq T(t) \leq T^{\max} \tag{7.3}$$

where $T^{\max}$ and $T^{\min}$ are the upper and lower limits of the temperature in the box respectively.

Equations (7.1)–(7.3) constitute the electricity consumption model of a single refrigerated container. Using a single refrigerated container as an optimization unit in a port scenario with more than a thousand containers would introduce hundreds of thousands or even millions of variables, greatly increasing the difficulty of optimization calculations. It is difficult to solve with the existing computing resources. Therefore, this chapter adopts the method of cluster equivalent modeling to describe the electricity consumption behavior and related constraints of the refrigerated container group. Several refrigerated container clusters (CR) are divided according to the type of goods, and refrigerated containers that are included in the same cluster have the same specific heat of goods, temperature setting value, and temperature allowable range. All refrigerated containers in the cluster can be equivalent to a large-capacity refrigerated container set, and the electricity consumption model of each CR can adopt the framework of a single refrigerated container electricity consumption model, with the quality and size taking the sum of all refrigerated container quality and size in the cluster.

7.3   Hierarchical Scheduling Modeling of Reefer Groups

7.3.1   Hierarchical Scheduling Architecture

The hierarchical scheduling architecture for reefer load groups is shown in Fig. 7.2 and consists of three layers: Port Dispatching Center (PDC), Reefer Aggregator (RFA), and Reefer (RF) load groups. By adopting a hierarchical scheduling strategy and establishing multiple small coordinated communication networks within the reefer load groups, the conventional management method of "collection-processing-control" loop entirely through the dispatching center is avoided, and a good balance between decentralized autonomy and centralized coordination is achieved from bottom to top. PDC-RFA-RF can achieve high-speed communication through 5G networks and fiber optic networks. The functions of each layer in the optimized scheduling process are as follows.

(1)   First level: PDC is responsible for supplying power to the basic load and refrigerated containers within its jurisdiction, and guides the power demand response of refrigerated containers by releasing time-of-use pricing signals, thus smoothing out load power fluctuations. PDC receives information about basic electricity prices, basic loads, and aggregated refrigerated container load demands during the optimization period, and calculates a new electricity price based on the total load and electricity price elasticity within the port area, and issues it to the load aggregator. Through the price signal guidance, refrigerated containers can be used as much as possible during periods of low load.

(2)   Second level: This layer is connected to the dispatching center and issues instructions to the reefer load group. After receiving the electricity price signal, RFA calculates the optimal power curve (i.e., pre-scheduling plan) for each CR based on the lowest cost as the optimization objective, according to the electricity price and CR model parameters, and issues it to the corresponding cluster. At the same time, RFA collects and integrates refrigerated container load information, aggregates the load demand curves of each CR into a total demand curve, and reports it to PDC.

(3)   Third level: This layer aggregates a certain number of refrigerated containers with similar electricity usage characteristics into a cluster, and adds a communication network between each refrigerated container in the cluster. An intelligent agent is assigned to each refrigerated container in the cluster, and during each scheduling cycle, each refrigerated container agent only communicates

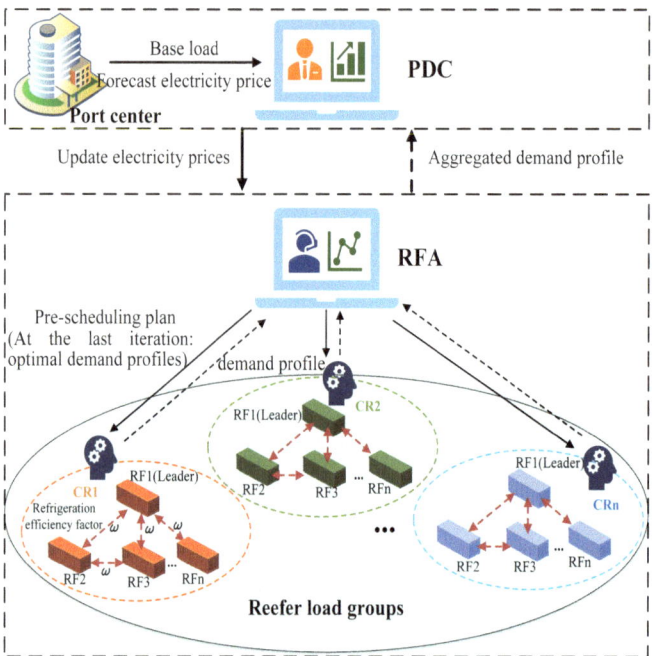

Fig. 7.2 The hierarchical dispatching architecture for reefers

with the adjacent two agents, and receives the pre-scheduled power instructions for the CR issued by the upper layer as the target for consistency control through the leading refrigerated container agent. By executing certain protocols and combining their own practical constraints, each refrigerated container can obtain its own power demand. Then, each CR uploads its actual demand curve to the upper layer.

7.3.2 Dynamic Model of PDC

The regulation goal of PDC is to reduce the peak-valley difference of system load. By utilizing dynamic time-of-use pricing mechanism, it achieves the transfer and reduction of cold box load during peak load periods. The forecasted electricity prices and the elasticity of electricity prices to cold box power demand during optimized periods are known, and the formula for updating electricity prices is shown in (7.4).

$$EP(t, n) = E\hat{P}(t) \cdot \left(1 + a \cdot \sum_{i=1}^{M} \sum_{j=1}^{N_i} P_{eR,ij}(t, n-1) \right) \cdot TSL(t, n) \qquad (7.4)$$

where $i = 1, 2, \ldots, M$, $j = 1, 2, \ldots, N_i$. M is the number of clusters. N_i is the number of reefers in CR_i. n is the number of iterations. $EP(t, n)$ is the electricity price at time t calculated by PDC at the nth iteration. $E\hat{P}$ is the predicted electricity price. a is the price elasticity factor, which indicates the change in unit price for every 1 kW change in electricity demand. TSL represents the load factor.

In order to define TSL, a high load threshold needs to be defined for the port grid. When the total load power of the port area exceeds this threshold, the TSL increases, thereby shifting the reefer load from peak to other time periods. TSL is calculated as shown in (7.5) and (7.6).

$$P_{total}(t, n - 1) = P_{load}(t) + \sum_{i=1}^{M} \sum_{j=1}^{N_i} P_{eR,ij}(t, n - 1) \tag{7.5}$$

$$TSL(t, n) = \begin{cases} 1, & P_{total}(t, n - 1) \leq P_{thres} \\ 1 + \rho \cdot \frac{P_{total}(t, n-1) - P_{thres}}{P_{thres}}, & P_{total}(t, n - 1) > P_{thres} \end{cases} \tag{7.6}$$

where P_{total} is the total load power in the port area. P_{load} is the base load in the port area except the reefer. P_{thres} is the threshold of high load in the port area. ρ is the penalty factor for overload.

In each iteration, the electricity price should be recalculated according to (7.4)–(7.6) until the electricity price converges.

7.3.3 RFA Decision Model

After collecting CR model parameters and electricity price information, RFA develops pre-scheduling strategy with the goal of lowest electricity cost for reefer, as shown in (7.7).

$$\min \left(\sum_{t=t_0}^{t_f} \sum_{i=1}^{M} P_{eR,i}(t) \cdot EP(t, n) \right) \tag{7.7}$$

where $P_{eR,i}$ is pre-scheduling planned power for CR_i.

To optimize scheduling in successive cycles, the temperature at the end of each CR scheduling cycle is set the same as the initial temperature, as shown in (7.8).

$$T_i(t_0) = T_i(t_f) \tag{7.8}$$

where $T_i(t_0)$ and $T_i(t_f)$ are respectively the equivalent temperature of CR_i at the beginning and end of the scheduling period.

7.4 Consensus Based Multi-agent Power Dynamic Distribution Model

7.4.1 Refrigeration Efficiency Factor of Reefers

After obtaining the CR pre-scheduling strategy, it is necessary to develop an efficient power allocation strategy to ensure that the reefers in the cluster fully responds to the RFA scheduling instructions. Due to the differences in the quality of goods in each reefer in the cluster, simply allocating the total power instruction equally to each reefer may result in some reefers with small refrigeration demands being over-cooled below the temperature lower limit, while reefers with larger refrigeration demands may be insufficiently cooled and exceed the temperature upper limit. On the other hand, power allocation is actually a process of anti-aggregation. In order to avoid the diversity of power consumption of reefer within the same cluster being undermined and its flexibility cannot be fully utilized, it is necessary to combine the reefer load model parameters and real-time operating status, judge the urgency and controllability of reefer power consumption at different times, formulate response tracking strategies, and achieve self-trend optimization operation of reefer load.

Consensus algorithm has the advantages of high computational efficiency, strong practicality, small amount of transmitted information, and "plug and play" characteristics when solving power allocation problems [17], and has gained some application research in the fields of power system economic dispatching [17, 18], automatic power generation control [19, 20], etc. Therefore, this chapter proposes the refrigeration efficiency factor of reefer, which is used as the consistency state variable among reefers, and adopts the "leader–follower" mode of power allocation algorithm to make reefers with higher temperature (i.e., larger refrigeration demands) in the same cluster bear a larger share of power. The jth reefer refrigeration efficiency factor ω_{ij} of CR_i is defined as (7.9).

$$\omega_{ij}(t) = \frac{\frac{ERR_{ij} \cdot P_{eR,ij}(t)}{m_{ij} \cdot c_{ij}}}{T_{ij}(t) - T_{ij}^{\min}} \tag{7.9}$$

In (7.9), the numerator represents the cooling rate, and the denominator represents the cooling margin of the reefer at time t. The uniformity of the cooling efficiency factor of reefers within the same cluster at each time point after total power allocation is used as the criterion for correct allocation. The reefers with higher cooling reserve have higher priority in refrigeration, and their cooling rates and corresponding power requirements are also higher.

7.4.2 Leader–Follower Consensus Algorithm for Refrigeration Efficiency of Multi-agent System

In the leader–follower refrigerating efficiency consensus (LREC) algorithm, the "leader–follower" mode is used as the basic framework to solve the centralized control problems in a distributed way to realize the autonomous response of loads. Each CR is regarded as a multi-agent system network, and each reefer represents an agent. Assume $B = [b_{jv}]$ the adjacency matrix of the multi-agent network, where $b_{jv} \geq 0$ represents the connection weight between agent j and agent v. The topology of the multi-agent network can be reflected by the Laplacian matrix $L_{jv} = [l]$ [21], which is defined as shown in (7.10).

$$l_{jj} = \sum_{v=1, v \neq j}^{N_i} b_{jv}, \quad l_{jv} = -b_{jv}, \quad \forall j \neq v \tag{7.10}$$

Refrigeration efficiency factor is selected as the consistent variable of each reefer in CR. Using the discrete time first-order consistency algorithm framework in literature [22], the refrigeration efficiency factor $\omega_{ij}[k + 1]$ of CR_i's jth follower reefer agent at the $k + 1$ iteration is related to the refrigeration efficiency factor of all reefers at the $k + 1$ iteration, as shown in (7.11).

$$\omega_{ij}[k + 1] = \sum_{v=1}^{N_i} d_{jv}[k] \omega_{iv}[k] \tag{7.11}$$

where $d_{jv}[k]$ represents the term of the row random matrix $D = d_{jv}[k] \in R^{N_i \times N_i}$ in the kth iteration, and its calculation is shown in (7.12).

$$d_{jv}[k] = \frac{|l_{jv}|}{\sum_{v=1}^{N_i} |l_{jv}|}, \quad j = 1, 2, \ldots, N_i \tag{7.12}$$

In order to keep the actual power demand of CR_i as consistent as possible with the pre-scheduling policy, $\Delta P_{error,i}(t)$ is defined to represent the power instruction difference of CR_i at time t, as shown in (7.13).

$$\Delta P_{error,i}(t) = P_{eR,i}(t) - \sum_{j=1}^{N_i} P_{eR,ij}(t) \tag{7.13}$$

The refrigeration efficiency factor updating rule of CR_i's leading agent is shown in (7.14).

$$\omega_{ij}[k + 1] = \sum_{v=1}^{N_i} d_{jv}[k] \omega_{iv}[k] + \mu_i \Delta P_{error,i}(t) \tag{7.14}$$

where μ_i is the power deviation adjustment factor for CR_i, and is a positive scalar that controls the convergence rate of LREC algorithm.

When LREC algorithm is used between reefers, some safety constraints need to be added to prevent the temperature or power of reefers from exceeding the limit. When the temperature of the jth reefer of CR_i reaches the lower limit at some point, the reefer stops refrigeration, $StR_{ij}(t) = 0$, and the corresponding power consumption $P_{eR,ij}$ is also zero. At this time, the reefer should exit from the network information topology, and the weight of the connection with the reefer agent j becomes zero, as shown in (7.15).

$$b_{jv} = 0, \quad v = 1, 2, \ldots, N_i \tag{7.15}$$

When the power consumption of the jth reefer of CR_i reaches its limit, it needs to be checked and corrected for safety, as shown in (7.16).

$$P_{eR,ij}(t) = \begin{cases} P_{eR,ij}^{min}(t), & P_{eR,ij}(t) < P_{eR,ij}^{min}(t) \\ P_{eR,ij}^{max}(t), & P_{eR,ij}(t) > P_{eR,ij}^{min}(t) \end{cases} \tag{7.16}$$

Similarly, at this time the connection weights with the reefer agent j all become zero. Equations (7.10)–(7.16) is a mathematical expression of the LREC algorithm. The flow chart of LREC algorithm is shown in Fig. 7.3. As can be seen from the figure, in each iteration, the leading cold-box agent needs to execute the whole process, while the follower agent only needs to carry out the steps of basic master–slave consistency algorithm and security constraint check in the small box shown in Fig. 7.3. When the security constraint of a reefer exceeds the limit, it needs to exit the multi-agent network immediately, and the corresponding network information connection weight also needs to be updated. When the difference between CR_i's actual power demand and the pre-scheduling policy $\Delta P_{error,i}(t)$ is less than the maximum allowable deviation ε_i, the algorithm iteration terminates.

The above analysis for the LREC algorithm is based on an ideal communication environment. In large-scale systems with a massive number of reefers, communication delay and noise may occur due to factors such as large amounts of interaction data, observation errors, and external interference in the communication network. To address these issues, the attenuation consistency gain function can be used to reduce the delay and noise assigned to the weights of corresponding edges of neighboring individuals, thereby utilizing the effective information of adjacent individuals for consensus calculation. Equations (7.11) and (7.14), which represent the updating rules of follower and dominant agents, can be modified into (7.17) and (7.18).

$$\omega_{ij}[k + 1] = \omega_{ij}[k] - c[k] \sum_{v=1}^{N_i} l_{jv}[k]\big(\omega_{iv}\big[k - \tau_{jv}[k]\big] + \eta_{jv}[k]\big) \tag{7.17}$$

$$\omega_{ij}[k + 1] = \omega_{ij}[k] - c[k] \sum_{v=1}^{N_i} l_{jv}[k]\big(\omega_{iv}\big[k - \tau_{jv}[k]\big] + \eta_{jv}[k]\big)$$

Fig. 7.3 The flow chart of LREC algorithm

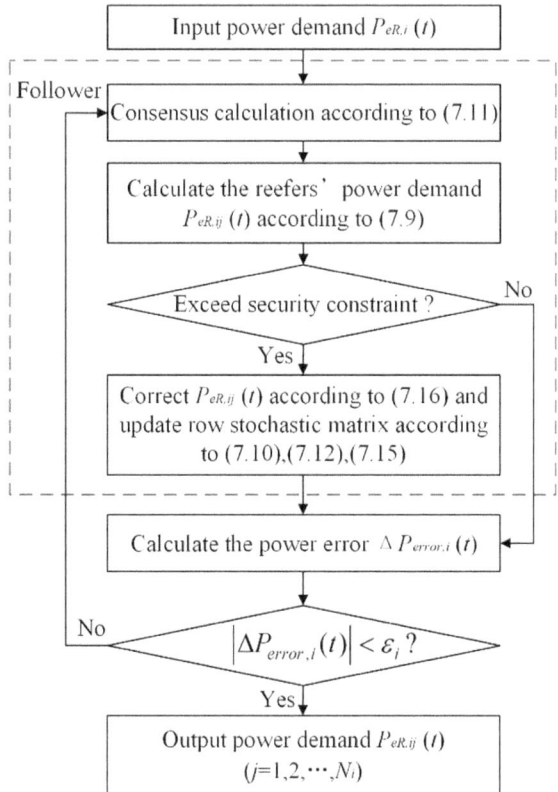

$$+ \mu_i \Delta P_{error,i}(t) \tag{7.18}$$

where $\tau_{jv}[k]$ is the communication delay of information transmitted from the v reefer agent to the j reefer agent. $\eta_{jv}[k]$ is the noise of channel transmission. $c[k]$ is the consistent gain function.

To ensure that the algorithm converges, $c[k]$ needs to satisfy the two conditions shown in (7.19) and (7.20) [23].

$$\sum_{k=0}^{\infty} c[k] = +\infty \tag{7.19}$$

$$\sum_{k=0}^{\infty} c^2[k] < +\infty \tag{7.20}$$

It has been proved in [23] that (7.19) is a convergence condition to ensure that the consistent state variables can converge at an appropriate rate. Equation (7.20) is a robust condition. When there is communication delay and noise in the system, this condition can also make the static error of the closed-loop system within a finite range. In other words, the robust condition ensures the convergence robustness of the algorithm after considering the effect of delay and noise.

7.4.3 Analysis of Power Deviation Adjustment Factor

The convergence rate of the LREC algorithm can be controlled by adjusting the power deviation adjustment factor μ_i. As can be seen from (7.14), the leader will increase or decrease the refrigeration efficiency factor according to the power instruction difference $\Delta P_{error,i}(t)$, and its updating range is affected by the adjustment factor. When the adjustment factor is too large, the updating amplitude of the consistency state variable of the dominant reefer agent is too large, which may lead to the algorithm non-convergent. However, when the adjustment factor is too small, the amplitude of updating the consistent state variable of the dominant reefer agent is small, and the corresponding convergence rate is slow. Therefore, it is necessary to select an appropriate adjustment factor to ensure that the LREC algorithm has better stability and faster convergence rate.

For CR_i containing N_i reefers, when the cooling efficiency factor of the leading reefer intelligent agent increases by $\mu \Delta P_{error,i}(t)$, the cooling efficiency factor of each reefer intelligent system increases by $\mu \Delta P_{error,i}(t)/N_i$ on average. According to (7.9), the total power demand increment of CR_i is shown in (7.21).

$$\Delta P_{eR,i}(t) = \sum_{j=1}^{N_i} \frac{\mu_i \Delta P_{error,i}(t) \cdot m_{ij} \cdot c_{ij} \cdot \left(T_{ij}(t) - T_{ij}^{\min}\right)}{N_i \cdot ERR_{ij}} \tag{7.21}$$

In LREC algorithm, taking the power instruction difference $\left|\Delta P_{error,i}(t)\right| < \varepsilon_i$ as the convergence criterion, the sufficient condition of algorithm convergence is $\left|\Delta P_{error,i}(t) - \Delta P_{eR,i}(t)\right| < \varepsilon_i$.

$$\begin{cases} \dfrac{N_i\left(\Delta P_{error,i}(t) - \varepsilon_i\right)}{\Delta P_{error,i}(t) \sum_{j=1}^{N_i} \frac{m_{ij}c_{ij}\left(T_{ij}(t)-T_{ij}^{\min}\right)}{ERR_{ij}}} < \mu_i < \dfrac{N_i\left(\Delta P_{error,i}(t) + \varepsilon_i\right)}{\Delta P_{error,i}(t) \sum_{j=1}^{N_i} \frac{m_{ij}c_{ij}\left(T_{ij}(t)-T_{ij}^{\min}\right)}{ERR_{ij}}}, \\ \qquad\qquad \Delta P_{error,i}(t) > 0 \\[4pt] \dfrac{N_i\left(\Delta P_{error,i}(t) + \varepsilon_i\right)}{\Delta P_{error,i}(t) \sum_{j=1}^{N_i} \frac{m_{ij}c_{ij}\left(T_{ij}(t)-T_{ij}^{\min}\right)}{ERR_{ij}}} < \mu_i < \dfrac{N_i\left(\Delta P_{error,i}(t) - \varepsilon_i\right)}{\Delta P_{error,i}(t) \sum_{j=1}^{N_i} \frac{m_{ij}c_{ij}\left(T_{ij}(t)-T_{ij}^{\min}\right)}{ERR_{ij}}}, \\ \qquad\qquad \Delta P_{error,i}(t) < 0 \end{cases} \tag{7.22}$$

It can be seen from (7.18) that the selection of μ_i depends primarily on the number of refrigerated cases N_i in CR_i, the characteristics and internal temperature of each refrigerated case, and the maximum permissible deviation ε_i. When power instructions for different periods are redistributed in different clusters, the value of μ_i can be updated according to the value range defined in (7.22).

7.5 Solution Methodology

Under the hierarchical scheduling architecture, PDC updates the electricity price information and sends it to RFA. RFA decides the pre-scheduling strategy for CR based on the electricity price information, and each refrigeration unit in the cluster calculates its own power consumption using the LREC algorithm to ensure that the actual power demand of CR remains consistent with the pre-scheduling plan within the constraint range as much as possible. The obtained load power curve is then aggregated by RFA and uploaded to PDC. PDC updates the electricity price based on the new power demand curve. The optimization process is shown in Fig. 7.4.

The optimization of refrigeration unit power consumption behavior and RFA pre-scheduling strategy is based on linear modeling, which is a mixed-integer linear programming problem. Therefore, each sub-problem can be solved quickly using the commercial solver Gurobi. The entire hierarchical optimization scheduling model can be built and solved on the Matlab platform.

7.6 Case Studies

7.6.1 Case Description

Taking Rizhao Port as an example, the proposed method is validated. The port's yard contains 3000 refrigerated containers. The model parameters used in this study are shown in Table 7.1. Twenty different temperature settings (-23 °C to $+14$ °C) and cargo with varying specific heat capacities (ranging from 1.46 to 4.06 kJ/kg.K) that allow for temperature variations are considered. Each refrigerated container loaded with the same cargo follows a log-normal distribution. The hysteresis width of the upper and lower temperature limits inside the container is 1 °C, and the initial temperature is equal to the set temperature. The upper limit of power consumption and cooling capacity of each refrigerated container at different temperature settings are shown in Fig. 7.5 [24]. The *ERR* at different temperature settings can be obtained based on the ratio of cooling capacity to upper limit of power consumption. The power consumption upper/lower limit ratio of the refrigerated containers is 9/1.

The high load threshold of the port is 80 MW, and the penalty factor for overloading is 0.5. The elasticity factor of electricity price a is 5×10^{-4} RMB/MW [25]. The scheduling period is 24 h, and each time interval Δt is 0.5 h. The simulation examples are run in the Matlab R2020b environment, and the computation platform parameters are as follows: CPU: Intel(R) Core TM i5; frequency: 3.1 GHz; memory: 16 GB.

Fig. 7.4 The flow chart of
hierarchical dispatch

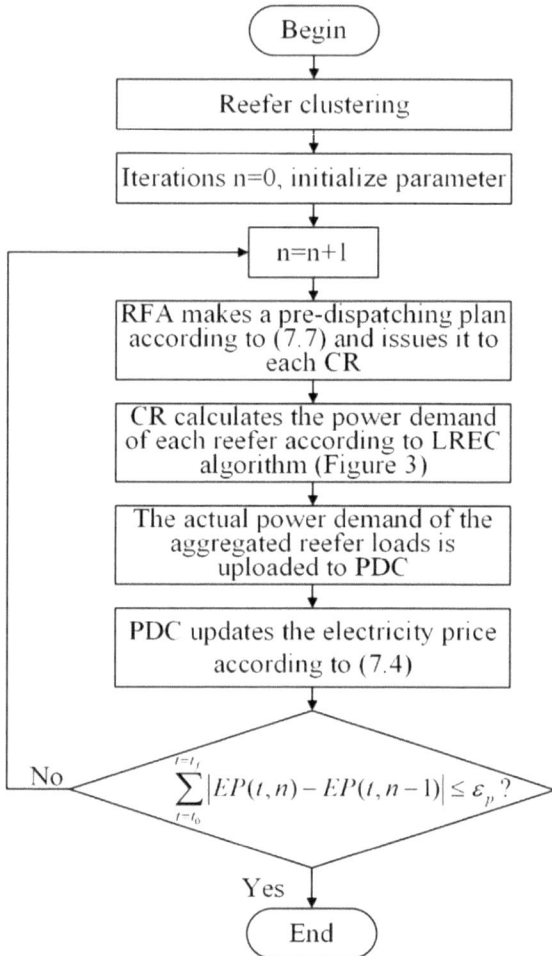

Table 7.1 Reefer parameters

	Reefer length (ft)	
	20	40
Surface area, A (m^2)	73.56	135.26
Cargo, m (kg)	17,880–21,000	22,300–26,240
k_t (W/m^2 K)	0.4	
c_p (kJ/kg K)	1.46–4.06	
Temperature set-point (°C)	−23 to +14	

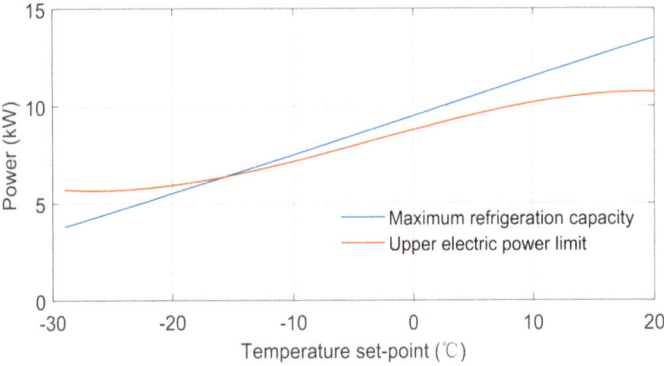

Fig. 7.5 Power limits of a reefer at different temperature set-points

7.6.2 Analysis of Scheduling Results

To verify the optimization effect of the proposed method, an unordered electricity usage scenario for refrigerated containers was added as a comparison. In the unordered electricity usage scenario, the electricity usage of the refrigerated containers is not affected by the electricity price and the port's basic load. When the internal temperature exceeds the upper limit, the refrigeration compressor of the refrigerated container starts and operates at maximum power. When the internal temperature is below the lower limit, the refrigerated container stops cooling. Figure 7.6 shows the comparison of the port's total load curve and the basic load curve under the layered optimization scheduling and the unordered electricity usage scenario. The evolution of electricity prices for each round and the load factor *TSL* are shown in Fig. 7.7. Compared with unordered electricity, the optimization method proposed in this chapter effectively shifts a large amount of refrigerated container load demand from peak hours (9:00–12:00, 19:00–23:00) to off-peak hours (3:00–7:00, 14:00–17:00) through electricity price guidance, effectively reducing the peak-valley difference of the port's load while reducing the electricity cost of refrigerated containers, which improves the system's operation. The entire optimization process only requires three iterations to converge. The predicted electricity price is used in the first round of iteration, and then adjusted appropriately according to (7.4).

The electricity cost of reefer optimized by the method in this chapter is 64,543 RMB, and the electricity cost is 73,793 RMB if the method of disorderly electricity consumption is adopted. The method proposed in this chapter can reduce the electricity cost of reefer by 12.5%.

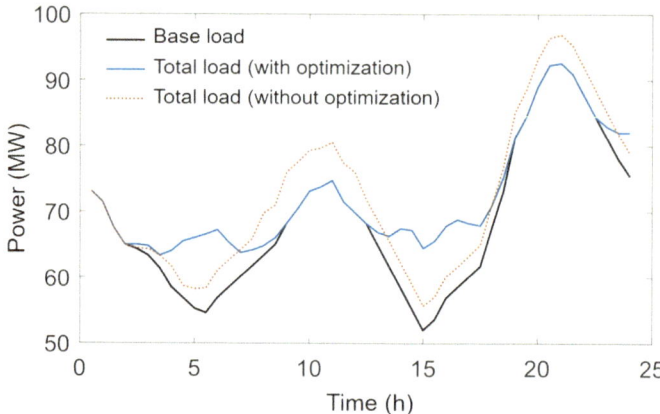

Fig. 7.6 Performance of hierarchical optimal dispatch

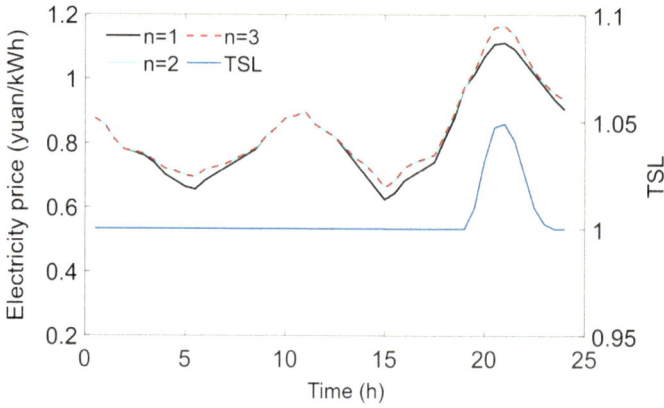

Fig. 7.7 Electricity price evolution process and loading factor

7.6.3 *Efficiency Analysis of Consensus Based Multi-agent Hierarchical Optimization*

The large-scale optimization and scheduling problem for cold storage load is a constrained mixed integer linear programming problem. If a single cold storage unit is used as an independent scheduling object, it involves 288,000 decision variables and 432,000 constraints in this case. It is actually impossible to solve this problem using a centralized optimization algorithm because it requires very long computation time. In Literature [16], an optimization method based on Multi-Agent Systems (MAS) is proposed. The MAS method can reduce the computational pressure by dividing this large-scale optimization problem into 3000 small-scale optimization problems.

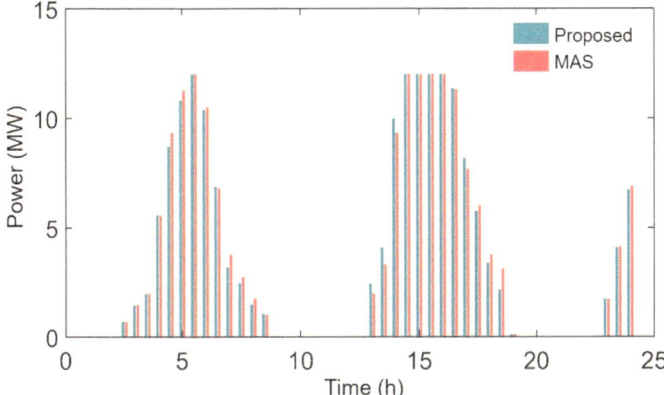

Fig. 7.8 Power consumed by the reefers with the proposed method and MAS

Table 7.2 The performance of the two methods		Proposed method	MAS method
	Electricity cost (RMB)	64,543	64,658
	Computing time (s)	733	3510

In this chapter, we compare our method with the MAS method to demonstrate the effectiveness of our method in improving optimization efficiency.

The optimal power demand for the port cold storage load obtained by our method and the MAS method are shown in Fig. 7.8. Table 7.2 compares the operating costs and optimization time of cold storage units under the two methods. It can be seen that the results obtained by our method are very similar to those obtained by the MAS method, and the optimized electricity cost differs by only 0.02%. However, the optimization time required by our method is reduced by 79.12% compared to the MAS method, and the computational efficiency is improved by about four times. This is because the improvement in computational efficiency by the MAS method depends on high-performance distributed parallel computing units. Therefore, when using the same computing device, our method has an advantage in computation time.

7.6.4   LREC Algorithm Analysis

(1)   Effect Analysis of LREC Power Dynamic Distribution

The consensus topology of information status in the CR intermediate reefer is shown in Fig. 7.2. The connection weight b_{ij} of information exchange is set to 1. In this section, CR_1 is taken as the research object, and other CR simulation analyses are similar. The convergence process of the consistency of the refrigeration efficiency factor of reefers in CR_1 at a certain moment randomly selected is shown in Fig. 7.9. It

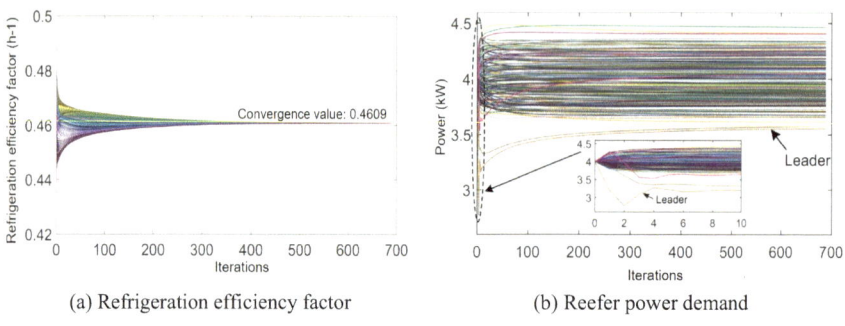

(a) Refrigeration efficiency factor   (b) Reefer power demand

Fig. 7.9 Convergence process of LREC algorithm in ideal communication environment

can be seen from Fig. 7.9a that the refrigeration efficiency factor of reefers gradually reaches consensus after continuous interaction of information. From Fig. 7.9b, it can be seen that when receiving the pre-scheduling instruction, the leader will respond first, and then other followers will follow the leader to respond to the pre-scheduling instruction.

To verify the convergence robustness of the algorithm in the presence of communication delay and noise, a non-ideal communication scenario under the same conditions was added for comparison. The communication noise follows a normal distribution $N[0, 4]$ and the probability distribution of the channel delay is as follows: the probability of a delay of 0 time unit is 0.8, the probability of a delay of 1 time unit is 0.1, and the probability of a delay of 2 time units is 0.1. The introduced consistency gain function $c[k]$ is shown in (7.23).

$$c[k] = 0.5\left[\frac{1}{0.05k+1} + \frac{\ln(0.05k+1)}{0.05k+1}\right] \tag{7.23}$$

The convergence process of the consistency of the CR_1 internal cooling box refrigeration efficiency factor under non-ideal communication conditions is shown in Fig. 7.10. In the presence of noise and delay, the refrigeration efficiency factor of the cooling box needs to iterate more times to reach consistency, and its convergence final value is 0.4602, with a deviation from the result under ideal communication environment of only 0.15%. It can be seen that the algorithm has good robustness to the interference of communication delay and noise.

The temperature variation inside the optimized reefers is shown in Fig. 7.11. The initial internal temperatures of reefers, ranging from CR_1 to CR_{20}, are shown from low to high in the figure. It can be seen that throughout the scheduling cycle, although the ambient temperature changes, the internal temperature of the reefers remains stable and within the allowable range of variation. To further verify the ability of the LREC power dynamic allocation algorithm to take into account the specific constraints of each reefer and to respond reasonably to the pre-scheduling strategy, the average allocation method is selected for simulation and comparative

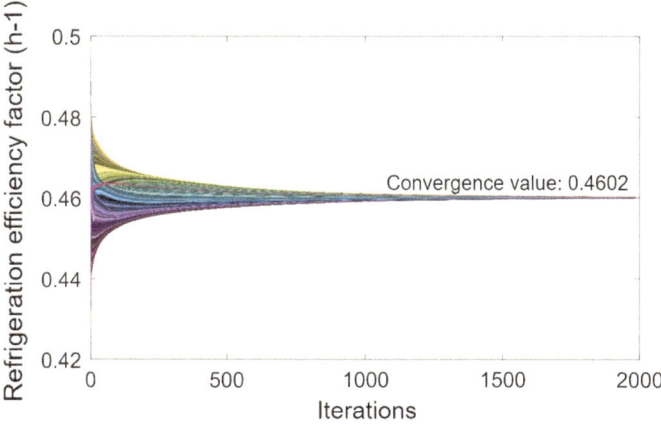

Fig. 7.10 Convergence process of Refrigerating Efficiency in non-ideal communication environment

analysis. Figure 7.12 shows the maximum and minimum temperatures of all refrigerators in CR_1 at different times using the two methods. It can be seen that the use of the power average allocation method, which does not consider the different characteristics and actual constraints of the refrigerators, can cause some refrigerators to exceed their internal temperature limits, while the LREC algorithm can ensure that the refrigerators always operate within the safe constraint range.

(2) **Analysis of the Convergence Rate of LREC Under Different Scales**

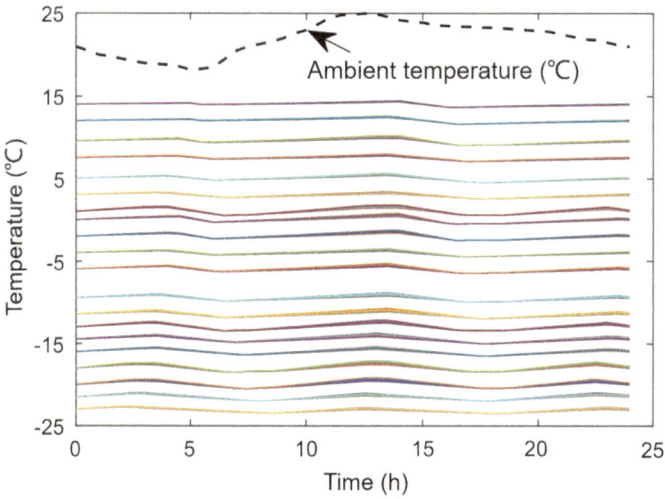

Fig. 7.11 Ambient temperature and internal temperature of reefers

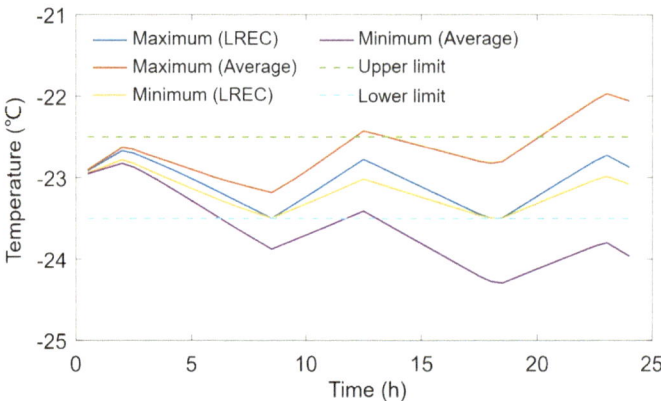

Fig. 7.12 Temperature of reefers in CR_1 with the LREC algorithm and average allocation

Generally speaking, as the number of reefers in the same cluster increases, the LREC algorithm needs to iterate more steps to achieve consensus. Here, we discuss the impact of different numbers of reefers on the convergence speed of consistency. Based on the above example, the reefer scale of CR_1 is set to (20, 40, 60, …, 200) in turn. The convergence speed of consistency within CR_1 under different scales is shown in Fig. 7.13. It can be seen that the median number of iterations only increases linearly. If the traditional centralized optimization method is used, the size of the solution space will exponentially increase with the increase of decision variables, and the optimization time will also increase exponentially with the number of reefers [26]. Therefore, the LREC algorithm can handle more reefers within a reasonable time.

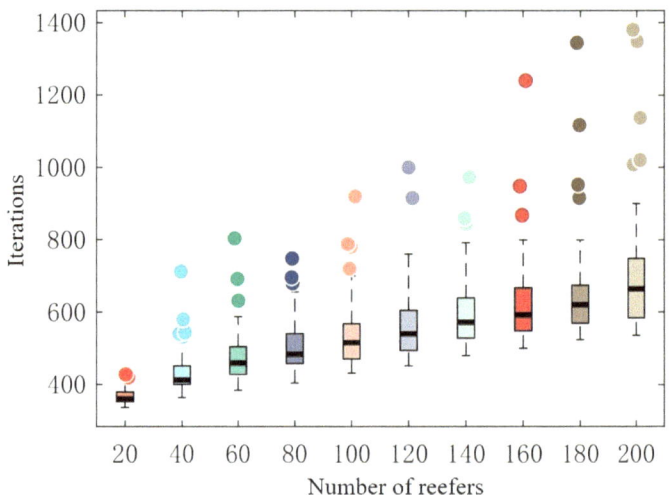

Fig. 7.13 Consensus convergence speed under different scales of CR_1

7.6.5 Method Accuracy Verification

To validate the accuracy of the proposed method, a simulation was conducted on a case of 1000 refrigerated containers using both the proposed method and a global optimization method. The refrigerated containers were still divided into 20 clusters. The implementation process of the global optimization method was as follows: after the PDC published the electricity price, the optimal power demand curve of all refrigerated containers was solved by RFA, with a single refrigerated container as the scheduling object, and then uploaded to the PDC. The same process was repeated until the electricity price converged. The optimal power demand of the refrigerated containers obtained by the proposed method and the global optimization method is shown in Fig. 7.14. It can be seen that the results obtained by the proposed method are basically consistent with those obtained by the global optimization method, indicating that the two methods are equivalent within the allowable error range and verifying the accuracy of the proposed method. The comparison of optimization effects is shown in Table 7.3. In the scenario of 1000 refrigerated containers, the proposed method can reduce the electricity cost by 10.9%, and compared with the case of 3000 refrigerated containers, the larger the scale of the container terminal, the more significant the cost-saving effect of using the proposed method.

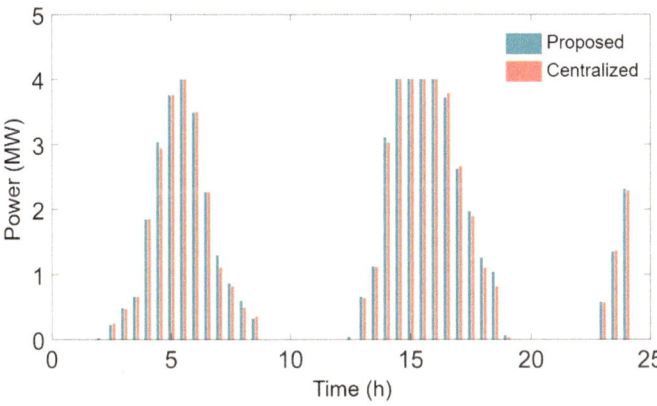

Fig. 7.14 Power consumed by the reefers with the proposed method and the global optimization algorithm

Table 7.3 Comparison of optimization effects

	Proposed method	Global optimization	Unregulated
Optimize time (s)	156	8720	–
Electricity cost (RMB)	21,033	20,928	23,611
Cost comparison	−10.9%	−11.4%	–

In this case, the electricity cost obtained by global optimization is 0.5% lower than that of the proposed algorithm. This small difference can be almost ignored, so the proposed method has good performance in reducing costs. Both methods were simulated on a personal computer. The proposed method converged to a solution after 3 iterations, requiring 156 s, while the global optimization method required 8720 s. In practical applications, the number of refrigerated containers stacked in a port is far more than 1000, and it is almost impossible to use the global optimization method on conventional computing devices.

7.7   Conclusion

This chapter addresses the difficult problem of optimizing the solution and calculation efficiency for large-scale refrigerated container load scheduling in ports. A refrigerated container cluster-layered iterative scheduling architecture and a multi-agent refrigeration efficiency collaborative optimization strategy are proposed. A pre-scheduling model is established for dynamic time-of-use electricity prices and iterative optimization of refrigerated container cluster power consumption. A pre-scheduling power dynamic algorithm for refrigeration efficiency consistency is proposed, achieving efficient optimization of large-scale refrigerated container loads. Using the container yard at Rizhao Port as an example, the proposed method reduced the electricity cost of the refrigerated container cluster by 12.5%, increased the optimization problem-solving speed by 4 times, and the deviation from global optimization was only 0.5%. In addition, the method's solution efficiency and accuracy are not affected by the number of refrigerated containers, and it has good robustness, providing technical support and reference solutions for energy conservation and emission reduction in large container ports.

References

1. Huang, Y., Huang, W., Wei, W., et al.: Logistics-energy collaborative optimization scheduling method for large seaport integrated energy system. Proc. CSEE **42**(17), 6184–6196 (2022)
2. Zhang, X., Shan, Q., Li, T., et al.: Smart port energy management strategy considering cold chain system. In: 2021 33rd Chinese Control and Decision Conference, Kunming (2021)
3. Fang, S., Zhao, C., Ding, Z., et al.: Port integrated energy systems toward carbon neutrality (I): typical topology and key problems. Proc. CSEE **43**(01), 114–135 (2023)
4. Fang, S., Zhao, C., Ding, Z., et al.: Port integrated energy systems toward carbon neutrality (II): flexible resources and key technologies in energy-transportation integration. Proc. CSEE **43**(03), 950–969 (2023)
5. The Port of Los Angeles: 2019 air emissions inventory (2021). https://kentico.portoflosang eles.org/getmedia/4696ff1aa441-4ee8-95ad-abe1d4cddf5e/2019_Air_Emissions_Inventory. Accessed 1 Oct 2020
6. Iris, C., Lam, S.: A review of energy efficiency in ports: operational strategies, technologies and energy management systems. Renew. Sustain. Energy Rev. **112**, 170–182 (2019)

7. German Insurance Association: The container faced (2016). https://www.containerhandbuch. de/chb_e/wild/index.html. Accessed 1 Oct 2016
8. Yang, Z., Ding, X., Lu, X., et al.: Inverter air conditioner load modeling and operational control for demand response. Power Syst. Prot. Control **49**(15), 132–140 (2021)
9. Wei, W., Wang, D., Jian, H., et al.: A hierarchical and distributed control strategy of thermostatically controlled appliances for city park based on load model prediction. Proc. CSEE **36**(08), 2049–2056 (2016)
10. Pu, Y., Chen, W., Zhang, R., et al.: Optimal operation strategy of port integrated energy system considering demand response. In: 2020 IEEE 4th Conference on Energy Internet and Energy System Integration, Wuhan (2020)
11. Song, T., Li, Y., Zhang, X., et al.: Integrated port energy system considering integrated demand response and energy interconnection. Int. J. Electr. Power Energy Syst. **117**, 105654 (2020)
12. Kermani, M., Shirdare, E., Parise, G., et al.: A comprehensive technoeconomic solution for demand control in ports: energy storage systems integration. IEEE Trans. Ind. Appl. **58**(2), 1592–1601 (2022)
13. Duin, J., Geerlings, H., Verbraeck, A., et al.: Cooling down: a simulation approach to reduce energy peaks of reefers at terminals. J. Clean. Prod. **193**, 72–86 (2018)
14. Heij, R.: Opportunities for peak shaving electricity consumption at container terminals. Master's thesis, Delft University, Delft (2015)
15. Palensky, P., Dietrich, D.: Demand side management: demand response, intelligent energy systems, and smart loads. IEEE Trans. Industr. Inf. **7**(3), 381–388 (2011)
16. Kanellos, F.D., Volanis, E., Hatziargyriou, N.: Power management method for large ports with multi-agent systems. IEEE Trans. Smart Grid **10**(2), 1259–1268 (2019)
17. Bian, X., Sun, M., Zhao, J., et al.: Distributed coordinative optimal dispatch and control of source and load based on consensus algorithm. Proc. CSEE **41**(04), 1334–1347 (2021)
18. Yang, S., Tan, S., Xu, J.: Consensus based approach for economic dispatch problem in a smart grid. IEEE Trans. Power Syst. **28**(4), 4416–4426 (2013)
19. Zhang, X., Yu, T.: Virtual generation tribe based collaborative consensus algorithm for dynamic generation dispatch of AGC in interconnected power grids. Proc. CSEE **35**(15), 3750–3759 (2015)
20. Xi, L., Zhang, L., Huang, Y., et al.: Multiple level automatic generation control based on the greedy strategy. Proc. CSEE **40**(16), 5204–5217 (2020)
21. Ji, L., Tang, Y., Liu, Q., et al.: On adaptive pinning consensus for dynamic multi-agent networks with general connected topology. In: Proceedings of 2016 International Joint Conference on Neural Networks, Vancouver (2016)
22. Zhang, Z., Chow, M.Y.: Convergence analysis of the incremental cost consensus algorithm under different communication network topologies in a smart grid. IEEE Trans. Power Syst. **27**(4), 1761–1768 (2012)
23. Liu, S., Xie, L., Zhang, H.: Distributed consensus for multi-agent systems with delays and noises in transmission channels. Automatica **47**(5), 920–934 (2011)
24. Gennitsaris, S.G., Kanellos, F.D.E.: Mission-aware and cost-effective distributed demand response system for extensively electrified large ports. IEEE Trans. Power Syst. **34**(6), 4341–4351 (2019)
25. Taylor, T.N., Schwarz, P.M., Cochell, J.E.: 24/7 hourly response to electricity real-time pricing with up to eight summers of experience. J. Regul. Econ. **27**(3), 235–262 (2005)
26. Deng, Q., Gao, J., Ge, D., et al.: Modern optimization theory and applications. Sci. Sin. Math. **50**(07), 899–968 (2020)

Chapter 8
Demand Side Response in Ports Considering the Discontinuity of the ToU Tariff

8.1 Introduction

Port power system dominated by conventional synchronous sources is gradually evolving to the power system equipped or even dominated by non-synchronous sources, such as wind power and photovoltaic power [1]. Non-synchronous sources are connected through power inverters which provide no mechanical inertia [2]. They are decoupled from the system frequency and cannot actively provide inertia support for the system under active power disturbance [3, 4]. Hence, frequency stability would be a key challenge for future power systems with low and variable inertia levels [5, 6]. The system will experience faster frequency drops and will have less time to respond, leading to a noticeable impact on its ability to recover from large disturbances such as sudden load variation.

Meanwhile, the fast development of flexible load and demand side response (DSR) is likely to make the future load more volatile. In particular, the time-of-use (ToU) tariff will provide a discontinued price signal at the time between peak and off-peak periods [7–9]. If all flexible load simultaneously responds to the price signal, the load will surge or plunge in a very short time. The load spike will, conversely, reinforce the sudden price jump, creating a "critical mass" effect. However, such sudden load variation can be fatal to the frequency stability of low inertia power systems, which do not have enough active power regulation capability.

At present, many scholars have studied the frequency stability problems. Literature [10] modeled the frequency response support capabilities of wind power from the field-measured data, and proposed a frequency-constrained stochastic planning method. In literature [11], a frequency-constrained stochastic unit commitment model was proposed considering the provision of synchronized and synthetic inertia, enhanced frequency response, primary frequency response and a dynamically-reduced largest power infeed. Literature [12] established the analytical formulation of system frequency nadir considering both the thermal generators and the renewable energy plants, and thus proposed a frequency-constrained unit commitment model.

W. Huang et al., *Energy Management of Integrated Energy System in Large Ports*,
Springer Series on Naval Architecture, Marine Engineering, Shipbuilding and Shipping 18, https://doi.org/10.1007/978-981-99-8795-5_8

Literature [13] proposed a system scheduling method on the premise of quantifying the impact of wind uncertainty on system inertia, which optimizes system operation by simultaneously scheduling energy production, standing/spinning reserves and inertia-dependent fast frequency response. The aforementioned studies have solved the frequency stability problems in power system planning, unit commitment, and system scheduling, but these works are applied to large grids, rather than microgrid systems with less inertia and more flexible loads.

For frequency stability problems in isolated microgrids, Literature [14] proposed an adaptive active power droop controller and voltage setpoint control scheme, which could enhance the primary frequency response and maintain the system frequency within the acceptable limits. Literature [15] presented a primary frequency control for the engine generators to regulate the frequency of isolated microgrids. Literature [16–18] also improved the frequency stability of isolated microgrids by augmenting primary frequency controls. Literature [19] proposed a hierarchical control strategy which divided the system frequency in three zones. This strategy ensured the frequency stability of stable zone in the isolated microgrid by scheduling precautionary zone and emergency zone. In literature [20], a novel adaptive droop control was designed for energy storage system in the hybrid AC/DC microgrid, which could provide DC bus voltage support and AC bus frequency support. Literature [21] proposed an approach of load frequency control for hybrid maritime microgrid system, and studied the system dynamics under random variation of wind, wave and sensitive load demand. Literature [22] proposed a multi-agent based multi-objective renewable energy management scheme, which considered consumer's needs for power reliability, cost saving, and green consumption. The scheme can reduce the steady frequency drop while ensuring economy. In literature [23], a dynamic event-triggered robust secondary frequency control scheme was presented, which provided a good balance between frequency stability performance and communication burden. Literature [24] established a cyber-attack model and presented a resilience-based frequency regulation scheme. It employed different control schemes to protect against the cyber-attack in the system. The previous works have considered some practical problems on frequency control of microgrids. However, the negative impact of DSR on frequency stability resulting from the simultaneous response to ToU tariff has not been investigated. This is a critical challenge for future power systems with many responsive loads but low inertia levels.

This chapter proposes a new frequency-based demand side response (FB-DSR) strategy. Due to the low inertia characteristic and the abundant DSR resources, the frequency stability problem of microgrids is particularly prominent. Hence, the proposed method is validated on a test multi-microgrid system based on real data collected from microgrids in Shanghai, China. The main contents of the chapter are summarized as follows:

(1) The frequency stability issue that arises from the DSR guided by ToU tariff is proposed and solved for the first time.

(2) The proposed strategy represents frequency dynamics through an analytical expression of exchange power. It achieves the conversion of frequency

constraints by modifying long-term DSR decisions by short-term power volatility suppression through two stages.

(3) Considering electricity charges and equipment depreciation, a day-ahead DSR optimization model is proposed, which ensures the economics of the strategy.

(4) Through the analysis of time–frequency domain, the whole process constraint of frequency involving inertia support, frequency nadir and quasi-steady-state is formulated. The power volatility is suppressed by means of the lowest cost, considering load shedding, RES curtailment and DSR control.

The remainder of this chapter is organized as follows. Section 8.2 discusses how the DSR guided by ToU tariff affects frequency stability, and introduces the scenarios prone to this problem. In Sect. 8.3, the FB-DSR strategy is proposed. Section 8.4 summarizes the case study. Conclusions are drawn in Sect. 8.5.

8.2 Problem Formulation

The total moment of inertia in the port microgrid system decreases with the introduction of renewable energy, and thus the ability to maintain frequency decreases. Meanwhile, the number of flexible loads such as electric vehicles and heat pumps keeps increasing. Their simultaneous responses to the ToU tariff may lead to frequency stability problems in the system operation.

As shown in Fig. 8.1, the price signal of the ToU tariff is discontinuous. When the price suddenly drops, if a large number of flexible loads respond at the same time, the load will surge in a very short time. Such sudden load variation will drag the system's frequency down and even develop to under frequency load shedding. To present such an undesired outcome, multiple constraints are required when the system's frequency is undergoing a dynamic process. The first stage of this process ($t_1 \sim t_2$ in Fig. 8.1) is referred to as system inertial response. The rate of change of frequency ($\Delta \dot{f}$) is mainly affected by the system inertia, and it is necessary to ensure that $\Delta \dot{f}$ does not exceed the limit. The second stage ($t_2 \sim t_3$ in Fig. 8.1) is referred to as primary frequency modulation. At this point, the governor starts to prevent the frequency from further reduction. In this stage, the frequency nadir of the dynamic process should be considered to make the minimum frequency (f^*) not lower than the set value of under frequency load shedding. The third stage (after t_3 in Fig. 8.1) is referred to as secondary frequency modulation. Since the governor cannot bring the frequency back to the original value, the automatic generation control (AGC) will start and use the reserve to bring the frequency back. At this moment, the quasi-steady-state frequency constraint must be satisfied, that is, the difference between the quasi-steady-state frequency and the initial value (Δf) meets the requirement. In general, throughout this frequency response process, the inertia support constraint, the frequency nadir constraint and the quasi-steady-state constraint should be satisfied at the same time. Otherwise, the frequency stability of system will be affected.

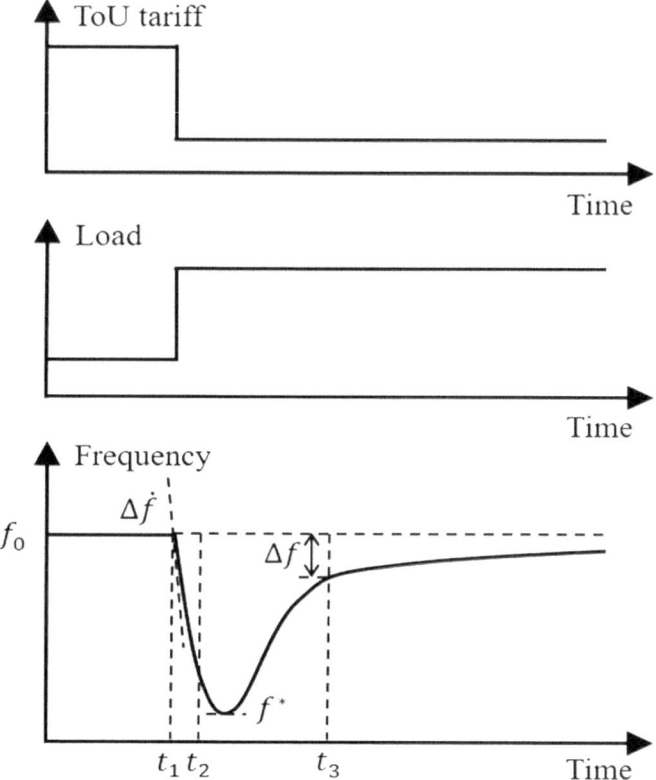

Fig. 8.1 A conventional diagram of load and frequency response for discontinued price signals

This is a problem that has not been fully discussed but could be critical in the future when flexible load like electric vehicles uptakes on a large scale. It is particularly urgent for microgrids with high penetrations of renewable energy and thus low and variable levels of inertia. With the development of interconnected microgrids and peer to peer (P2P) trading based on ToU tariff, unregulated demand response may lead to serious frequency issues. Therefore, this chapter will study the FB-DSR strategy under the condition of interconnected microgrids with P2P trading based on ToU tariff.

8.3 The FB-DSR Strategy

The mismatch between the power consumption and the ToU tariff leads to high electricity costs. However, the exact match will trigger simultaneous response and potential frequency issues. This section proposes an FB-DSR strategy to address the trade-off between customers' economic benefits and systems' frequency stability.

Section 8.3.1 introduces the day-ahead complete-period DSR optimization method, which runs iteratively every hour. The conversion of frequency to power and the short-term power volatility suppression method is given in Sect. 8.3.2.

8.3.1 Day-Ahead Complete-Period DSR Optimization

(1) Equipment Depreciation Model

The service life of power regulating equipment will be compromised when they participate in DSR. The depreciation of energy storage equipment is related to the charging and discharging cycles during the equipment operation. A complete cycle with $x\%$ depths of charge means the storage equipment's SOC decreases from 100 to $x\%$ and then increases from x to 100%. The equipment life depreciation of a complete cycle is defined as $D_{100\sim x}^{comp}$. Therefore, the life depreciation during the incomplete cycle can be calculated according to the rain-flow cycle counting algorithm [25]. Equation (8.1) calculates the life depreciation of the equipment when its SOC changes from x to y%, where $n_{x\sim y}$ means the number of the cycles the equipment completed during the operation period.

$$D_{x\sim y}^{incomp} = n_{x\sim y} \left| D_{100\sim x}^{comp} - D_{100\sim y}^{comp} \right| \tag{8.1}$$

Combining the calculation formula of $D_{100\sim x}^{comp}$ [25], the relation between equipment depreciation and their SOC can be obtained, as shown in (8.2).

$$D(k, h) = 0.5 \left| \frac{SOC(k, h+1) - b_k}{a_k} - \frac{SOC(k, h) - b_k}{a_k} \right| \tag{8.2}$$

where a and b are the parameters determined by the type of equipment.

(2) Demand Side Response Model

In microgrids, the heat pump group is often used as flexible load [26]. Generally, the heat demand is met through centralized heat production, which is more efficient. Large water storage tanks will be equipped for heat storage, which can work with the battery for demand response. The heat pump group generally includes base-load heat pumps and flexible regulating heat pumps. The former has high rated power and bears the base load of heat demand. The latter has low rated power and is able to adjust the switch state rapidly with the change of the users' heat demand. The heat output power of heat pump is defined in (8.3).

$$Q_{hpi}(h) = C_{hp} \cdot P_{hpi}(h) \tag{8.3}$$

where $Q_{hpi}(h)$ is the heat output power of heat pump i at hour h, C_{hp} is the energy efficiency ratio coefficient of heat pump, $P_{hpi}(h)$ is the electric power of heat pump

i at hour h. The heat generated by the heat pump group is transmitted to the heat storage tank through the heat exchanger by heat circulating working medium. The heat storage tank supplies the heat demand of each user. The first-order equivalent thermal parameter model is used to simulate water temperature changes in the water tank, as defined in (8.4) and (8.5).

$$V \cdot \rho_W \cdot c_W \cdot \frac{dT_t(h)}{dh} = \sum_i Q_{hpi}(h) - Q_{load}(h) - \frac{T_t(h) - T_{en}(h)}{R_t} \qquad (8.4)$$

$$T^{min} \le T_t(h) \le T^{max} \qquad (8.5)$$

where V is water storage volume, ρ_W is water density, c_W is the specific heat capacity of water, $T_t(h)$ is water temperature of heat storage tank at hour h, $T_{en}(h)$ is environment temperature at hour h, $Q_{load}(h)$ is the forecast result of heat demand at hour h, R_t is thermal resistance of heat storage tank, T^{min} and T^{max} are the minimum/maximum temperatures of the stored water respectively.

The energy stored in the heat storage tank, which has been converted into electricity, is described in (8.6).

$$S_{tank}(h) = V \cdot \rho_w \cdot c_w \cdot \frac{[T_t(h) - T^{min}]}{C_{hp}} \qquad (8.6)$$

where $S_{tank}(h)$ is the energy stored in the tank at hour h. Analogous to the SOC of battery, the SOC of heat storage tank is established in (8.7), which is described by the water temperature.

$$SOC_{tank}(h) = \frac{[T_t(h) - T^{min}]}{[T^{max} - T^{min}]} \qquad (8.7)$$

where $SOC_{tank}(h)$ is the SOC of heat storage tank energy storage system at hour h. The heat pump group and heat storage tank are able to be regarded as a load and storage coordination power control unit, which can cooperate with the battery to realize the DSR.

(3) Day-Ahead Complete-Period Optimization Model

There is likely to be a mismatch between the power consumption and the ToU tariff because of the intermittence of RES and the energy usage habit of users. The DSR optimization can shift electricity from low price periods to high price periods by controlling the heat pump group and battery. Besides, the model proposed in this chapter considers the depreciation expense of storage equipment during the power regulation. The power of heat pump group and battery is adjusted every hour at this stage. The forecast results of load and RES are also based on one-hour timescale. Furthermore, since both photovoltaic and user loads have obvious daily patterns, the energy storage equipment take 24 h as an operation cycle, that is, the SOC after 24 h should be equal to the initial SOC. The optimization model is shown as follows.

$$\min : \sum_{h=1}^{N} [C_{elec}(h) P_{line}(h) + C_{batt} D_{batt}(h) + C_{hs} D_{hs}(h)] \tag{8.8}$$

$$P_{line}(h) = P_{load}(h) + P_{hp}(h) - P_{wind}(h) - P_{pv}(h) + P_{batt}^{cha}(h) - P_{batt}^{dis}(h) \tag{8.9}$$

where $C_{elec}(h)$ is the ToU tariff at hour h, $P_{line}(h)$ is the optimization result of exchange power at hour h, C_{batt} is the battery cost, $D_{batt}(h)$ is the battery depreciation at hour h, C_{hs} is the heat storage system cost, $D_{hs}(h)$ is the heat storage system depreciation at hour h, $P_{load}(h)$ is the forecast result of electricity demand at hour h, $P_{wind}(h)$ is the forecast result of wind power at hour h, $P_{pv}(h)$ is the forecast result of photovoltaics power at hour h, $P_{batt}^{cha}(h)$ and $P_{batt}^{dis}(h)$ are the charging and discharging power of the battery respectively.

Objective function (8.8) minimizes the daily electricity cost and the equipment depreciation expense. The exchange power is determined by load, heat pump group, RES and battery, as shown in (8.9). The model also includes energy storage operation constraints, upper and lower power constraints and security constraints, which can be found in [27].

After solving the complete-period DSR optimization model, the optimal operating status of the heat pump group and battery for the next 24 h is obtained. Then, the results of the first hour are taken as the reference power of heat pump group and battery for the next hour. The adjustments of heat pump group and battery power are based on these reference power at the stage of short-term volatility suppression. In addition, the SOC of the energy storage equipment at the end of the first hour is the target SOC of short-term volatility suppression during the adjustment of the next hour, and the minute-level target SOC is obtained by linearization.

8.3.2 Short-Term Power Volatility Suppression

As the amount of flexible load increases, the system frequency deviation may be too large when all flexible load simultaneously responds. Hence, this subsection discusses the relationship between frequency and short-term power volatility in detail, and proposes a method to judge whether the short-term volatility needs to be suppressed after the complete-period DSR optimization. In addition, the optimal short-term power suppression method is given.

(1) Frequency Stability Constraints

When analyzing the power volatility, the multi-microgrid systems can be simplified as Fig. 8.2. If the Microgrid A conducts demand response, a sudden change power ΔP occurs on the tie-line and will have an impact on other microgrids. According to the analysis in Sect. 8.2, ΔP_B, ΔP_C, and ΔP_N need to satisfy the inertia support constraint, frequency nadir constraint and quasi-steady-state constraint at the same time, so as to ensure the frequency stability.

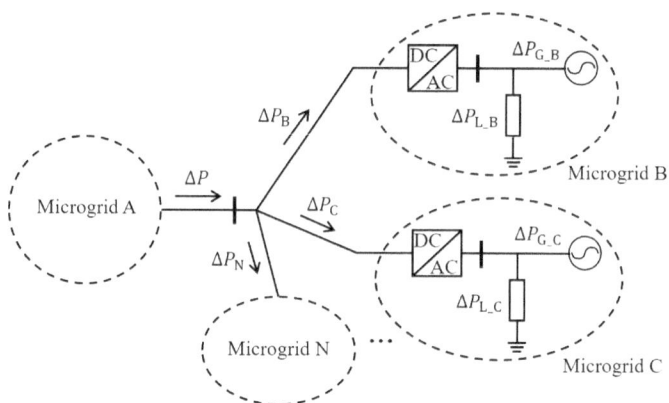

Fig. 8.2 The equivalent of multi-microgrid systems

After the sudden disturbances, the system first relies on inertia to suppress the frequency change. The initial rate of change of frequency should meet the following requirements.

$$\left| \Delta \dot{f} \right| = \left| \frac{\Delta P}{2H} f_0 \right| \leq \Delta \dot{f}_{max} \tag{8.10a}$$

where $\Delta \dot{f}$ is the rate of change of frequency, f_0 is the system nominal frequency, H is the system inertia. The above equation is the inertia support constraint, from which the maximum power mutation value, ΔP_1^{max}, can be deduced.

$$\Delta P_1^{max} = \left| \frac{2H \Delta \dot{f}_{max}}{f_0} \right| \tag{8.10b}$$

where $\Delta \dot{f}_{max}$ is the maximum value of rate of change of frequency. At the end of the inertia support stage, the governors begin to respond to the sudden change of frequency to prevent its further changes. The frequency nadir constraint should be considered to ensure that the maximum frequency offset value does not exceed the set value of under frequency load shedding.

The dynamic model of the governor is equivalent to the first-order model as follows.

$$\Delta P_G = \frac{K_G}{1 + T_G s} \Delta \omega \tag{8.11a}$$

The first-order frequency response model of the system considering the governors is shown in Fig. 8.3, and its transfer function can be written as (8.11b). Note that Microgrid A is connected to multiple microgrids. Hence, parameter equivalent aggregation method is used to merge the parameters of connected microgrids. The parameter aggregation formulas are shown in (8.11c), (8.11d) and (8.11e).

Fig. 8.3 The first-order frequency response model

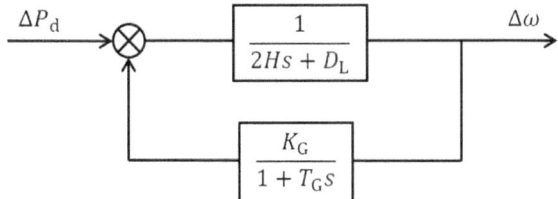

$$\frac{\Delta\omega}{\Delta P_\mathrm{d}} = \frac{1}{2Hs+D_\mathrm{L}+\frac{K_\mathrm{G}}{1+T_\mathrm{G}s}} \tag{8.11b}$$

where D_L is the load damping constant. Note that Microgrid A is connected to multiple microgrids. Hence, parameter equivalent aggregation method is used to merge the parameters of connected microgrids. The parameter aggregation formulas are shown in (8.11c), (8.11d) and (8.11e).

$$D_\mathrm{L} = \frac{\sum_{i=\mathrm{B}}^{N} P_{\mathrm{N}i}\cdot D_i}{\sum_{i=\mathrm{B}}^{N} P_{\mathrm{N}i}} \tag{8.11c}$$

$$K_\mathrm{G} = \frac{\sum_{i=\mathrm{B}}^{N} P_{\mathrm{N}i}\cdot K_i}{\sum_{i=\mathrm{B}}^{N} P_{\mathrm{N}i}} \tag{8.11d}$$

$$T_\mathrm{G} = \frac{\sum_{i=\mathrm{B}}^{N} P_{\mathrm{N}i}\cdot T_i}{\sum_{i=\mathrm{B}}^{N} P_{\mathrm{N}i}} \tag{8.11e}$$

where N is the number of power optimization periods. The above Eq. (8.11b) can be converted to

$$\frac{\Delta\omega}{\Delta P_\mathrm{d}} = \frac{1}{2HT_\mathrm{G}}\frac{1+T_\mathrm{G}s}{s^2+2\varepsilon\omega_\mathrm{n}s+\omega_\mathrm{n}^2} \tag{8.11f}$$

where:

$$\omega_\mathrm{n}^2 = \frac{K_\mathrm{G}+D_\mathrm{L}}{2HT_\mathrm{G}} \tag{8.11g}$$

$$\varepsilon = \frac{1}{2}\frac{2H+D_\mathrm{L}T_\mathrm{G}}{\sqrt{2HT_\mathrm{G}(K_\mathrm{G}+D_\mathrm{L})}} \tag{8.11h}$$

where in (8.11g), ω_n is the nominal angular velocity. For sudden disturbances, we assume a step function, i.e., $\Delta P_\mathrm{d}(s) = -\Delta P/s$. Taking t as the index of the minute, the time-domain response can be derived as

$$\Delta\omega(t) = -\frac{\Delta P}{2HT_\mathrm{G}\omega_\mathrm{n}^2} - \frac{\Delta P}{2H\omega_r}e^{-\varepsilon\omega_\mathrm{n}t}$$
$$\times\left(\sin(\omega_r t) - \frac{1}{\omega_\mathrm{n}T_\mathrm{G}}\sin(\omega_r t + \phi)\right) \tag{8.11i}$$

where:

$$\omega_r = \omega_n \sqrt{1 - \varepsilon^2} \qquad (8.11j)$$

$$\phi = \arcsin\left(\sqrt{1 - \varepsilon^2}\right) \qquad (8.11k)$$

In the extreme points of $\Delta\omega(t)$, $\frac{d\Delta\omega}{dt} = 0$, so the maximum frequency offset time can be obtained as follows

$$t^* = \frac{1}{\omega_r} \arctan\left(\frac{\omega_r}{\varepsilon\omega_n - \frac{1}{T_G}}\right) \qquad (8.11l)$$

Substituting (8.11l) into (8.11i), the maximum frequency offset $\Delta\omega(t^*)$ can be found.

$$\Delta\omega(t^*) = \chi \Delta P \qquad (8.11m)$$

where,

$$\chi = -\frac{1}{K+D_L}\left(1 + e^{-\varepsilon\omega_n t^*}\sqrt{\frac{T_G K}{2H}}\right) \qquad (8.11n)$$

The minimum frequency is calculated as

$$f^* = f_0 + f_0\Delta\omega(t^*) \qquad (8.11o)$$

where f^* is the minimum frequency. This frequency should not be lower than the set value of under frequency load shedding f_{shed}, that is, $f^* \geq f_{shed}$. Therefore, the maximum power mutation value obtained from the frequency nadir constraint is as follows,

$$\Delta P_2^{max} = \left|\frac{f_0 - f_{shed}}{f_0 \chi}\right| \qquad (8.11p)$$

where ΔP_2^{max} is the maximum power mutation value derived from frequency nadir constraint. After the primary frequency control participated by governors, the sudden change power, ΔP, is compensated by generation reduction, ΔP_G, and load increase, ΔP_L. The formulas are shown below.

$$\Delta P + \Delta P_G - \Delta P_L = 0 \qquad (8.12a)$$

$$\Delta P_G = -K_G \cdot \Delta f \qquad (8.12b)$$

$$\Delta P_L = -D_L \cdot \Delta f \qquad (8.12c)$$

By solving (8.12a), (8.12b) and (8.12c), the frequency deviation is obtained as shown in (8.12d).

$$\Delta f = \frac{\Delta P}{K_G + D_L} \tag{8.12d}$$

The frequency at this point is called the quasi-steady-state frequency, and it should be within the interval $[f^{min}, f^{max}]$. Therefore, according to the quasi-steady-state constraint, the upper and lower limits of power mutation value can be obtained as follows.

$$\Delta \overline{P_3^{max}} = (f^{max} - f_0) \cdot (K_G + D_L) \tag{8.12e}$$

$$\Delta \underline{P_3^{max}} = (f^{min} - f_0) \cdot (K_G + D_L) \tag{8.12f}$$

where $\Delta \overline{P_3^{max}}$, $\Delta \underline{P_3^{max}}$ are the upper/lower limits of the maximum power mutation value derived from quasi-steady-state constraint.

(2) Construction of the Suppression Power

Define the allowable range of sudden change power considering the frequency stability constraints to be $[\Delta P^{min}, \Delta P^{max}]$, and its calculation formulas are shown as follows.

$$\Delta P^{max} = \max\{\Delta P_1^{max}, \Delta P_2^{max}, \Delta \overline{P_3^{max}}\} \tag{8.13a}$$

$$\Delta P^{min} = \min\{-\Delta P_1^{max}, -\Delta P_2^{max}, \Delta \underline{P_3^{max}}\} \tag{8.13b}$$

where ΔP_1^{max} is the maximum power mutation value derived from inertia support constraint. The allowable range of power exchanged between Microgrid A and the system can be calculated. The calculation formulas are shown below.

$$P_{s\_line}^{max}(t) = P_{s\_line}(t-1) - \Delta P^{min} \tag{8.14a}$$

$$P_{s\_line}^{min}(t) = P_{s\_line}(t-1) - \Delta P^{max} \tag{8.14b}$$

where P_{s_line} is the exchange power after using the real-time data, ΔP^{max} and ΔP^{min} are the upper/lower power fluctuation limits of the microgrid respectively. After the complete-period power optimization, the initial exchanged power is determined by the user load, the real-time output of RES and the reference power of heat pump group and battery, which is shown in (8.15).

$$P_{s\_line}(t) = P_{s\_load}(t) - P_{s\_wind}(t) - P_{s\_pv}(t) + P_{hp}^{ref} + P_{batt}^{ref} \tag{8.15}$$

where $P_{s\_load}(t)$, $P_{s\_wind}(t)$, $P_{s\_pv}(t)$ are the real-time electricity demand/wind power/photovoltaics at time t, P_{batt}^{ref}, P_{hp}^{ref} are the reference power of battery/heat pump group respectively. Influenced by the sudden changes of the ToU tariff, the initial exchanged power may have significant short-term volatility. The portion of this power that exceeds the allowable range is the short-term suppression power that needs to be processed at this minute.

$$P_{s\_line}^{flu}(t) = \begin{cases} P_{s\_line}^{max}(t) - P_{s\_line}(t), & \text{if } P_{s\_line}(t) > P_{s\_line}^{max}(t) \\ P_{s\_line}^{min}(t) - P_{s\_line}(t), & \text{if } P_{s\_line}(t) < P_{s\_line}^{min}(t) \\ 0, & \text{otherwise} \end{cases} \tag{8.16}$$

where $P_{s\_line}^{flu}$ is the fluctuation power that need to be stabilized.

(3) The Suppression Approach of Power Volatility

To ensure frequency stability, power volatility can be suppressed by load shedding, RES curtailment and DSR control. The optimal suppression model of short-term power volatility is established as follows.

$$\min : C_{shed} \cdot P_{shed} + C_{curt} \cdot P_{curt} + C_{dsr} \cdot P_{dsr} \tag{8.17}$$

s.t.

$$P_{s\_line}^{flu} = -P_{shed} + P_{curt} + P_{dsr} \tag{8.18}$$

$$P_{shed} \geq 0 \tag{8.19}$$

$$P_{curt} \geq 0 \tag{8.20}$$

$$C_{curt} = C_{elec}(h) \tag{8.21}$$

$$C_{dsr} = C_{elec}(h) - C_{elec}(h+1) \tag{8.22}$$

where C_{shed} is the load shedding cost, C_{curt} is the renewable energy sources (RES) cost, C_{dsr} is the demand side response control cost, P_{shed} is the load shedding power, P_{curt} is the RES curtailment power, P_{dsr} is the demand side response control power.

Objective function (8.17) minimizes the cost of load shedding, RES curtailment and DSR control. The sum of the power suppressed by these three measures equals to $P_{s\_line}^{flu}$, as shown in (8.18). Note that load shedding can only reduce power, and RES curtailment can only reduce generation. The cost of load shedding is determined by the importance of load. The cost of RES curtailment is consistent with the electricity price, and the cost of DSR control is consistent with the difference of ToU tariff.

After solving the optimal suppression model, load shedding and RES curtailment are implemented. The battery, together with the power control unit composed of the heat pumps and heat storage tank, participates in the DSR control. The DSR suppression task is distributed according to the energy storage capacity of battery and heat storage tank, so as to coordinate the output of battery and heat pump group. The base proportion that the battery should undertake is shown in (8.23).

$$\alpha_0 = \frac{S_{\text{batt}}^{\text{max}}}{\left(S_{\text{tank}}^{\text{max}} + S_{\text{batt}}^{\text{max}}\right)} \tag{8.23}$$

where α_0 is the ratio of battery energy storage, $S_{\text{batt}}^{\text{max}}$, $S_{\text{tank}}^{\text{max}}$ are the maximum capacity of the battery/heat storage tank respectively.

The proportion of the task assignment is dynamically adjusted based on α_0. The fuzzy control method is utilized, and the deviation between the actual state and target state of each energy storage equipment is taken as the reference, which is calculated in (8.24) and (8.25). When energy consumption is required, that is, $P_{s\_line}^{\text{flu}}(t) > 0$, if $deSOC_{\text{batt}}(t)$ is higher than $deSOC_{\text{tank}}(t)$, the suppression proportion undertaken by the battery will be reduced, otherwise the proportion will be increased. When energy discharging is required, the opposite is true. Under different circumstances, the suppression proportion of battery is shown in Fig. 8.4.

$$deSOC_{\text{tank}}(t) = SOC_{\text{tank}}(t) - SOC_{\text{tank}}^{\text{tar}}(t) \tag{8.24}$$

$$deSOC_{\text{batt}}(t) = SOC_{\text{batt}}(t) - SOC_{\text{batt}}^{\text{tar}}(t) \tag{8.25}$$

where $deSOC_{\text{batt}}(t)$, $deSOC_{\text{tank}}(t)$ are the deviation between the actual state and the target state of battery/heat storage tank energy storage system at time t respectively. $SOC_{\text{batt}}^{\text{tar}}(t)$, $SOC_{\text{tank}}^{\text{tar}}(t)$ are the target soc of battery/heat storage tank energy storage system at time t respectively. $SOC_{\text{batt}}(t)$ is the SOC of heat storage tank energy storage system at time t.

After considering taking part in short-term volatility suppression, the control power of battery at this time is shown as follows.

$$P_{s\_\text{batt}}^{\text{flu}}(t) = \alpha(t) \cdot P_{\text{dsr}}(t) \tag{8.26}$$

$$P_{s\_\text{batt}}^{\text{tar}}(t) = P_{\text{batt}}^{\text{ref}} + P_{s\_\text{batt}}^{\text{flu}}(t) \tag{8.27}$$

where $P_{s\_\text{batt}}^{\text{flu}}$ is the fluctuation power that need battery to stabilize, $P_{s\_\text{batt}}^{\text{tar}}$ is the target power of the battery. The actual power of the battery is also limited by the rated power.

The heat pump group reduces the short-term power volatility by changing the switch state of heat pump and adjusting the consumed power. The target power of heat pump group is the sum of the reference power and the remaining short-term suppression power. The difference between the target power and the actual power

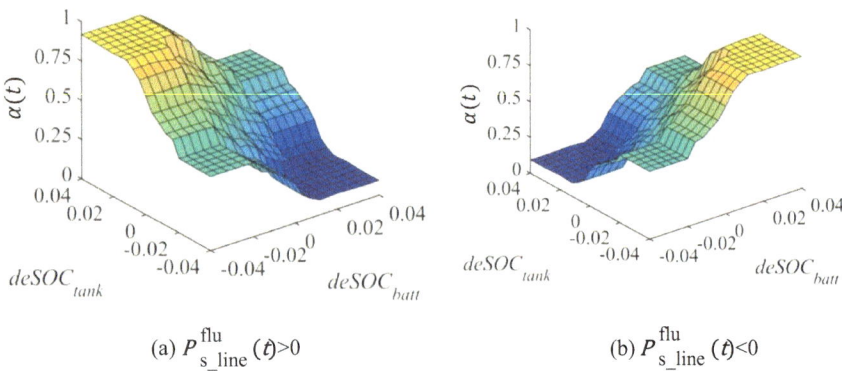

(a) $P^{\text{flu}}_{\text{s\_line}}(t)>0$

(b) $P^{\text{flu}}_{\text{s\_line}}(t)<0$

Fig. 8.4 The suppression ratio of battery during the short-term power volatility suppression

at the previous moment is the on–off control target of the heat pump group at this moment. The formulas are shown as follows.

$$P^{\text{tar}}_{\text{s\_hp}}(t) = P^{\text{ref}}_{\text{hp}} + P_{\text{dsr}}(t) - \left[P^{\text{act}}_{\text{s\_batt}}(t) - P^{\text{ref}}_{\text{batt}} \right] \tag{8.28}$$

$$P^{\text{oc}}_{\text{s\_hp}}(t) = P^{\text{tar}}_{\text{s\_hp}}(t) - P^{\text{act}}_{\text{s\_hp}}(t-1) \tag{8.29}$$

where $P^{\text{tar}}_{\text{s\_hp}}$ is the target power of heat pump group. $P^{\text{act}}_{\text{s\_batt}}$ is the actual power of battery, and $P^{\text{oc}}_{\text{s\_hp}}$ is the on–off control target of heat pump group.

The heat pump group includes two types: the base-load heat pump and the flexible regulating heat pump. When controlling the heat pump group to meet the on–off control target, the following rules are considered in this chapter: (1) avoid the flexible regulating heat pumps all on or off; (2) average the on–off times of the same type heat pump; (3) turn on or off as few heat pumps as possible. When rule 1 contradicts rule 3, follow rule 1 first. These rules will help to extend the overall service life of heat pumps.

8.4   Case Studies

8.4.1   Case Description

The case study of the proposed FB-DSR strategy is based on a test multi-microgrid system. This system consists of three AC microgrids, which are connected by DC transmission line. The data is collected from microgrids in Shanghai, China. Microgrid A shall perform peak shaving and valley filling through DSR, and its frequency

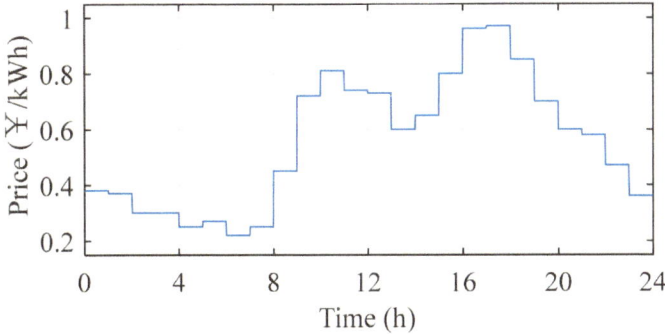

Fig. 8.5 An example of typical ToU tariff for microgrids

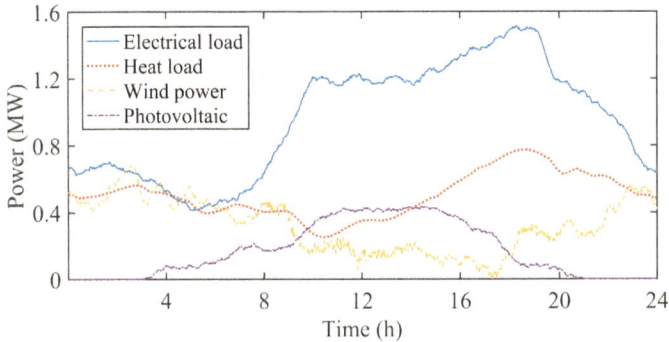

Fig. 8.6 The actual load and power generation of Microgrid A

shall be kept within the allowable range. Microgrid A uses the heat pump group as flexible loads, which contains 20 base-load heat pumps and 25 flexible regulating heat pumps. In addition, the maximum power of the battery and the SOC range of the hybrid energy storage system are limited during complete-period DSR optimization, so as to satisfy the requirement of short-term suppression. The ToU tariff [28], which is obtained from net load of the multi-microgrid system, is shown in Fig. 8.5 and the load shedding cost is ¥27.08/kWh [29]. The actual load and power generation of Microgrid A is shown in Fig. 8.6. Table 8.1 summarizes the equipment parameters [30]. Other parameters for case simulation are shown in Table 8.2.

8.4.2 The Results of the Proposed Strategy

Figure 8.7 shows the simulation results of the proposed strategy, and Fig. 8.8 demonstrates the regulating power of the battery and heat pump group. Note that, the

Table 8.1 Equipment parameters for the case study

Parameter		Value
Heat pump group	Base-load heat pump power	40 kW
	Flexible regulating heat pump power	8 kW
	Energy efficiency ratio coefficient	350%
Heat storage tank	Volume	107 m^3
	The maximum temperature	80 °C
	The minimum temperature	40 °C
	Thermal resistance	2 °C/kWh
	Heat storage system cost	1.2e + 5¥
Battery	Capacity	1000 kWh
	Rated power	300 kW
	Charge–discharge efficiency	92%
	Battery cost	1.655e + 6¥

Table 8.2 Simulation parameters for the case study

Parameter	Value
The maximum power of battery in complete-period suppression	280 kW
The SOC range of hybrid energy storage system in complete-period suppression	0.05–0.95
System nominal frequency	50 Hz
The maximum rate of change of frequency	0.2 Hz/s
The set value of under frequency load shedding	49.8 Hz
The allowable range of frequency	±0.1 Hz
System inertia	8 s
Load damping constant	1.5
Frequency regulation factor of generator	20
Generator time constant	7 s

exchange power fluctuates according to the ToU tariff by adjusting the power of controllable devices.

The SOC target and actual SOC of battery and heat storage tank are shown in Fig. 8.9. After participating in the short-term volatility suppression, the actual SOC is still able to follow the target SOC obtained by the complete-period DSR optimization. This indicates that the short-term suppression strategy proposed in this chapter will not have a significant impact on the DSR result. Figure 8.10 shows the on–off times of heat pumps. It can be seen that the on–off times of the same type heat pumps are equal, which is conducive to extending the overall service life of the heat pump group.

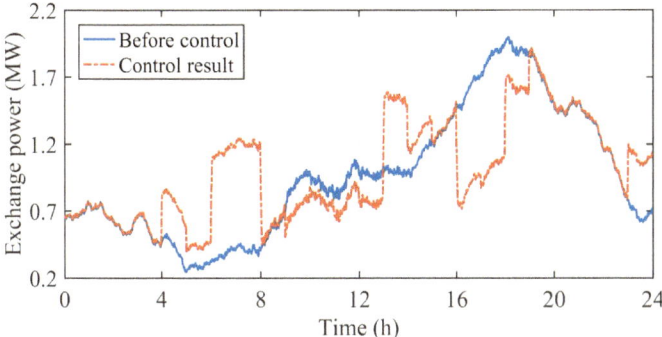

Fig. 8.7 The exchange power obtained by the proposed strategy and for previous situation

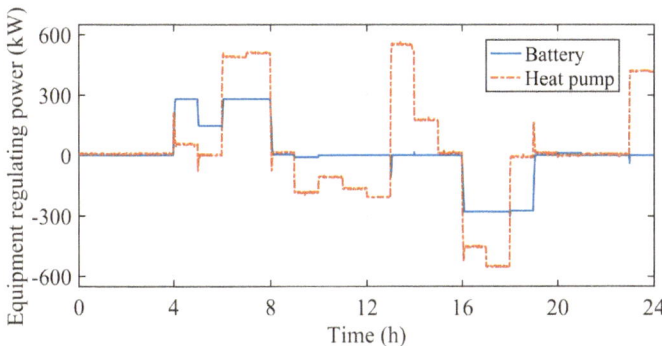

Fig. 8.8 The equipment regulating power obtained by the proposed strategy

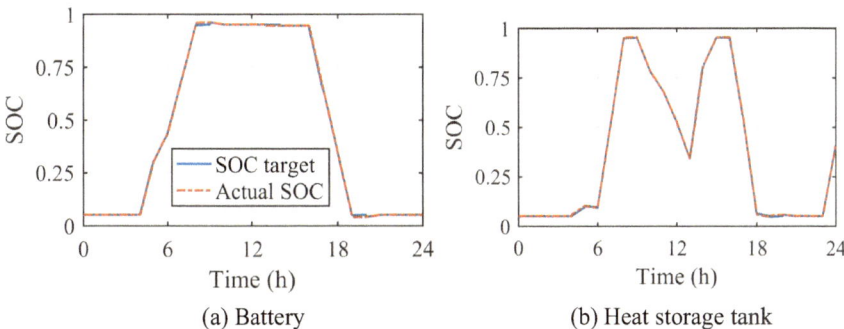

(a) Battery   (b) Heat storage tank

Fig. 8.9 The SOC of each equipment obtained by the proposed strategy

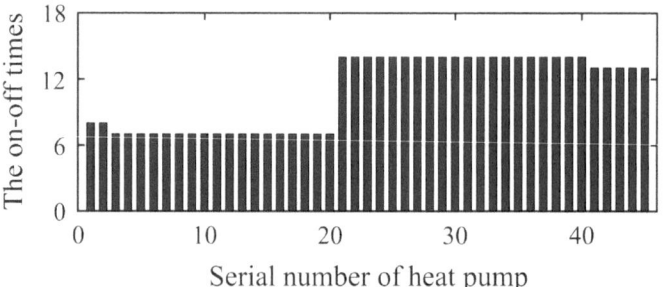

Fig. 8.10   The on–off times of heat pumps obtained by the proposed strategy

8.4.3   Comparative Studies

In addition to the proposed FB-DSR strategy (S_1), the common DSR strategy that ignores frequency stability (S_2) is also utilized in this chapter. We have magnified the simulation results of the two strategies at thirteen o'clock, as shown in Fig. 8.11. It can be seen that as the ToU tariff suddenly decreases, the load power increases immediately. Under the control of S_2, the sudden increase of power results in the 1.8% decrease in the frequency. While, under the control of S_1, the exchange power gradually increases, so that the frequency is always maintained within the allowable range. The generators start AGC to increase output and eventually recover the power volatility. The differential power of load under two strategies is compensated by controlling the power of heat pump and battery. In the actual operation process, when the frequency goes beyond the lower limit, it will maintain the frequency stability by cutting the load. Therefore, under the control of S_2, under frequency load shedding will occur when the frequency is in the shaded part of the figure.

Table 8.3 presents the results comparison for the two strategies. The maximum absolute value (MAV) of the 1-min exchange power volatility is used to measure the short-term volatility. The MAV of S_1 and S_2 is 313.02 kW and 845.77 kW, respectively, which proves that S_1 is effective in short-term volatility suppression. The electricity cost and the depreciation expense of equipment are almost equal for the two strategies. The power volatility suppression cost of S_1 is much lower than S_2. This is because when the power fluctuates greatly, S_2 can only adopt load shedding and RES curtailment to maintain frequency stability, while S_1 can adopt DSR control. In general, S_1 is superior to S_2 in economy and security.

8.4.4   Short-Term Volatility Suppression Effect

This subsection analyses the short-term volatility suppression effect of the proposed strategy. The short-term suppression power of each equipment is presented in Fig. 8.12 and the frequency deviation in Fig. 8.13. It can be seen from Figs. 8.5

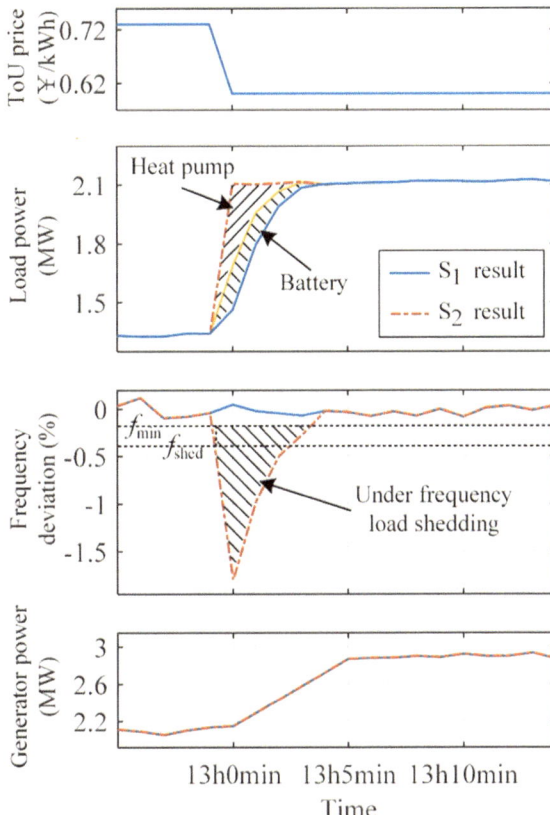

Fig. 8.11 The comparison of results obtained by the proposed strategy (S_1) nd the common DSR strategy (S_2) at thirteen o'clock

Table 8.3 Comparison of results obtained by the proposed strategy (S_1) and the common DSR strategy (S_2)

	S_1	S_2
MAV (kW)	313.02 (−62.99%)	845.77
Electricity cost (¥)	14,016.70 (−0.0062%)	14,017.57
Depreciation expense (¥)	345.69 (+0.37%)	344.41
Power volatility suppression cost (¥)	2.68 (−99.77%)	1141.72
Total cost (¥)	14,365.07 (−7.34%)	15,503.7

and 8.7 that the sudden change in price will cause the sudden change of exchange power, but will not destabilize the frequency. The proposed strategy can adjust the power of battery and heat pump group when the short-term volatility of exchange power is too large, so that the frequency can be kept within the allowable range. For example, at eight o'clock, the price of electricity jumps 80% and the load power of Microgrid A drops rapidly. At this moment, the battery and the heat pump group

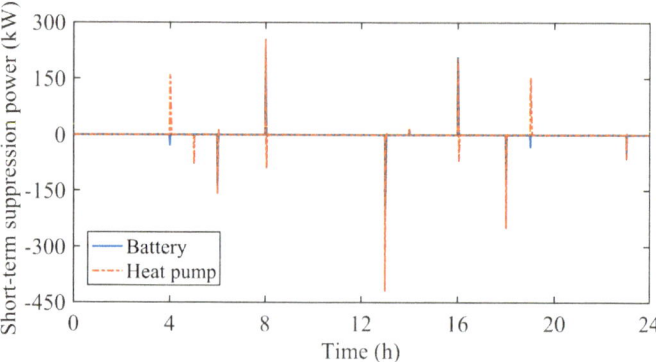

Fig. 8.12 The short-term suppression power of each equipment obtained by the proposed strategy

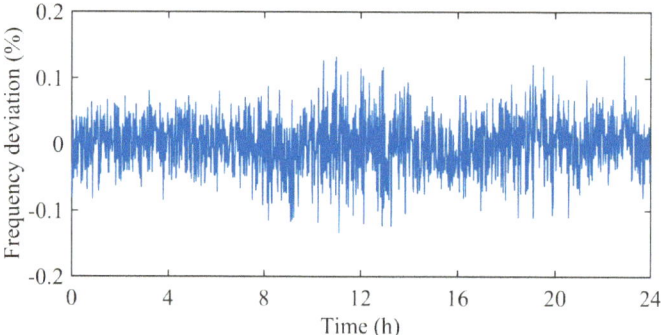

Fig. 8.13 Frequency deviation of Microgrid A obtained by the proposed strategy

consume additional power based on the reference power determined by the complete-period DSR optimization, to alleviate the short-term volatility of power and maintain the frequency within the allowable range.

8.5   Conclusion

The DSR guided by the ToU tariff can change the electricity consumption behaviour of users and promote the balance between energy supply and demand under the condition of high renewable energy penetration. However, with the increase of flexible load, the discontinuity of the ToU tariff could trigger sudden and drastic variations of loads, threatening the frequency stability of low-inertia system. This chapter proposes a FB-DSR strategy, which integrates time-domain optimization and frequency-domain control and realizes the whole process constraint of frequency. By dynamically distributing the regulating power of each equipment, it adjusts the complete-period power fluctuation, as well as the drastic short-term volatility. The strategy reduces the electricity cost, and circumvents frequency stability issues associated with DSR. The equipment depreciation due to power regulation is also considered. Case simulation demonstrates that the proposed strategy can improve the acceptability of power system to DSR. The MAV of 1-min exchange power volatility decreases by 62.99%, which keeps the system frequency within the allowable range and reduces the loss caused by frequency instability. The overall operating cost is reduced by 7.34%.

References

1. Mayer, M.J., Gróf, G.: Extensive comparison of physical models for photovoltaic power forecasting. Appl. Energy **283**(12), 116239 (2020)
2. Zhang, C., Lu, X., Ren, G., et al.: Optimal allocation of onshore wind power in China based on cluster analysis. Appl. Energy **285**, 116482 (2021)
3. Howlader, A.M., Sadoyama, S., Roose, L.R., et al.: Active power control to mitigate voltage and frequency deviations for the smart grid using smart PV inverters. Appl. Energy **258**, 114000 (2020)
4. Magdy, G., Ali, H., Xu, D.: A new synthetic inertia system based on electric vehicles to support the frequency stability of low-inertia modern power grids. J. Clean. Prod. **297**(6), 126595 (2021)
5. Qazi, H., Wall, P. Escudero, M.V., et al.: Impacts of fault ride through behavior of wind farms on a low inertia system. IEEE Trans. Power Syst. **37**(4), 3190–3198 (2020)
6. Badesa, L., Teng, F., Strbac, G.: Optimal portfolio of distinct frequency response services in low-inertia systems. IEEE Trans. Power Syst. **35**(6), 4459–4469 (2020)
7. Badesa, L., Strbac, G., Magill, M., et al.: Ancillary services in Great Britain during the COVID-19 lockdown: a glimpse of the carbon-free future. Appl. Energy **285**(1), 116500 (2021)
8. Bejan, I., Jensen, C.L., Andersen, L.M., et al.: Inducing flexibility of household electricity demand: the overlooked costs of reacting to dynamic incentives. Appl. Energy **284**(15), 116283 (2021)
9. Wang, X., Huang, W., Wei, W., et al.: Day-ahead optimal economic dispatching of integrated port energy systems considering hydrogen. IEEE Trans. Ind. Appl. **58**(2), 2619–2629 (2021)
10. Meinrenken, C.J., Mehmani, A.: Concurrent optimization of thermal and electric storage in commercial buildings to reduce operating cost and demand peaks under time-of-use tariffs. Appl. Energy **254**, 113620 (2019)
11. Li, H., Qiao, Y., Lu, Z., et al.: Frequency-constrained stochastic planning towards a high renewable target considering frequency response support from wind power. IEEE Trans. Power Syst. **36**(5), 4632–4644 (2021)
12. Badesa, L., Teng, F., Strbac, G.: Simultaneous scheduling of multiple frequency services in stochastic unit commitment. IEEE Trans. Power Syst. **34**(5), 3858–3868 (2019)
13. Zhang, Z., Du, E., Teng, T.F.: Modeling frequency dynamics in unit commitment with a high share of renewable energy. IEEE Trans. Power Syst. **35**(6), 4383–4395 (2020)
14. Teng, F., Trovato, V., Strbac, G.: Stochastic scheduling with inertia-dependent fast frequency response requirements. IEEE Trans. Power Syst. **31**(2), 1557–1566 (2016)
15. Alghamdi, B., Cañizares, C.A.: Frequency regulation in isolated microgrids through optimal droop gain and voltage control. IEEE Trans. Smart Grid **12**(2), 988–998 (2021)
16. Davari, M., Gao, W., Jiang, Z., et al.: An optimal primary frequency control based on adaptive dynamic programming for islanded modernized microgrids. IEEE Trans. Autom. Sci. Eng. **18**(3), 1109–1121 (2021)
17. Hussain, A., Shireen, W.: Model for frequency dynamics in an islanded microgrid and primary frequency control based on disturbance compensation. IEEE Access **9**, 52784–52795 (2021)

18. Schneider, K.P., Radhakrishnan, N., Tang, Y., et al.: Improving primary frequency response to support networked microgrid operations. IEEE Trans. Power Syst. **34**(1), 659–667 (2019)
19. Mendieta, W., Cañizares, C.A.: Primary frequency control in isolated microgrids using thermostatically controllable loads. IEEE Trans. Smart Grid **12**(1), 93–105 (2021)
20. Zhao, Z., Yang, P., Guerrero, J.M., et al.: Multiple-time-scales hierarchical frequency stability control strategy of medium-voltage isolated microgrid. IEEE Trans. Power Electron. **31**(8), 5974–5991 (2016)
21. Li, X., Dong, C., Jiang, W., et al.: An improved coordination control for a novel hybrid AC/ DC microgrid architecture with combined energy storage system. Appl. Energy **292**, 116824 (2021)
22. Al, A., Smsh, B., Dcd, A., et al.: Double stage controller optimization for load frequency stabilization in hybrid wind-ocean wave energy based maritime microgrid system. Appl. Energy **282**, 116171 (2020)
23. Xiong, L., Li, P., Wang, Z., et al.: Multi-agent based multi objective renewable energy management for diversified community power consumers. Appl. Energy **259**, 114140 (2020)
24. Yang, C., Yao, W., Fang, J., et al.: Dynamic event-triggered robust secondary frequency control for islanded AC microgrid. Appl. Energy **242**, 821–836 (2019)
25. Mishra, D.K., Ray, P.K., Li, L., et al.: Resilient control based frequency regulation scheme of isolated microgrids considering cyber attack and parameter uncertainties. Appl. Energy **306**, 118054 (2022)
26. Ke, X., Lu, N., Jin, C., et al.: Control and size energy storage systems for managing energy imbalance of variable generation resources. IEEE Trans. Sustain. Energy **6**(1), 70–78 (2015)
27. Huang, W., Wang, X., Tai, N., et al.: Energy coupling conversion model and cascade utilization method for microgrid with heat and power system. Proc. CSEE **40**(21), 6804–6814 (2020)
28. Wang, X., Huang, W., Tai, N., et al.: Two-stage full-data processing for microgrid planning with high penetrations of renewable energy sources. IEEE Trans. Sustain. Energy **12**(4), 2042–2052 (2021)
29. Shanghai Municipal Development & Reform Commission. http://fgw.sh.gov.cn/cxxxgk/201 90531/0025-36105.html. Accessed 24 March 2021
30. How do we price in VoLL? https://www.cornwall-insight.com/newsroom/all-news/black-out-what-black-out-the-curious-case-of-unexciting-imbalance-prices-on-9-august. Accessed 24 June 2021
31. Wang, X., Huang, W., Tai, N., et al.: A tie-line power smoothing strategy for microgrid with heat and power system using source-load-storage coordination control. Trans. China Electrotech. Soc. **35**(13), 2817–2829 (2020)

Chapter 9
Energy Cascade Utilization of Electric-Thermal Port Microgrids

9.1 Introduction

Electric-thermal port microgrid is one of the typical applications of port integrated energy systems. Based on electrical and thermal demands, it integrates the supply, conversion, and storage equipment in electric and thermal energy flows, coordinates and optimizes protection and control methods, so as to achieve economical and reliable operation [1–4]. With the increasingly diverse energy needs of industrial, commercial, and residential users supplied by microgrids, there exist more complex multi-energy couplings, such as energy cascade utilization, which poses difficulties for energy optimization management of electric thermal microgrids [5, 6]. Energy cascade utilization is an effective method to improve energy utilization efficiency and supply quality. It is an important direction in current research on energy optimization management of electric-thermal port microgrids [7–9].

Currently, there has been a large number of researches on energy management in electric-thermal microgrids [10–12]. Literature [13] considered the coupling constraints of heating, cooling, and electricity, and used a clustering algorithm to obtain the optimal energy storage configuration for the integrated energy system to improve the system's economic performance. Literature [14] took into account the thermal storage characteristics of the heating network and established a system optimization and scheduling model to improve the consumption level of wind and solar energy in the region. Literature [15, 16] established a hybrid energy storage system model and proposed a coordinated operation strategy for electric and thermal energy storage. Literature [17] constructed an electric, thermal, and gas coupling network model and proposed an optimization strategy based on demand-side response, achieving optimized peak-shaving and valley-filling operation of the system. Literature [18] considered the dual uncertainties of renewable energy and load terminals and established a stochastic optimization model for cogeneration. The above literatures have provided relatively comprehensive models for the devices involved in electric-thermal microgrids, such as the sources, networks, loads, and storage. These

© The Author(s) 2023
W. Huang et al., *Energy Management of Integrated Energy System in Large Ports*,
Springer Series on Naval Architecture, Marine Engineering, Shipbuilding and Shipping
18, https://doi.org/10.1007/978-981-99-8795-5_9

researches enable the optimization configuration and operation of the system, and lay a research foundation for the coordinated complementarity of heterogeneous energy sources and the mutual substitution of different-grade energy sources.

Literature [19, 20] constructed a hierarchical optimization model for regional integrated energy systems considering multi-objective optimization in terms of economics, environmental protection, and energy efficiency. Literature [21, 22] proposed an energy management method for microgrids with energy storage devices, which enhanced system reliability while simultaneously reducing operating costs. Literature [23–25] established an economic dispatch model for energy systems with multiple energy supply sources, with the objective of minimizing operating costs. Literature [26] proposed a weighting method based on the Analytic Hierarchy Process to establish a comprehensive evaluation index for integrated energy systems from economic, reliability, environmental, and energy consumption aspects. Currently, scholars have suggested the utilization of flexible and controllable gas turbines in microgrid energy management based on the current production situation, which further improves energy utilization efficiency and is more in line with the development trend of actual production [27–29]. Literature [30] proposed a comprehensive energy system multi-energy collaborative optimization model that considers energy cascade utilization for industrial parks. This model followed the idea of "matching quality and cascade utilization" to achieve the comprehensive optimization of equipment operating parameters and industrial production processes.

The research on the comprehensive optimization and evaluation of multi-energy synergies in integrated energy systems has been increasingly matured. However, the energy utilization structure for energy cascading utilization is still relatively simple, lacking analysis of the substitution relationship between heterogeneous energy sources and energy sources with different grades, and the economic operating potential of the electric-thermal port microgrid has not been fully explored. In this chapter, different grades of energy sources in the electric-thermal microgrid are considered, and an energy cascade utilization flow structure based on the cascade utilization principle of gas turbines is constructed. An optimization model for energy cascade utilization in the electric-thermal microgrid is proposed with the objective of minimizing daily operating costs. On the Matlab platform, the system optimization operating model is established and the optimal operating scheme is solved. The effectiveness of the proposed strategy is verified through a case study of the "China-Italy Green Energy Experimental Center" at a university in southwest Shanghai.

9.2 Electric-Thermal Port Microgrids

9.2.1 Structure of Electric-Thermal Port Microgrids

An electric-thermal microgrid includes multiple forms of energy such as cold, heat, electricity, and gas. In addition to supplying electricity to user loads, it also includes steam loads, high/medium-temperature hot water loads, and chilled water loads. A typical microgrid structure that includes industrial, commercial, and residential users is shown in Fig. 9.1. Among them, the steam load corresponds to the demand of industrial users; in addition to meeting the heating needs of commercial and residential users, high-temperature hot water also satisfies the hot water needs of some industrial users; medium-temperature hot water meets the domestic water needs of commercial and residential users; and chilled water supplies the cooling load needs of all users.

Electric thermal microgrids can be divided into micro-power grids and micro-thermal grids (including cooling and heating), which couples different forms of energy through integrated energy stations. The integrated energy station includes various energy equipment such as gas turbines, absorption heat pumps, absorption refrigeration, and energy storage, consuming natural gas and interacting with the power grid through transmission lines to supply or store cold, heat, and electrical energy. Wind power generation equipment is connected to the micro-power grid, and its output has certain fluctuations and randomness. Solar thermal power plants use solar thermal co-generation to supply electricity to the micro-power grid and to supply heat to the micro-thermal grid.

Considering the diverse energy needs of users in the electric thermal microgrid, in addition to electrical loads, there are also steam loads, high-temperature hot water loads, and medium-temperature hot water loads. Different heat load requires different

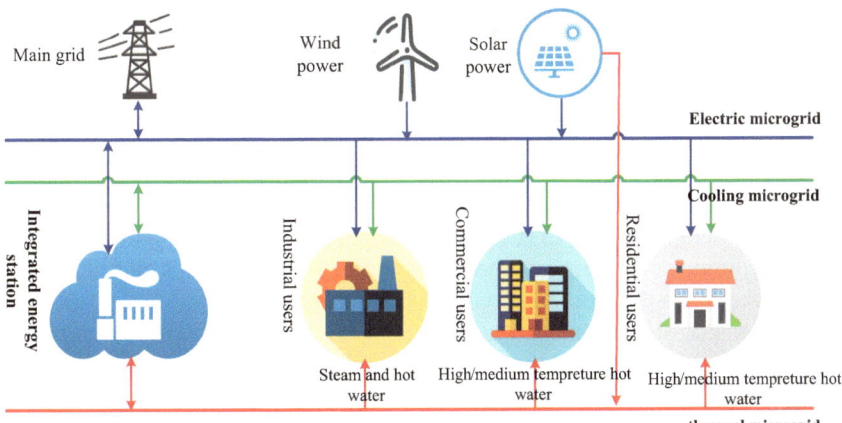

Fig. 9.1 The structure and energy flow of electric-thermal microgrid

water supply temperatures. Multiple forms of energy flows require the configuration of various types of energy coupling and conversion equipment in the electric thermal microgrid, making the operating mode of the electric thermal microgrid more complex than that of traditional distribution networks. However, considering energy cascade optimization, multiple energy flows and the abundance of controllable energy coupling equipment also make it more flexible for mutual support among various of energy forms. When purchasing electricity prices, natural gas prices, and renewable energy generation change, the adjustment and control methods of electric-thermal microgrids are also more diverse. Therefore, the principle of energy cascade utilization has broader optimization space for electric-thermal microgrids with diversified load demands.

9.2.2 Cascaded Utilization of the Electric-Thermal Microgrids

The cascaded utilization of electric-thermal microgrid follows the principles of "electric-thermal complementarity, temperature matching, and cascaded utilization." Thermal energy is divided into different grades based on temperature, and the higher the temperature, the higher the thermal energy grade. Furthermore, according to the temperature requirements of different thermal loads, efforts are made to meet the heat utilization of temperature matching as much as possible. In cascaded utilization, higher-grade thermal energy is recycled, recovered and gradually converted to lower-grade thermal energy, thereby achieving high-efficiency energy utilization. Additionally, the advantages of "electric-thermal complementarity" are leveraged in the electric-thermal microgrid, and the reliability and efficiency of cascaded utilization are improved based on the deep coupling of electric-thermal energy flow.

The energy cascading utilization method in gas turbines is shown in Fig. 9.2, where the extracted air heat and exhaust heat are utilized separately. The exhaust heat is recovered by the waste heat boiler to produce low-temperature hot water of around 34 °C. Since the value of storage and transfer low-temperature hot water are low, a portion of the waste heat is used as the low-temperature heat source of the absorption heat pump, and the remaining portion is used to heat the boiler feedwater or space heating load. The extracted air heat is high-temperature steam, and a portion of it drives the absorption heat pump to heat a large amount of low-temperature hot water to mid-temperature of 75 °C, and another portion is heated to high-temperature of 120 °C through the peak heater heat exchanger to supply steam load and produce cold water through the absorption refrigeration.

Based on the above cascaded utilization process and considering the electric-thermal coupling relationship, the energy equipment and loads of the electric-thermal microgrid are shown in Fig. 9.3. The sources of power and heat supply include gas turbines, waste heat boilers, gas boilers, absorption heat pumps, absorption refrigeration, peak heaters, electric heat pumps, electric refrigeration, and electricity

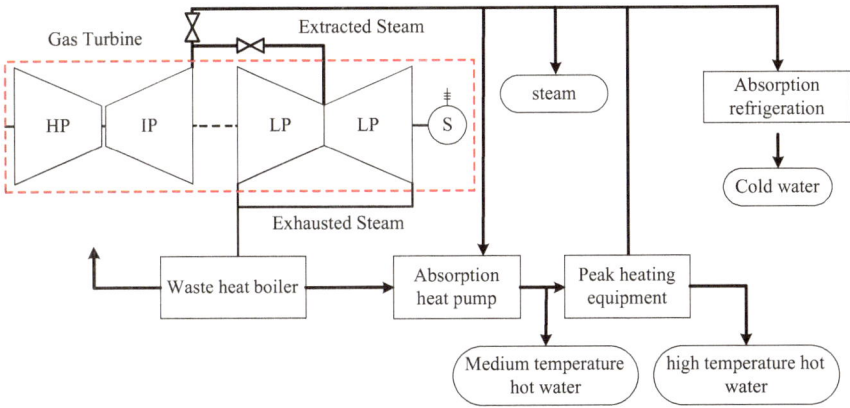

Fig. 9.2 The energy cascade utilization method of gas turbine

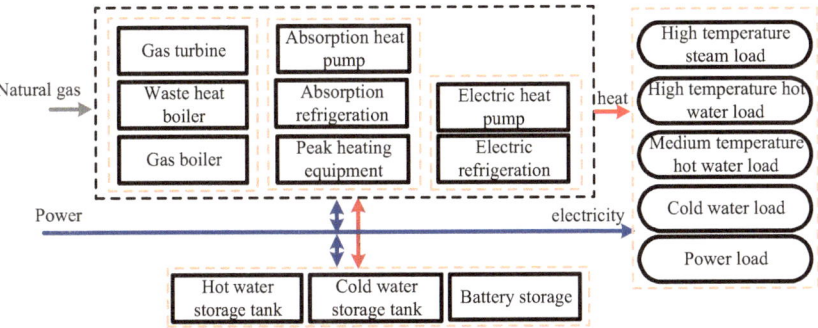

Fig. 9.3 The source-demand-storage relationship in energy cascade utilization

purchased from the main power grid. Cold, heat, and electricity storage include cold water storage tanks, hot water storage tanks, and battery storage. The loads include high-temperature steam loads, high-temperature hot water loads, mid-temperature hot water loads, and cold water loads.

9.3 Energy Flow Analysis of Cascaded Utilization in Electric-Thermal Port Microgrids

Considering the electric-thermal coupling relationship at different thermal energy levels and utilizing the advantages of multi-energy complementarity, the energy flow structure of electric-thermal coupling cascaded utilization is shown in Fig. 9.4. From the energy perspective, it can be divided into electric power bus, steam bus, low-temperature hot water bus, medium-temperature hot water bus, and high-temperature

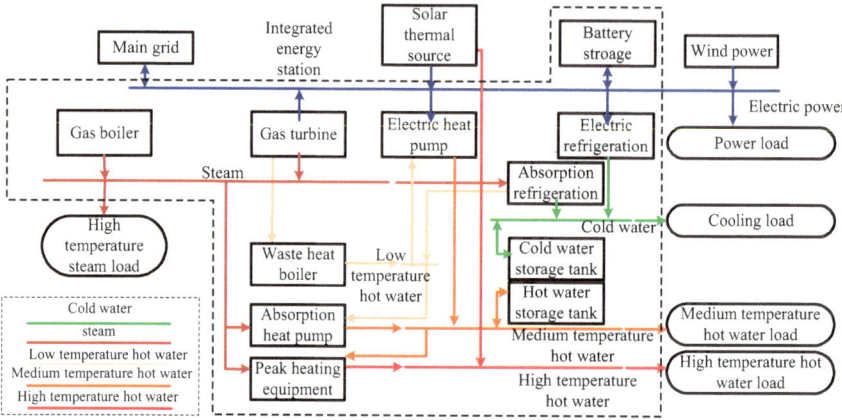

Fig. 9.4 The energy flow structure in heat and power cascade utilization

hot water bus. The electric power bus is connected to gas turbine power generation, solar thermal power generation, wind power generation, and interacts with the main power grid and battery energy storage, supplying electric loads, electric heat pumps, and electric refrigeration equipment. The steam bus is connected to gas turbine exhaust and gas boilers, supplying absorption heat pumps, peak heating equipment, absorption refrigeration equipment and high-temperature steam loads. The low-temperature hot water bus is connected with waste heat boilers and absorption refrigeration, supplying absorption heat pumps and electric heat pumps. The low-temperature hot water in medium-temperature hot water bus is heated to supply the peak heating equipment and medium-temperature hot water loads. The high-temperature hot water bus is connected with peak heating equipment and solar thermal energy, satisfying high-temperature hot water loads.

The energy flow structure of the cascade utilization of the electric-thermal microgrid includes electricity, steam, high-temperature hot water, medium-temperature hot water, low-temperature hot water, and cold water. The next section will analyze the cascade utilization model from three aspects: energy flow coupling relationship, energy grade conversion model, and energy supply-demand relationship.

9.3.1 The Coupling Relationship of Energy Flow

Different from traditional microgrids, there are multiple forms of energy flow in the electric-thermal microgrid, and there are also various coupling relationships between them. Taking full advantage of these relationships and achieving heterogeneous energy flow coupling and mutual aid in different operating conditions is an effective way to improve energy utilization efficiency and save operating costs. The energy coupling models used in this section are introduced as follows.

(1) Gas Turbine

The back-pressure gas turbine is selected as the modeling object, which has high efficiency in power generation. Throughout the operation, the power generation, gas consumption, and exhaust ratios are fixed. The waste gas from the gas turbine is recovered by the waste heat boiler while filtering out pollutants to reduce emissions. The cogeneration model of gas turbine is shown in Eq. (9.1).

$$
\begin{cases}
P_{GT}^t = \eta_{GT} \cdot \lambda F_{GT}^t \\
H_{GT\_L}^t = \eta_{exh} \cdot (1 - \eta_{GT}) \cdot \lambda F_{GT}^t \\
H_{GT\_S}^t = \eta_{ext} \cdot (1 - \eta_{GT}) \cdot \lambda F_{GT}^t
\end{cases}
\tag{9.1}
$$

In (9.1), P_{GT} is the power output of the gas turbine, λ is the lower heating value of natural gas, F_{GT} is the gas turbine's air intake flow rate, and η_{GT} is the gas turbine's power generation efficiency. H_{GT_L} is the recovered heat power from the waste gas, and η_{exh} is the waste heat recovery efficiency. H_{GT_S} is the steam power generated by the gas turbine's exhaust gas, and η_{ext} is the exhaust coefficient. The superscript t denotes time.

(2) Gas Boiler

The gas boiler also consumes natural gas, but lacks the power generation process. The combustion process generates high temperature steam directly for heating. The heating model is shown in Eq. (9.2).

$$
H_{GB}(t) = \eta_{GB} \cdot \lambda F_{GB}(t)
\tag{9.2}
$$

where F_{GB} is the intake amount of the gas boiler, H_{GB} is the heat energy contained in the steam generated by the gas boiler, and η_{GB} is the efficiency of the gas boiler.

(3) Electric Heat Pumps and Electric Refrigeration

Electric heat pumps and electric refrigeration have similar working principles, and their models can be represented by coefficients of energy efficiency ratio, as shown in Eqs. (9.3) and (9.4).

$$
H_{HP\_M}^t = C_{HP} \cdot P_{HP}^t
\tag{9.3}
$$

$$
H_{RE\_C}^t = C_{RE} \cdot P_{RE}^t
\tag{9.4}
$$

where H_{HP_M} represents the heating power of the electric heat pump, P_{HP} represents the power consumption of the electric heat pump, and C_{HP} represents the energy efficiency ratio of the electric heat pump. H_{RE_C} represents the cooling power of the electric refrigeration, P_{RE} represents the power consumption of the electric refrigeration, and C_{RE} represents the energy efficiency ratio of the electric refrigeration.

(4) Absorption Heat Pump and Absorption Refrigeration

The working principles of absorption heat pump and absorption refrigeration are similar. The absorption heat pump consumes a small amount of high temperature heat energy to produce abundant medium temperature heat energy, while the absorption refrigeration consumes high temperature heat energy to produce cold energy, accompanied by a large amount of low temperature hot water by-product. Their mathematical models are shown in Eqs. (9.5) and (9.6).

$$H_{AHP\_M}^t = C_{AHP} \cdot H_{AHP\_S}^t \qquad (9.5)$$

$$\begin{cases} H_{ACH\_C}^t = C_{ACH} \cdot H_{ACH\_S}^t \\ H_{ACH\_L}^t = k_{ACH} \cdot C_{ACH} \cdot H_{ACH\_S}^t \end{cases} \qquad (9.6)$$

In the above equations, H_{AHP_M} represents the heating power of the absorption heat pump, C_{AHP} represents the energy efficiency ratio of the absorption heat pump, and H_{AHP_S} represents the steam heat consumption of the absorption heat pump. H_{ACH_C} represents the cooling power of the absorption refrigeration, C_{ACH} represents the energy efficiency ratio of the absorption refrigeration, H_{ACH_L} represents the low temperature heat energy produced by the absorption refrigeration, and k_{ACH} is a proportional constant based on the characteristics of the absorption refrigeration equipment.

(5) Peak Heater

The peak heater can use steam to heat hot water to a high temperature. Its mathematical model is similar to that of a heat exchanger, as shown in Eq. (9.7).

$$H_{PH\_H}^t = C_{PH} H_{PH\_S}^t \qquad (9.7)$$

In (9.7), H_{PH_H} represents the heat energy transferred by the peak heater, C_{PH} represents the heat transfer efficiency, and H_{PH_S} represents the steam heat energy consumed by the peak heater.

(6) Solar Thermal Equipment

Solar thermal power generation uses concentrated solar light to generate high-temperature steam. A part of it is used for power generation through a steam turbine, and the remaining part is used for heating. Its mathematical model is shown in Eq. (9.8).

$$\begin{cases} P_{PT} = \eta_{ST} x_p E_{solar} \\ H_{PT\_H} = \eta_{ex} \big((1 - x_p) + \eta_{WH} (1 - \eta_{ST}) x_p \big) E_{solar} \end{cases} \qquad (9.8)$$

In (9.8), P_{PT} represents the solar thermal power generation capacity, η_{ST} represents the steam turbine power generation efficiency, E_{solar} represents the heat energy provided by the solar collector system, and x_p is the proportion coefficient of thermal energy entering the steam turbine. H_{PT_H} represents the solar thermal heating capacity, η_{ex} represents the efficiency of the heat exchanger, and η_{WH} represents the recovery efficiency of the steam turbine waste heat.

(7) Battery Energy Storage

The battery energy storage model is represented by the state of charge, as shown in Eq. (9.9).

$$SOC_{ess}^t = \begin{cases} (1 - \sigma_{ess})SOC_{ess}^{t-1} + \eta_{chr}\frac{P_{ess}^{t-1}\cdot\Delta t}{E_{ess,max}} \\ (1 - \sigma_{ess})SOC_{ess}^{t-1} + \frac{1}{\eta_{dch}}\frac{P_{ess}^{t-1}\cdot\Delta t}{E_{ess,max}} \end{cases} \tag{9.9}$$

where SOC_{ess} represents the state of charge of the battery energy storage, P_{ess} represents the charging/discharging power (positive for charging and negative for discharging), $E_{ess,max}$ represents the maximum capacity, η_{chr} represents the charging efficiency, η_{dch} represents the discharging efficiency, σ_{ess} represents the self-discharge coefficient, and Δt represents the time interval.

(8) Thermal Energy Storage Using Cold and Hot Water

Cold and hot water energy storage is stored at a constant temperature, and the change in water storage reflects the energy storage status. Its mathematical model is shown in Eq. (9.10).

$$\begin{cases} Q_M^t = (1 - \sigma_M)Q_M^{t-1} + H_{tank\_M}^{t-1}\Delta t \\ Q_C^t = (1 - \sigma_C)Q_C^{t-1} + H_{tank\_C}^{t-1}\Delta t \end{cases} \tag{9.10}$$

where Q_M represents the thermal energy stored in the water tank, σ_M is the heat self-loss coefficient, and H_{tank_M} is the thermal power of the water tank (positive for input and negative for output); Q_C represents the cold energy stored in the water tank, σ_C is the cold self-loss coefficient, and H_{tank_C} is the cooling power of the water tank (positive for input and negative for output).

9.3.2 Energy Grade Conversion Model

Electric energy and thermal energy, as well as different grades of thermal energy, can be transformed from high to low grade by equipment such as electric heat pumps, absorption heat pumps, and peak heaters. During the conversion process, a certain proportion of the high-grade energy input on the driving side of the device is converted to the heating side, raising the grade of thermal energy on the heating side. The process of grade conversion and thermal energy transfer in the energy flow is shown in Fig. 9.5. Based on the energy conservation law and the definition of specific enthalpy, the multi-energy coupling relationships between different grades of thermal energy are analyzed as shown in Eqs. (9.11–9.13).

Fig. 9.5 Multi-grade energy conversion diagram

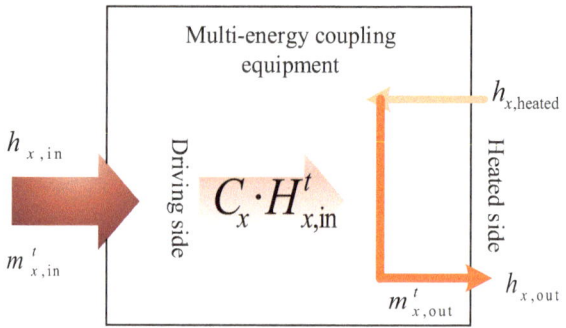

$$\begin{cases} H^t_{x,\text{out}} = C_x \cdot H^t_{x,\text{in}} + H^t_{x,\text{heated}} \\ m^t_{x,\text{in}} \cdot (h_{\text{in}} - h_{\text{base}}) \cdot \Delta t = H^t_{x,\text{in}} \\ m^t_{x,\text{out}} \cdot (h_{\text{heated}} - h_{\text{base}}) \cdot \Delta t = H^t_{x,\text{heated}} \\ m^t_{x,\text{out}} \cdot (h_{\text{out}} - h_{\text{base}}) \cdot \Delta t = H^t_{x,\text{out}} \end{cases} \tag{9.11}$$

$$R_x = \frac{H^t_{x,\text{heated}}}{H^t_{x,\text{in}}} = \frac{C_x \cdot (h_{x,\text{heated}} - h_{\text{base}})}{h_{x,\text{out}} - h_{x,\text{heated}}} \tag{9.12}$$

$$\begin{cases} H^t_{x,\text{out}} = (C_x + R_x) \cdot H^t_{x,\text{in}} \\ H^t_{x,\text{heated}} = R_x \cdot H^t_{x,\text{in}} \end{cases} \tag{9.13}$$

In Eq. (9.11), x represents the type of energy equipment, $H_{x,\text{out}}$ is the output thermal energy, C_x is the energy efficiency ratio constant, $H_{x,\text{in}}$ is the input energy on the driving side, and $H_{x,\text{heated}}$ is the input thermal energy on the heated side. $m_{x,\text{in}}$ and $m_{x,\text{out}}$ are the working fluid flow rates on the driving and heated sides, respectively, and h_{in}, h_{heated}, and h_{out} are the specific enthalpies of input on the driving side, input on the heated side, and output on the heated side, respectively. h_{base} is the specific enthalpy of water at normal temperature and is used as a reference.

In Eq. (9.12), R_x is the ratio of input thermal energy on the heated side to the input energy on the driving side. It can be seen that this proportionality factor depends on the energy efficiency ratio and the design input-output specific enthalpy of the equipment and can be treated as a constant for a specific equipment. Equation (9.13) analyzes the coupling and conversion relationship between different thermal energy grades based on the proportionality factor R_x. The input-output mathematical models of electric heat pumps, absorption heat pumps, and peak heating equipment are rewritten based on the original mathematical model.

9.3.3 Energy Supply and Demand Analysis

Based on the energy flow structure shown in Fig. 9.4, the energy flow coupling relationships and the energy grade conversion model described above, the energy supply and demand on different energy buses are analyzed from the perspective of energy cascade utilization. The energy sources on each bus are shown in Eq. (9.14).

$$
\begin{cases}
P_{E\_supply}^t = P_{grid}^t + P_{GT}^t + P_{PT}^t + P_{wind}^t \\
H_{S\_supply}^t = H_{GB\_S}^t + H_{GT\_S}^t \\
H_{H\_supply}^t = H_{PT\_H}^t + H_{PH\_H,out}^t \\
H_{M\_supply}^t = H_{HP\_M,out}^t + H_{AHP\_M,out}^t \\
H_{L\_supply}^t = H_{GT\_L}^t + H_{ACH\_L}^t \\
H_{C\_supply}^t = H_{RE\_C}^t + H_{ACH\_C}^t
\end{cases}
\tag{9.14}
$$

where $P_{E\_supply}, H_{S\_supply}, H_{H\_supply}, H_{M\_supply}, H_{L\_supply}, H_{C\_supply}$ are the total energy sources of the power, steam, high-temperature hot water, medium-temperature hot water, low-temperature hot water, and chilled water buses, respectively.

From the equation above, it can be seen that electricity and steam, the two high-grade energy sources, are directly supplied by power generation equipment or natural gas; high-temperature hot water is not only directly supplied by solar thermal energy but also by utilizing the surplus steam and medium-temperature hot water energy through the peak heating equipment. Medium-temperature hot water is supplied by consuming electricity through electric heat pumps and steam and low-temperature hot water through absorption heat pumps. Low-temperature hot water is supplied by the waste heat recovery power of gas turbines and the consumption of steam through absorption refrigeration. Cold water is supplied by consuming electricity through electric refrigeration and steam through absorption refrigeration. During the energy supply process, surplus high-grade energy can supply low-grade energy, and low-grade heat energy can also be heated into high-grade heat energy.

The energy consumption of each bus is shown in Eq. (9.15), which takes into account the energy interaction between electricity, cooling, thermal storage, and their corresponding buses.

$$
\begin{cases}
P_{E\_load}^t = P_{EL}^t + P_{HP}^t + P_{RE}^t + P_{ess}^t \\
H_{S\_load}^t = H_{SL}^t + H_{ACH\_S}^t + H_{AHP\_S,in}^t + H_{PH\_S,in}^t \\
H_{H\_load}^t = H_{HL}^t \\
H_{M\_load}^t = H_{ML}^t + H_{PH\_M,heated}^t + H_{tank\_M}^t \\
H_{L\_load}^t = H_{HP\_L,heated}^t + H_{AHP\_L,heated}^t \\
H_{C\_load}^t = H_{CL}^t + H_{tank\_C}^t
\end{cases}
\tag{9.15}
$$

where $P_{E\_load}, H_{S\_load}, H_{H\_load}, H_{M\_load}, H_{L\_load}, H_{C\_load}$ represent the total energy consumption of the power, steam, high-temperature hot water, medium-temperature hot water, low-temperature hot water, and chilled water buses respectively. $P_{EL}, H_{SL}, H_{HL}, H_{ML}$, and H_{CL} represent the loads of electric power, steam, high-temperature hot water, medium-temperature hot water, and chilled water, respectively.

As shown in the above equation, the total energy consumption of each energy bus including power, steam, high-temperature hot water, medium-temperature hot water, low-temperature hot water, and cold water, is considered while taking into account the energy interaction with the corresponding energy storage system. PEL, HSL, HHL, HML, HCL represent the electrical, steam, high-temperature hot water, medium-temperature hot water, and cold water loads, respectively.

As can be seen from the above equation, the two high-grade energy sources, electric power and steam, are not only consumed by the corresponding loads, but also consumed by energy coupling devices to supply low-grade energy demands. The surplus of medium-temperature and low-temperature hot water can also supply higher-grade energy demands through energy coupling devices. The energy supply and demand under the cascade utilization mode is not limited to a single form of energy, but achieves complementarity between different grades of energy.

9.4 Optimization Strategy for Cascaded Utilization of Electric-Thermal Microgrids

9.4.1 Objective Function

The objective of the optimization strategy takes into account the minimum daily operating cost, which is composed of the costs of purchasing natural gas and electricity, as well as equipment operating costs, as shown in Eq. (9.16).

$$C_{all} = C_{ng} + C_{grid} + C_{device} \tag{9.16}$$

Natural gas is consumed by gas turbines and gas boilers, and the cost of purchasing natural gas is shown in Eq. (9.17). Here, i represents the internal number of the same type of equipment.

$$C_{ng} = \sum_i \sum_t F_{GT,i}^t + \sum_i \sum_t F_{GB,i}^t \tag{9.17}$$

When the microgrid is in parallel operation with the main grid, electricity is purchased and sold from the main grid based on time-of-use pricing. The cost of purchasing and selling electricity from the grid is shown in Eq. (9.18). Here, p^t_{buy} represents the time-of-use purchase price of electricity.

$$C_{grid} = \sum_t p_{buy}^t \cdot P_{buy}^t \tag{9.18}$$

Equipment operating costs can be divided into energy equipment maintenance costs and battery energy storage depreciation costs. Equipment maintenance costs are defined by the cost per unit of equipment power. Battery energy storage depreciation costs are related to the amount of charged and discharged electricity, and the depreciation is assumed to be linear with increasing charged and discharged electricity. Therefore, the equipment operating cost is shown in Eq. (9.19).

$$C_{\text{device}} = \left(\sum_x \sum_i \sum_t p_x \cdot P_{x,i}^t + \sum_t c_{\text{bat}} \frac{P_{\text{ess}}^t}{Q_{\text{ess,max}}} \right) \cdot \Delta t \qquad (9.19)$$

where x represents the cost per unit output power of different energy equipment, c_{ess} represents the replacement cost of the battery energy storage system, and $Q_{\text{ess,max}}$ represents the total charge and discharge amount of the battery over its entire lifecycle.

9.4.2 Constraints

Energy supply and demand must be balanced on each bus during operation, and the constraints on each bus are shown in Eq. (9.20). Here, the storage of large amounts of low-temperature hot water is uneconomical, and the study do not consider factors such as space heating demand in the plant area and boiler return water heating. Therefore, the constraint is set to ensure that supply is greater than demand.

$$\begin{cases} P_{E\_\text{supply}}^t = P_{E\_\text{load}}^t \\ H_{S\_\text{supply}}^t = H_{S\_\text{load}}^t \\ H_{H\_\text{supply}}^t = H_{H\_\text{load}}^t \\ H_{M\_\text{supply}}^t = H_{H\_\text{load}}^t \\ H_{L\_\text{supply}}^t \geq H_{L\_\text{load}}^t \\ H_{C\_\text{supply}}^t = H_{C\_\text{load}}^t \end{cases} \qquad (9.20)$$

In addition, the operation of energy equipment must satisfy constraints on maximum and minimum power and ramp rate, as shown in Eq. (9.21).

$$\begin{cases} P_{x,\min} \leq P_{x,i}^t \leq P_{x,\max} \\ -D_x \cdot \Delta t \leq P_{x,i}^t - P_{x,i}^t \leq B_x \cdot \Delta t \end{cases} \qquad (9.21)$$

where $P_{x,\min}$ and $P_{x,\max}$ represent the minimum and maximum operating power of different energy equipment, while D_x and B_x represent the downward and upward ramp rates of different energy equipment, respectively.

9.4.3 Solution Methodology

This chapter studies a linearized modeling approach, which belongs to the mixed-integer linear programming problem, and uses the commercial solver Gurobi for solving. The energy flow mathematical model and constraints of the integrated energy system are constructed using the Matlab platform and Yalmip toolbox. During the simulation process, the day is divided into 96 time nodes, and the scheduling plan for each device is developed with the objective of minimizing daily operating costs.

During the solution process, there is some randomness in the solver's selection of output devices for multiple devices of the same type due to the same constraint definition. In order to make the device selection more distinctive, Eq. (9.22) is used to process the operating and maintenance costs of devices of the same type, with the operating costs of devices of the same type increasing sequentially, thus prioritizing devices with lower numbers. Here, x represents the device type, i represents the device number, and e_r represents the cost increment. The cost increment is defined to be extremely small (10^{-4} in this study) and only serves to differentiate devices, so its numerical value is negligible in the final solution.

$$p_{x,i+1} = p_x \cdot (1 + i \cdot e_r) \tag{9.22}$$

9.5   Case Studies

9.5.1   Case Description

The simulation case is based on the structural topology and parameters of the "China-Italy Green Energy Experimental Center" at a university in southwest Shanghai, which corresponds to an electric-thermal microgrid scenario that includes industrial, commercial, and residential users. The simulation analysis is performed with a 15-min interval, which satisfies the time scale requirement for optimizing scheduling considering the dynamic characteristics of multiple energy flows and the response capabilities of each device. The example includes devices such as gas turbines, gas boilers, absorption heat pumps, absorption refrigeration, electric heat pumps, electric refrigeration, peak heating equipment, battery energy storage, hot water storage tanks, and chilled water storage tanks. These devices cooperate to supply users with electric power, steam, high- temperature and medium-temperature hot water, and chilled water loads. The key parameters of each device are shown in Table 9.1, while the maintenance costs, time-of-use electricity prices, and natural gas purchase costs are shown in Table 9.2. The microgrid load and renewable energy daily output predictions are shown in Fig. 9.6, where wind power and electric load are in electric power, while the rest are in thermal power.

9.5.2   Results Analysis

Based on the simulation case described above, the comprehensive energy station's thermal-electric dispatch is shown in Figs. 9.7, 9.8, 9.9, 9.10, and 9.11, where energy input to the bus is represented by positive values, and energy acquired from the bus is represented by negative values. Based on the electricity and steam dispatch plan

Table 9.1   Equipment parameter

Equipment	Paramteter	Value
Battery energy storage	Capacity (kWh)	3000
	Maximum chaging/discharging power (kW)	400
	Self-discharge coefficient σ	0.0025
	Charging efficiency η_{chr}	0.98
	Discarging efficiency η_{dch}	0.98
	Initial state of charge	0.7
Hot water storage tanks	Water storage capacity (t)	25
	Maximum water flow rate (t/h)	5
	Self-dissipation rate	0.0001
Cold water storage tanks	Water storage capacity (t)	25
	Maximum water flow rate (t/h)	5
	Self-dissipation rate	0.0001
Gas turbine	Maximum operating power (kW)	1000
	Minimum operating power (kW)	50
	Generation efficiency η_{GT}	0.33
	Extraction ratio coefficient a η_{ext}	0.6
	waste heat recovery efficiency η_{exh}	0.3
Gas boiler	Maximum heating capacity (kW)	2000
	Gas boiler efficiency η_{GB}	0.9
Absorption heat pump	Maximum heat input (kW)	1000
	Energy efficiency ratio COP_{AHP}	2.6
	Steam utilization efficiency η_{AHP}	0.916
Absorption refrigeration	Maximum heat intake volume (kW)	1000
	Energy efficiency ratio COP_{ACH}	1.8
	Steam utilization efficiency η_{ACH}	0.916
	Cold water ratio k_{RE}	1.5
Electric heat pump	Maximum power consumption (kW)	600
	Energy efficiency ratio COP_{HP}	3
Electric refrigeration	Maximum power consumption (kW)	600
	Energy efficiency ratio COP_{RE}	3
Peak heating equipment	Steam utilization efficiency η_{PH}	0.896
Solar thermal steam turbine	Maximum power generation (kW)	500
	Minimum power generation (kW)	400
	Generation efficiency η_{ST}	0.4
	Waste heat recovery efficiency η_{WH}	0.2
	Heat exchanger heat transfer efficiency η_{ex_PT}	0.8

Table 9.2 Operating cost parameter

Cost type	Cost calculation basis	Value
Maintenance cost (RMB/kWh)	Charging and discharging electricity for battery storage	0.005
	Gas turbine generation	0.0063
	Heat generation of gas boiler	0.03
	Heat intake of absorption heat pump	0.0008
	Heat intake of absorption refrigeration	0.0008
	Power consumption of electric heat pump	0.001
	Power consumption of electric refrigeration	0.001
Peak and valley electricity price (RMB/kWh)	Peak electricity price period (10:00–15:00; 18:00–21:00)	1.3902
	Flat electricity price period (7:00–10:00; 18:00–23:00)	0.8645
	Valley electricity price period (23:00–7:00 next day)	0.3648
Natural gas price (RMB/m^3)	–	3.4

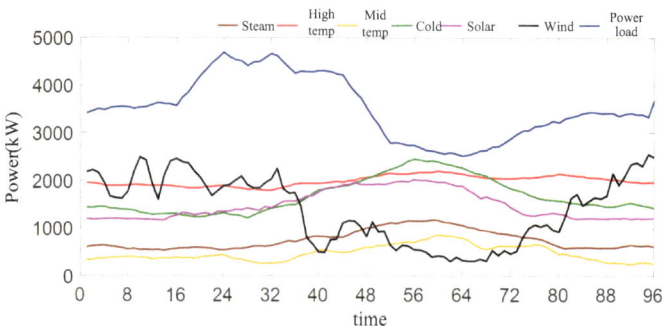

Fig. 9.6 Renewable energy output and load prediction of microgrid with heat and power system

results, Analysis are conducted from the perspective of electricity prices during peak and off-peak hours:

(1)   23:00 to 7:00 is the off-peak electricity pricing period. It is economically efficient to directly purchase electricity from the grid, so the gas turbine is not started. The power supply side is composed of solar-thermal power generation, wind power, and the main power grid, while the power load side is composed of electric heat pumps, electric refrigeration, and uncontrollable electric loads. During this period, only the gas boiler supplies steam, and the steam load side is composed of absorption refrigeration, peak heating equipment, and steam loads. The reason for using absorption refrigeration during this period is that the electricity price is low, and the low-temperature hot water produced by the

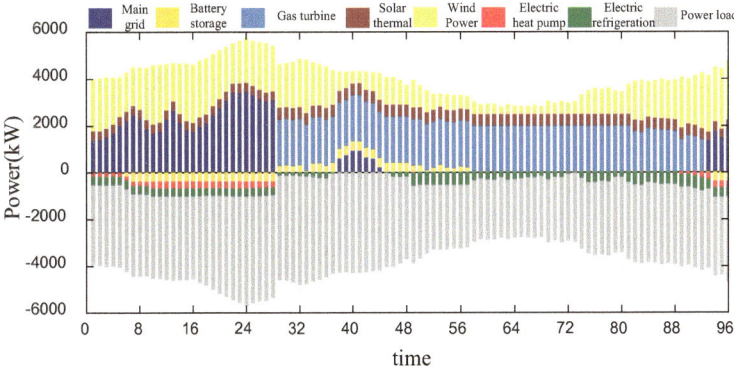

Fig. 9.7 Power dispatch plan

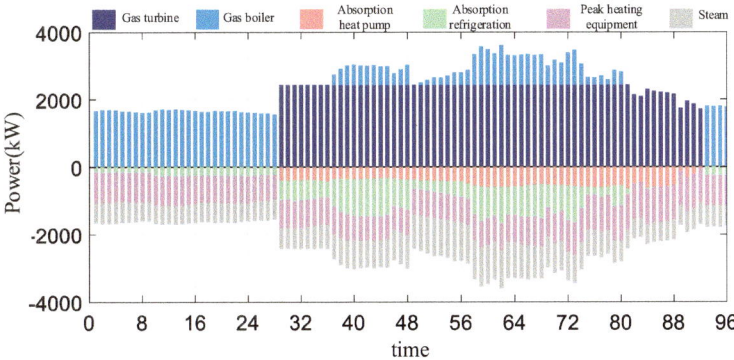

Fig. 9.8 Steam dispatch plan

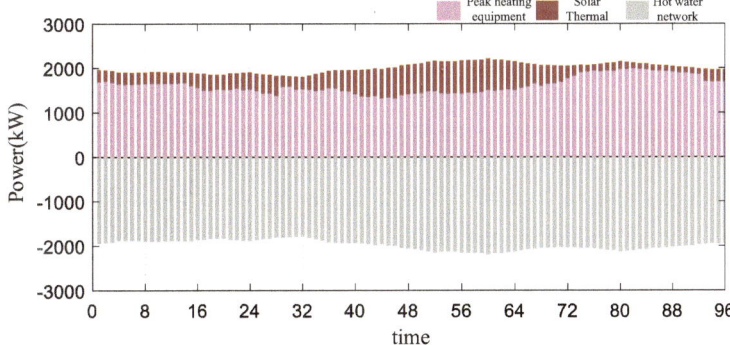

Fig. 9.9 High-temperature hot water dispatch plan

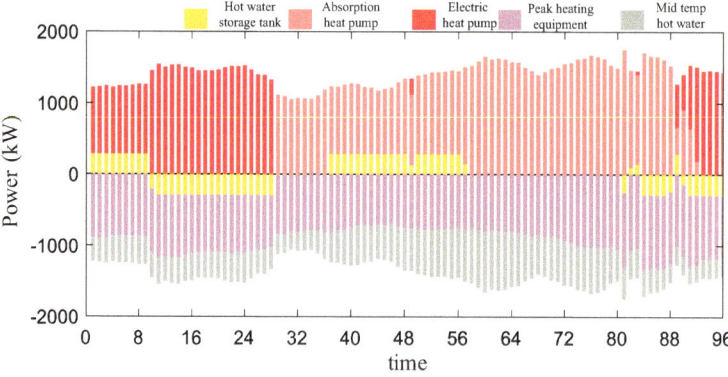

Fig. 9.10 Medium-temperature hot water dispatch plan

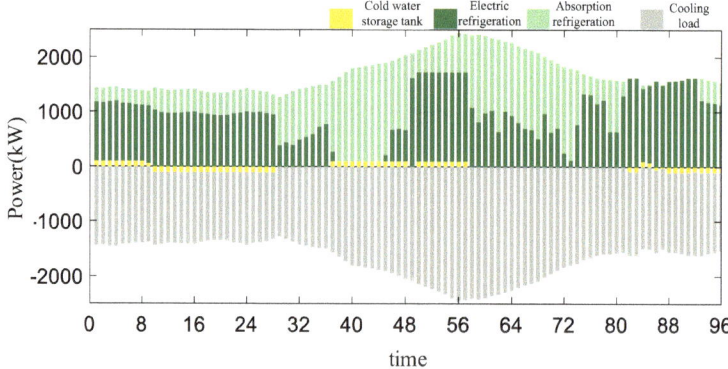

Fig. 9.11 Cold water dispatch plan

absorption refrigeration can be heated and utilized by the electric heat pump, which has good economic benefits. The battery energy storage is fully charged at maximum power before the end of the off-peak period.

(2) 7:00 to 23:00 is the peak and flat electricity pricing period, and the power supply by the gas turbine is more economical. During this period, the power supply side is mainly composed of gas turbines, solar-thermal power generation, and wind power. Note that at around 10:00, due to the intermittent decrease in wind power output, both the gas turbine and the battery energy storage reach maximum output, and electricity is purchased from the grid to meet the load demand. The power load side is mainly composed of electric refrigeration and uncontrollable electric loads, and electric heat pumps are only selectively used. The steam supply side is provided by the gas turbine and gas boiler, and the steam load side is composed of absorption heat pumps, absorption refrigeration, peak heating equipment, and steam loads. Due to the large amount of steam

provided by the gas turbine combined heat and power and the high electricity price, absorption heat pumps and absorption refrigeration have better economic benefits. During the period from 7:00 to 15:00, due to high electric load and the decrease in wind power output, the battery energy storage is discharged to assist in peak shaving until the state of charge reaches the lower limit.

Based on the cold and hot water scheduling plan results in Figs. 9.9, 9.10, and 9.11, analysis are conducted from the perspective of the energy supply composition of cold and hot loads:

(1) For high-temperature hot water supply, since the solar thermal power station does not participate in scheduling control, the steam turbine's electricity generation is relatively stable, and the remaining solar thermal energy is used for heating. Therefore, the peak heating equipment adjusts the heat production according to the heating supply of the solar thermal power station. Its steam consumption is a rigid demand, and the steam production of gas turbines and gas boilers needs to be prioritized.

(2) For medium-temperature hot water supply, it is supplied by electric heat pumps during the valley electricity price period and by absorption heat pumps during the peak and flat electricity price periods. Electric heat pumps are used for heating when steam supply is tight. Due to energy self-loss, the heat storage water tank first releases water during the valley electricity price period and then heats up through electric heat pumps before the end of the valley electricity price period, storing water to the maximum capacity and releasing heat during the peak electricity price period.

(3) For cold water supply, electric refrigeration is mainly used, and the proportion of absorption refrigeration supply increases during the peak and flat electricity price periods. This is because the energy efficiency ratio of absorption refrigeration is lower than that of electric refrigeration, but considering that absorption refrigeration can provide low-temperature hot water while cooling, it can be used for further heating by electric heat pumps during the valley electricity price period. When steam supply is sufficient during the peak and flat electricity price periods, the comprehensive benefits of consuming steam refrigeration by absorption refrigeration are higher. The operation mode of the cold storage water tank is similar to that of the hot storage water tank. It stores water to the maximum capacity before the end of the valley electricity price period and releases cold during the peak electricity price period.

9.5.3 Economic Analysis of Diverse Energy Supply Structures

The energy supply of electric-thermal microgrid is realized by electric-thermal coupling cascade utilization (structure 1), cascade utilization without electric-thermal coupling (structure 2) and traditional tri-generation supply structure (structure 3), and the daily cost of the three energy supply structures is shown in Table 9.3 below. The daily operating cost includes natural gas costs, electricity purchase costs and maintenance costs. At the same time, because the equipment used in the three structures is not identical, the total cost also includes the equipment cost converted to the daily price.

Structure 2 only adopts the energy cascade utilization strategy without considering the coupling between electric and thermal energy, so it does not use electric heat pumps, electric refrigeration and other electric-thermal coupling equipment. In terms of energy supply, absorption heat pumps provide all medium temperature hot water, and absorption refrigeration provides all cold water. The power and steam dispatch plan for structure 2 is shown in Figs. 9.12 and 9.13. Gas turbines and gas boilers operate around the clock, providing steam for absorption heat pumps, absorption refrigeration and steam loads. The steam supply is mainly gas boilers during the valley electricity price period, and gas turbines are mainly used during the peak electricity price period. During the valley tariff period, the purchase of electricity from the grid is reduced because the gas turbine runs to generate electricity. The operation mode of battery energy storage is consistent with structure 1, that charging during the valley electricity price period and discharging during the peak electricity price period.

Structure 3 adopts the traditional tri-generation supply mode without dividing the thermal energy grade. Its specific supply strategy is referred to in literature [31]. The power and steam dispatch plan of Structure 3 is shown in Figs. 9.14 and 9.15. Since the thermal energy grade is not divided, the load of medium-temperature hot water and hot water in the heat network are unified as high-temperature heat load, and the hot water is supplied through the heat network. In terms of supply strategy, during the off-peak electricity price period, the steam generated by gas turbines and gas boilers is supplied to the peak heater and high-temperature steam load. The peak heater produces high-temperature hot water to meet the hot water load of the heat network, and electric refrigeration provides all the required cold water. During the peak and flat electricity price period, gas turbines and gas boilers operate at high loads, providing

Table 9.3 Daily cost comparison

Energy supply structure	Structure 1	Structure 2	Structure 3
Natural gas cost (RMB)	39,871	46,644	49,642
Electricity purchasing cost (RMB)	7897	4572	6183
Maintenance cost (RMB)	838	1428	1327
Daily operation cost (RMB)	48,606	52,644	57,152
comparison	–	+8.31%	+17.58%
Device cost (RMB)	1522	1295	1388
Total cost (RMB)	50,128	53,939	58,540
Comparison	–	+7.60%	+16.78%

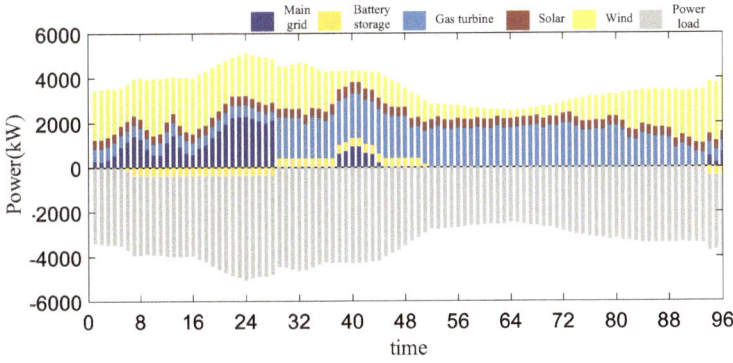

Fig. 9.12 Power dispatch plan of structure 2

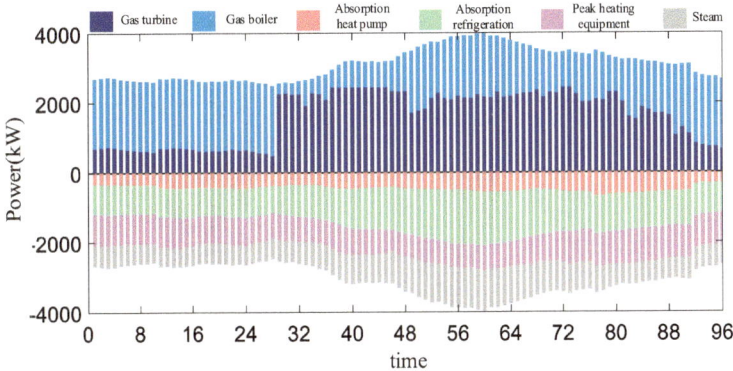

Fig. 9.13 Steam dispatch plan of structure 2

the steam required for absorption refrigeration, peak heater, and high-temperature steam load. Electric refrigeration is turned on after the uncontrollable electrical load in the microgrid decreases, consuming surplus electricity and reducing the steam demand of absorption refrigeration, thus balancing the power and steam demand of the microgrid.

Based on the information presented in the table and figures, the following conclusions can be drawn:

(1) The cascade utilization strategy of electric-heat coupling can effectively improve the economic performance of electric-thermal microgrids. As microgrids have demand for both heating and cooling, the benefit of using natural gas for cogeneration is significant, hence natural gas expenses are relatively high. Structure 1 consumes more electricity at lower rates and thus saves on natural gas expenses,

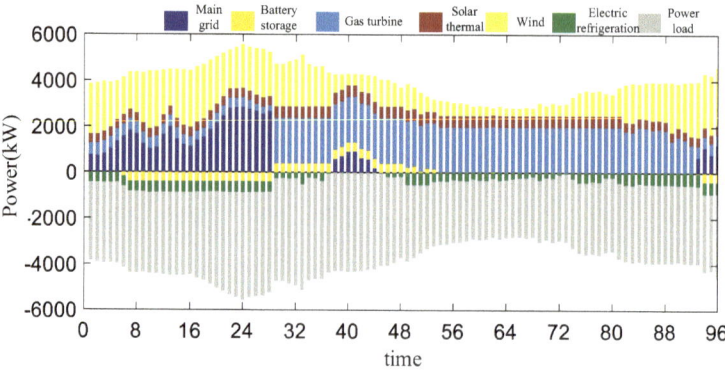

Fig. 9.14 Power dispatch plan of structure 3

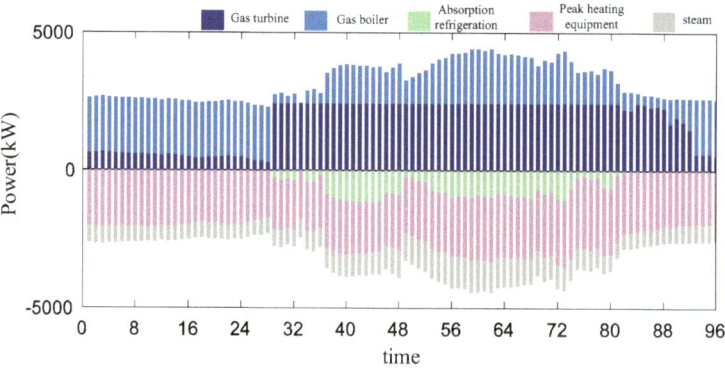

Fig. 9.15 Steam dispatch plan of structure 3

resulting in lower overall energy costs. Furthermore, the high-efficiency equipment such as absorption heat pumps, electric heat pumps, and electric refrigeration are utilized to minimize maintenance costs. In terms of daily operating costs, structures 2 and 3 have increased by 8.31% and 17.58% respectively compared to structure 1.

(2) As structure 1 utilizes the most equipment, it has the highest daily equipment cost. However, as the prices of the various equipment types are relatively low among the three structures, it has not had a significant impact on the total cost. The total cost of structures 2 and 3 have increased by 7.60% and 16.78% respectively compared to structure 1.

(3) The lack of electric-thermal coupling equipment negatively impacts the flexibility of electric-thermal combined microgrid operations. In structure 2, due to the single type of heating and cooling supply equipment, the operation of absorption heat pumps and absorption refrigeration needs to be carried out based on

load requirements, while the gas turbine and gas boiler need to operate at high load throughout the day, leaving no room for optimization.

(4) Without considering the cascade utilization of energy, the operating efficiency of electric-thermal combined microgrid is low. In structure 3, direct heat exchange is used for heating, resulting in low efficiency, and a large amount of steam is consumed for heating and cooling supply in the microgrid, causing high natural gas expenses. Without distinguishing the heat quality requirements of microgrid users, the actual heating efficiency of the system is low, and it cannot achieve efficient and targeted high-quality heating.

9.6 Conclusion

This chapter aims to address the issue of the comprehensive optimization and utilization of various forms of energy in electric-thermal microgrids. Based on the principle of cascade utilization, a cascade utilization energy flow structure for electric-thermal coupling conversion is constructed to achieve energy utilization and supply with matched thermal energy grade, and to enhance the operational economy and flexibility of cold and heat supply equipment by leveraging the complementary advantages of electric and thermal energy. For the source-load-storage of electric-thermal microgrids, an energy flow supply and demand model with different energy buses is established. Based on the operational constraints of the energy cascade optimization utilization model and with the objective of minimizing daily operating costs, this mixed integer optimization problem is solved via the establishment of constraints and objectives in Matlab. By coordinating and scheduling various energy sources and coupling devices, the optimized economic operation of the electric-thermal microgrid system is achieved. The method used in this study fully considers the principles of mutual coupling and complementary optimization of various forms of energy, making effective use of energy with different grades in the production process, providing good engineering application prospects. Future research will further improve the energy optimization and scheduling capabilities of electric-thermal microgrid systems through rolling optimization or by establishing optimization models that take into account uncertain factors.

References

1. Yang, X., Su, J., Lyu, Z., et al.: Overview on micro-grid technology. Proceed. CSEE **34**(1), 57–70 (2014)
2. Teng, Y., Sun, P., Luo, H.: Autonomous optimization operation model for multi–source micro-grid considering electrothermal hybrid energy storage. Proceed. CSEE **39**(18), 5316–5324 (2019)
3. Kienzle, F., Ahcin, P., Göran, A.: Valuing investments in multi-energy conversion, storage, and demand-side management systems under uncertainty. IEEE Trans. Sustain. Energy **2**(02), 194–202 (2011)

4. Jiang, S., Fan, C., Huang, N., et al.: A fault location method for DC lines connected with DAB terminal in power electronic transformer. IEEE Trans. Power Deliv. **34**(1), 301–311 (2019)
5. Li, Z., Zhang, F., Liang, J., et al.: Optimization on microgrid with combined heat and power system. Proceed. CSEE **35**(14), 3569–3576 (2015)
6. Chen, Y., Wei, W., Liu, F., et al.: A multi-lateral trading model for coupled gas-heat-power energy networks. Appl. Energy **200**, 180–191 (2017)
7. Jin, H.: Efficient cascade utilization and total energy system. Science China Press **62**(23), 2589–2593 (2017)
8. Jin, H., Zhang, G., Gao, L., et al.: The development and prospect of total energy system. J. Mech. Eng. **45**(03), 39–48 (2009)
9. Yongping, Y., Chenxu, Z., Gang, X., et al.: A heat integration system based on boiler-turbine coupling for large scale coal-fired power plants. Proceed. CSEE **35**(2), 375–382 (2015)
10. Tabar, V.S., Jirdehi, M.A., Hemmati, R.: Energy management in microgrid based on the multi objective stochastic programming incorporating portable renewable energy resource as demand response option. Energy **118**, 827–839 (2017)
11. Liu, X., Wu, H.: A control strategy and operation optimization of combined cooling heating and power system considering solar comprehensive utilization. Automat. Electric Power Syst. **39**(12), 1–6 (2015)
12. Pei, W., Deng, W., Shen, Z., et al.: Energy coordination and optimization of hybrid microgrid based on renewable energy and CHP supply. Automat. Electric Power Syst. **38**(16), 9–15 (2014)
13. Zheng, G., Li, H., Zhao, B., et al.: Comprehensive optimization of electrical/thermal energy storage equipments for integrated energy system near user side based on energy supply and demand balance. Power Syst. Protect. Control **46**(16), 8–18 (2018)
14. Wang, W., Yang, L., Wang, L., et al.: Optimal dispatch of integrated electricity-heat energy system considering heat storage characteristics of heating network. Automat. Electric Power Syst. **42**(21), 45–55 (2018)
15. Fan, F., Huang, W., Tai, N., et al.: A conditional depreciation balancing strategy for the equitable operation of extended hybrid energy storage systems. Appl. Energy **228**, 1937–1952 (2018)
16. Fan, F., Tai, N., Zheng, X., et al.: Equalization strategy for multi-battery energy storage systems using maximum consistency tracking algorithm of the conditional depreciation. IEEE Trans. Energy Convers. **33**(3), 1242–1254 (2018)
17. Ai, X., Chen, Z., Sun, Y., et al.: (2019) Study on integrated DLC coordination optimization of electric-thermal-gas coupling system considering demand response. Power Syst. Tech. **04**, 1160–1171 (2019)
18. Xu, L., Yang, G., Xu, Z., et al.: Combined scheduling of electricity and heat in a microgrid with volatile wind power. Automat. Electric Pow. Syst. **35**(9), 53–60 (2011)
19. Hao, R., Ai, Q., Zhu, Y., et al.: Hierarchical optimal dispatch based on energy hub for regional integrated energy system. Automat. Electric Pow. Syst. **37**(06), 171–178 (2017)
20. Zhou, C., Zheng, J., Jin, Z., et al.: Multi-objective optimal design of integrated energy system for park-level microgrid. Power Sys. Tech. **42**(06), 1687–1697 (2018)
21. Ju, C., Wang, P., Goel, L., et al.: A two-layer energy management system for microgrids with hybrid energy storage considering degradation costs. IEEE Trans. Smart Grid **9**(6), 6047–6057 (2018)
22. Ross, M., Abbey, C., Bouffard, F., et al.: Microgrid economic dispatch with energy storage systems. IEEE Trans. Smart Grid **9**(4), 3039–3047 (2018)
23. Jin, X., Mu, Y., Jia, H., et al.: Optimal scheduling method for a combined cooling, heating and power building microgrid considering virtual storage system at demand side. Proceed. CSEE **37**(2), 581–590 (2017)
24. Liu, D., Ma, H., Wang, B., et al.: Operational optimization of regional integrated energy system with CCHP and energy storage system. Automat. Elect. Power Syst. **42**(04), 9 (2018)
25. Wei, Z., Zhang, S., Sun, G., et al.: Carbon trading based low–carbon economic operation for integrated electricity and natural gas energy system. Automat. Elect. Power Syst. **40**(15), 9–16 (2016)

26. Zhang, S., Lyu, S.: Evaluation method of park-level integrated energy system for microgrid. Power Syst. Tech. **42**(08), 2431–2439 (2018)
27. Comodi, G., Renzi, M., Cioccolanti, L., et al.: Hybrid system with micro gas turbine and PV (photovoltaic) plant: guidelines for sizing and management strategies. Energy **89**, 226–235 (2015)
28. Zhao, X., Fu, L., Sun, T., et al.: Technology research on deep utilization of flue gas waste heat for 9F grade gas-steam combined cycle back pressure heat supply unit. Heat. Ventilating & Air Cond. **47**(08), 83–87 (2017)
29. Li, Y., Wang, W., Ma, Y.F., et al.: Study of new cascade heating system with multi-heat sources based on exhausted steam waste heat utilization in power plant. Appl. Therm. Eng. **136**, 475–483 (2018)
30. Xu, H., Dong, S., He, Z., et al.: Multi-energy cooperative optimization of integrated energy system in plant considering stepped utilization of energy. Automat. Electric Pow. Syst. **42**(14), 123–130 (2018)
31. Wu, M., Luo, Z., Ji, Y., et al.: Optimal dynamic dispatch for combined cooling heating and power microgrid based on model predictive control. Proceed. CSEE **37**(24), 7174–7184 (2017)

Chapter 10
Optimal Coordination Operation of Port Integrated Energy Systems

10.1 Introduction

Seaport is the significant hub of maritime industry, which undertakes nearly 90% global trades [1]. The increasing trade has led to high energy consumption and carbon emissions in the past few decades [2, 3]. It is estimated that 3–5% total global greenhouse gas (GHG) emission comes from maritime transportation [4]. This data will rise to 18% by 2025 if no measure being adopted. To solve the urgent environmental problem, seaports begin to introduce energy technologies at the aims of energy structure transition and restricting fuel use. These technical measures generally include shore-side power supply (cold-ironing) for docking ships, electrification of loading/unloading equipment and the integration of multiple energy sources (e.g., renewable energy, natural gas, heating and cooling) [5–8], which undoubtedly will contribute to the coupling between logistic system (LS) and energy system (ES). On one hand, the LS (ships and QCs) can be regarded as the load of ES, so it is constrained by both logistic operation and energy supply capacities. On the other hand, the behavior of LS impacts not only the transportation efficiency, but also the distribution of energy flows, then further influences the economic benefits.

Conventionally, LS and ES operate independently. The LS only focuses on transportation efficiency, while the ES is completely dominated by LS. However, the uncooperative operating mode may lead to mismatch between energy demand and supply in the case of coupling. In the spatial dimension, the berthing position of ships determines the electrical distance from nodes that connected with multiple energy sources in the electricity networks. Inappropriate docking position will cause more energy loss on the transmission line, or the energy cannot even be delivered to the ship. In the perspective of time dimension, due to the intermittent nature of renewable energy, its generation may not match the temporal distribution of ships/QCs loads. Then, the energy efficiency and economic benefits decline. Therefore, an effective coordination method is needed for the two systems, which minimizes the total cost without operating constraint violations.

© The Author(s) 2023
W. Huang et al., *Energy Management of Integrated Energy System in Large Ports*,
Springer Series on Naval Architecture, Marine Engineering, Shipbuilding and Shipping
18, https://doi.org/10.1007/978-981-99-8795-5_10

Since the coordination between LS and ES is an interdisciplinary issue, the energy community and the seaport logistic community have conducted separate studies on the two systems for decades. In the area of seaport logistic engineering, two fundamental problems are berth allocation and QC assignment. These research, e.g., [9–13], emphasized only on the berth allocation problems, while others integrated berth allocation and QC assignment together [14–19]. These works only concerned about logistic operation and usually ignored the energy consumption. Therefore, their schemes may not be optimal (may be even unacceptable) from the viewpoints of energy dispatching.

In the terms of port energy management, literature [20] analyzed the significance of good energy management principles and electrical distribution architecture for seaport business from a conceptual level. Literature [21] discussed ways to reduce GHG emissions by integrating renewable energy into port microgrids. Demand response (DR), as an effective tool for peak-shaving and filling load valleys, has been applied in green ports to enhance the flexibility of power grid operation. Literature [22] proposed to reduce reefers' peak power demand through adjusting charging time. The authors in [23] suggested an evaluated drive system for port cranes to optimize energy absorptions. They also discussed the installation of ultracapacitors and flywheels into port cranes to reduce the maximum power demands [24]. For large seaports, thousands of electricity loads cause heavy computational burden and curse of dimensionality. An multiagent framework was proposed in [25–28] to meet this challenge. Many power loads are aggregated into various agents to obtain less data scale and higher computing efficiency. In addition, some research such as [29] and [30] discussed the application of integrated energy systems (IES) into seaports. The above research solely analyze the energy management under given power demands and do not consider the operational characteristics of energy loads (e.g., logistic systems).

In recent years, researchers are beginning to recognize the interdependency between the LS and ES brought by the deeper electrification of seaports. Fang et al. [31] discussed the structure of future seaport microgrid with the indication of the connections between seaports and ships expanding from logistic-side to the electric-side. Their works are mostly reviews and did not give the specific coordination strategy. Peng et al. [32] proposed a cooperated optimization of berth and shore-side power allocation. Their method can obtain a balance between economic costs and environmental benefits but is only about shore-side power without modeling power grids. Although these research have made remarkable progresses in the interdisciplinary area, there are still some gaps to fill.

In this chapter, an optimal operation of port logistic-energy systems is proposed to minimize total costs including logistic operating and energy service costs. The discrete logistics operating model is adopted, which integrates the concepts of space and time to describe the time-space behaviors of ships and QCs. The ES is established by using Dist-Flow model and Weymouth flow equation, which can also describe the temporal and spatial characteristics of power and natural gas flows. Then the decisions of berth allocation, QC assignment, unit commitment and energy flows are determined simultaneously under energy-logistic dual-constraints. This not only maximizes the utility of dock sources to ensure transportation efficiency, but also matches energy demand and supply through temporal and spatial shifting of ship/QC loads, thereby improving energy efficiency and reducing carbon emission.

10.2 Structure of Port Integrated Energy Systems (PIES)

Figure 10.1 illustrates a general structure of PIES coupling LS and ES. The ES consists of energy equipment, electricity and natural gas network. Electricity and natural are purchased from main grid and natural gas wells, and then are transmitted via electric lines and gas pipelines. Various energy technologies including gas turbine (GT), power to gas (P2G), energy storage system (ESS) and photovoltaic (PV) are integrated in the two energy supply networks. The ES feed all the energy demands from docked ship and QCs, fixed electricity and natural gas loads.

The LS comprises ships and QCs, as Fig. 10.2 shows. In the respect of logistic transportation, QCs are assigned for docked ships to load and unload containers on ships. The arrival ships anchor to wait or dock at free berth according to berth allocation plan. After loading/unloading tasks are completed, the ships leave the seaport. In the terms of energy consumption, electrified QCs and docked ships are support by electrical networks. The QCs are directly connected with certain electrical nodes while the docked ships are linked via shore-side power interfaces (cold-ironing).

Note that the berthing time of ships determines the start and end time of ships power demand. The QCs are allocated when ships dock at berth, so QCs also generate power demands during ship berthing period. Different berthing time leads to the energy demands of ships and QCs at different time intervals. On the other hand, because of the node-to-node connection between ships/QCs and electricity network, the different berthing position of ships and the QCs running state can cause different

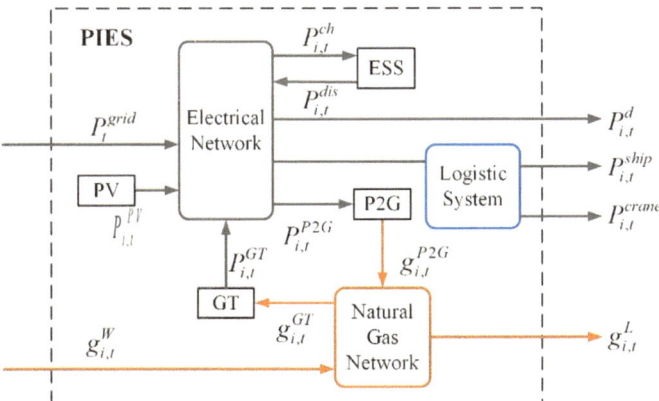

Fig. 10.1 The structure of port integrated energy systems

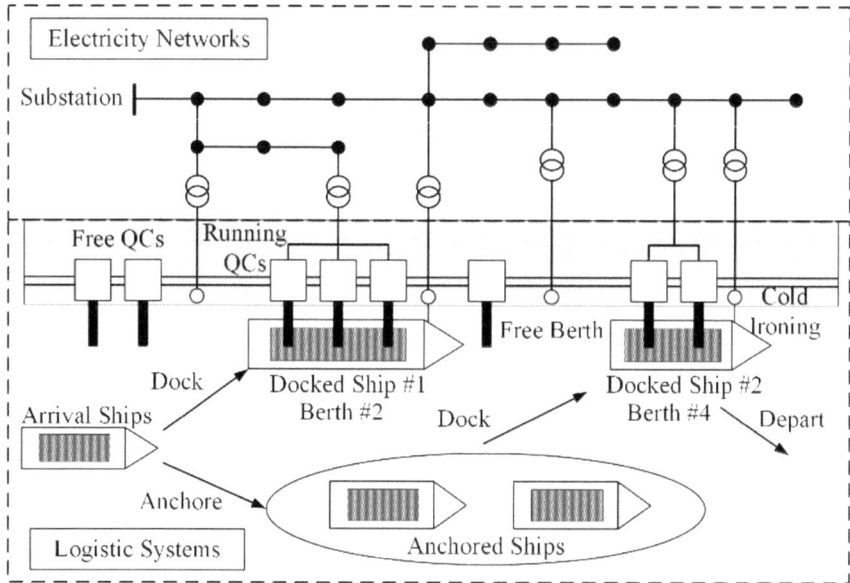

Fig. 10.2 The connection between electricity networks and logistic systems

spatial distribution of energy flows. Therefore, a proper dispatching scheme of ships and QCs can realize the temporal and spatial shifting of power loads, at the same time to ensure logistic transportation efficiency, so as to minimize the total costs.

10.3   PIES Formulation

10.3.1   Logistics System

(1)   Berth Allocation of Ships

For the berth allocation problem, there are generally two decisions: berthing position and berthing time. In this work, discrete berth is adopted to correspond one berth to one shore-side power interface. The effects of the berthing positions on the QCs handling efficiency are ignored so that the berthing time are basically determined if the handling containers and the number of assigned QCs are known.

Note that ships are the cause of QCs assignment and power demands so that the docking status of ships needs to be defined firstly in Eq. (10.1):

$$X_{bst} = \begin{cases} 1 & if b = B_s and t \in \left(t_s^1, t_s^2\right) \\ 0 & otherwise \end{cases}, \tag{10.1}$$

The docking status X_{bst} is a three-dimensional binary variable that combines berthing position B_s, ship number s and berthing duration together. The value of X_{bst} is 1 if the ship docks at berth b at time interval t, otherwise, it equals to 0. Equation (10.1) has two implications, (1) ensure that the berthing position do not change in the whole scheduling period once it is allocated; (2) the value of X_{bst} remains 1 only during docking period.

The relationship between arrival time, docking start time and departure time are constrained by Eqs. (10.2–10.3). The docking start time should be greater than or equal to ship arrival time. The departure time should be greater than docking start time and less than or equal to latest departure time.

$$t_{0,s} \leq t_{1,s}, \forall s \tag{10.2}$$

$$t_{1,s} + 1 \leq t_{2,s} \leq t_s^{latest}, \tag{10.3}$$

The above two equations restrict berthing time of ships. Based on the docking status X_{bst}, berthing position constraints can be formulated as follows:

$$\sum_{b \in B} X_{bst} \leq 1, \quad \forall s, t \tag{10.4}$$

$$\sum_{s \in S} X_{bst} \leq 1, \quad \forall b, t \tag{10.5}$$

$$B_s = \sum_{b \in B} b\sigma_{bs}, \quad \forall s, t \tag{10.6}$$

$$\sum_{b \in B} \sigma_{bs} = 1, \quad \forall s, t \tag{10.7}$$

$$\sum_{b \in Berth_s} \sigma_{bs} = 1, \quad \forall s, t \tag{10.8}$$

$$1 \leq B_s \leq B_{max} \tag{10.9}$$

Constraints (10.4) and (10.5) ensure that one ship is assigned to one berth. Constraint (10.6) establishes the relationship between integer variable and binary variable of berthing position. The binary variable σ_{bs} means that the ship s docks at berth b when its value is equal to 1. Constrain (10.7) is similar to (10.4) and (10.5) to ensure that each ship is assigned to one berth. Preferred and alternative berthing position are considered in constraint (10.8). Each ship can only dock at preferred or alternative berths. The set $Berth_s$ represents all the possible berthing position for ship s. The preferred berth is usually the first choice of arrival ships. However, the ships may deviate from their preferred berths and are arranged at the other alternative berths in order to coordinate with energy systems. In addition, all the ships must dock within permissible berthing position (10.9).

(2) QC Assignment

The QCs are used to load/unload containers on ships and directly influence the berthing duration of ships. A time-variant QC assignments model is considered in this book in order to increase the flexibility of logistic operation. That means the QCs allocated to a ship are not fixed from a period to another period of time. The QC assignment constraints are formulated as follows:

$$\sum_{s \in S} \omega_{qst} \leq 1, \quad \forall q, t \tag{10.10}$$

$$Q_{st} = \sum_{q \in Q} \omega_{qst}, \quad \forall s, t \tag{10.11}$$

$$\sum_{s \in \mathcal{S}} Q_{st} \leq Q^{max} \tag{10.12}$$

$$Q_s^{min} \sum_{b \in \mathcal{B}} X_{bst} \leq Q_{st} \leq Q_s^{max} \sum_{b \in \mathcal{B}} X_{bst} \quad \forall s, t \tag{10.13}$$

$$\sum_t \eta Q_{st} \geq TEU_s, \quad \forall s, t \tag{10.14}$$

The running status of QC is also represented by a three-dimensional binary variable ω_{qst}. The value of ω_{qst} is equal to 1 when the QC q serves for the ship at time interval t. Constraint (10.10) guarantees that a specific QC is capable of serving only one ship at most during each time interval. The number of allocated QCs is calculated by constraint (10.11). Because of the limited resources of QCs, the sum of assigned QCs cannot be larger than the total number of available QCs (10.12). The ship length and loading/unloading requirements restrict the minimum and maximum number of QCs that can be allocated to each ship (10.13). Constraint (10.14) ensures that enough QCs are assigned to handle the containers on ships during docking time.

In addition to number-related constraints, the following practical operating constraints that can be seen from Fig. 10.2 should also be considered:

$$-1 \leq \omega_{(q+1)st} - \omega_{qst} + \omega_{q'st} \leq 1, \quad \forall q' \leq q - 1, s, t \tag{10.15}$$

$$B_{qt} = \begin{cases} B_s \text{ if } \omega_{qst} = 1 \\ 0 \text{ if } \omega_{qst} = 0 \end{cases} \quad \forall q, s, t \tag{10.16}$$

$$\xi_{qt} = \begin{cases} 1 \text{ if } B_{qt} \geq 1 \\ 0 \text{ if } B_{qt} = 0 \end{cases} \quad \forall q, t \tag{10.17}$$

$$\xi_{(q+1)t} \left(B_{(q+1)t} - B_{qt} \right) \geq 0, \quad \forall q, t \tag{10.18}$$

Constraint (10.15) ensures that all the QCs serving for one ship are adjacent. That means there should be no free QCs positioned between the running QCs. Each running QC is stationed at the same location as the ship it serves, while free QCs are assumed at the zeroth position that actually does not exist (10.16). This is because free QCs are not involved in loading/unloading containers and do not generate power demands, so they are not needed to be considered. Equation (10.17) indicates the relationship between the location of QC and its running status. All QCs are installed in a row thus they cannot cross each other.

Equations (10.1–10.18) limit the behaviors of ships and QCs within allowed range to ensure normal logistics transportation operations. However, the decision of berth allocation and QC assignment should also consider energy systems constraints.

10.3.2 Energy System

(1) Energy Technologies

Constraints (10.19–10.21) show the charge/discharge power limitations of ESS. The ESS hourly energy change is given in (10.22). Constraints (10.23) give the state limitation of ESS. The stored energy at the end of period should be equal to the initial stored energy to ensure that ESS can operate continuously (10.24).

$$v_t^{ch} P_{min}^{ch} \leq P_t^{ch} \leq v_t^{ch} P_{max}^{ch} \tag{10.19}$$

$$v_t^{dis} P_{min}^{dis} \leq P_t^{dis} \leq v_t^{dis} P_{max}^{dis} \tag{10.20}$$

$$v_t^{ch} + v_t^{dis} = 1 \tag{10.21}$$

$$S_t^{ESS} = S_{t-1}^{ESS} + \eta^{ch} P_t^{ch} - \frac{P_t^{dis}}{\eta^{dis}} \tag{10.22}$$

$$S_{min}^{ESS} \leq S_t^{ESS} \leq S_{max}^{ESS} \tag{10.23}$$

$$S_{t=1}^{ESS} = S_{t=T}^{ESS} \tag{10.24}$$

The upper/lower bound of GT generation is imposed in (10.25). Constraint (10.26) shows the up/down ramp limit. Constraints (10.27) and (10.28) indicate the minimum up- and down-time. Constraints (10.29) and (10.30) present the logical relation of start-up and shut-down binary variables. The natural gas consumption is calculated by (10.31).

$$I_t^{GT} P_{min}^{GT} \leq P_t^{GT} \leq I_t^{GT} P_{max}^{GT} \tag{10.25}$$

$$-D_{max}^{GT} \leq P_t^{GT} - P_{t-1}^{GT} \leq U_{max}^{GT} \tag{10.26}$$

$$T_{on}^{GT}\left(I_t^{GT} - I_{t-1}^{GT}\right) + \sum_{j=t-T_{on}^{GT}}^{t-1} I_j^{GT} \geq 0 \tag{10.27}$$

$$T_{off}^{GT}\left(I_{t-1}^{GT} - I_t^{GT}\right) + \sum_{t'=t-T_{off}^{GT}}^{t-1} I_{t'}^{GT} \geq 0 \tag{10.28}$$

$$x_t^{GT} + y_t^{GT} = I_t^{GT} - I_{t-1}^{GT} \tag{10.29}$$

$$x_t^{GT} + y_t^{GT} \leq 1 \tag{10.30}$$

$$g_t^{GT} = \frac{h_0^{GT}\left(P_t^{GT}\right)^2 + h_1^{GT} P_t^{GT} + h_2^{GT}}{HHV_g} \tag{10.31}$$

The electricity consumption and natural gas production of P2G are shown in (10.32) and (10.33).

$$0 \leq P_t^{P2G} \leq P_{max}^{P2G} \tag{10.32}$$

$$g_t^{P2G} = \frac{\phi P_t^{P2G} \eta^{P2G}}{HHV_g} \tag{10.33}$$

(2)   Electrical Network

The electrical network of seaport operates as a radial network. The DistFlow model proposed in [33] is employed to represent power flows in the electrical network:

$$P_{ij} + P_j^{in} - r_{ij} I_{ij}^2 = \sum_{k \in \Theta^E(j)} P_{jk} + P_j^L \tag{10.34}$$

$$Q_{ij} + Q_j^{in} - x_{ij} I_{ij}^2 = \sum_{k \in \Theta^E(j)} Q_{jk} + Q_j^L \tag{10.35}$$

$$V_j^2 = V_i^2 - 2\left(r_{ij} P_{ij} + x_{ij} Q_{ij}\right) + \left(r_{ij}^2 + x_{ij}^2\right) I_{ij}^2 \tag{10.36}$$

$$I_{ij} = \sqrt{\frac{P_{ij}^2 + Q_{ij}^2}{V_i^2}}, \; P_{ij} \geq 0, \; Q_{ij} \geq 0 \tag{10.37}$$

$$0 \leq I_{ij} \leq I_{ij}^r, \; V_i^f \leq V_i \leq V_i^r \tag{10.38}$$

Equations (10.34) and (10.35) show the nodal active and reactive power balance. Equation (10.36) presents the voltage drop at each electric line. Equation (10.37) relates voltage, current, active power, and reactive power. Constraint (10.38) imposes bounds on line currents and nodal voltages.

(3)   Natural Gas Network

Constraint (10.39) represents the nodal natural gas balance. Constraint (10.40) limits the supply of natural gas wells. Constraint (10.41) and (10.42) restricts nodal gas pressure and natural gas flow through pipelines. The natural gas flow is caused by gas pressure difference between nodes and can be formulated by Weymouth flow Eq. (10.43).

$$\sum_{i\in\Omega^G(m)} g_{lm} + g_j^{in} = \sum_{k\in\Theta^G(m)} g_{mn} + g_j^L \tag{10.39}$$

$$\underline{G}_m^W \leq g_m^W \leq \overline{G}_m^W \tag{10.40}$$

$$\underline{\pi}_m \leq \pi_m \leq \overline{\pi}_m \tag{10.41}$$

$$\underline{G}_{mn} \leq g_{mn} \leq \overline{G}_{mn} \tag{10.42}$$

$$g_{mn} = K_{mn}\mathrm{sgn}(\pi_m - \pi_n)\sqrt{\left|\pi_m^2 - \pi_n^2\right|} \tag{10.43}$$

10.3.3 The Nexus Between Logistics System and Energy System

As Fig. 10.2 shows, QCs and ships are connected with certain electrical nodes of electricity networks. The nodal power loads of QCs and ships can be formulated as Eqs. (10.44) and (10.45) using rated power P^{crane}/P^{ship} and status ω_{qst}/X_{bst}, where $\Phi(i)$ and $\Gamma(i)$ are the set of QCs and berths, respectively, that are linked with electrical node i.

$$P_{i,t}^{crane} = \sum_{q\in\Phi(i)}\sum_{s\in S} \omega_{qst} P_{crane} \tag{10.44}$$

$$P_{i,t}^{ship} = \sum_{s\in S} X_{bst} P_s^{ship} b \in \Gamma(i) \tag{10.45}$$

Note that the nodal electricity demands of electricity networks originate from the fixed power loads, ESS charging, P2G, QCs and ships, so can be calculated by Eq. (10.46) as follows:

$$P_{i,t}^L = P_{i,t}^d + P_{i,t}^{ch} + P_{i,t}^{P2G} + P_{i,t}^{ship} + P_{i,t}^{crane} \tag{10.46}$$

The above three equations achieve the coupling between LS and ES. Equations (10.44) and (10.45) include binary variables ω_{qst}/X_{bst} that represent ships/QCs status and are restricted by logistic operation constraints. Then the two variables are added to ES through Eq. (10.46), so as to also comply with energy dispatching constraints. In this regard, berth allocation of ships and QC assignment are restricted by logistic and energy dual-constraints. Thus, the two systems restrict and influence each other as a whole, and can be dispatched uniformly.

10.3.4 Coordinated Optimization of PIES

In this study, PIES is dispatched by seaport control center and is allowed to purchase energy from the main power grid and natural gas wells at certain prices. The control center is eligible to dispatch ships, QCs and energy flows. The objective is to minimize the total costs, including the logistic operating costs F_L and the energy service costs F_E as follows:

$$\min F_L + F_E$$

s.t.

$$
\begin{aligned}
&\text{Logistic system constraints } (10.1 - 10.18)\\
&\text{Energy system constraints } (10.19 - 10.43)\\
&\text{Coupling constraints } (10.44 - 10.46)
\end{aligned}
\qquad (10.47)
$$

where the logistic operating cost F_L is given by

$$F_L = \sum_{s \in S} \left[c_s^d \left(\sigma_{bs}^{pref} - \sigma_{bs} \right) + c_s^w \left(t_{1,s} - t_{0,s} \right) + c_s^b \left(t_{2,s} - t_{1,s} \right) \right] \qquad (10.48)$$

The first term is penalty cost of deviation from preferred berth, where $\sigma_{bs}^{pref} - \sigma_{bs}$ is equal to 0 when the ship dock at preferred berth, otherwise, it equals to 1. The second and third terms are anchoring cost and docking cost respectively.

The energy service cost F_E is expressed in Eq. (10.49), includes start-up and shutdown cost of GT, energy purchase cost and carbon emissions penalty. The carbon emission comes from energy consumption and auxiliary engines of anchored ships.

$$
\begin{aligned}
F_E = &\sum_{t \in T} \sum_{o \in N_{GT}} \left[x_{o,t}^{GT} S_i^{on} + y_{o,t}^{GT} S_i^{off} \right] + \sum_{t \in T} \left(c_t^{grid} P_t^{grid} + \sum_{j \in g^W} c^{gas} g_{j,t}^W \right) \\
&+ c^{CO_2} \sum_{t \in T} \left(\alpha^{CO_2} P_t^{grid} + \sum_{j \in g^W} \beta^{CO_2} g_{j,t}^W \right) \\
&+ c^{CO_2} \sum_{s \in S} \left[\left(t_{1,s} - t_{0,s} \right) \left(\gamma_s^0 \left(P_s^{ship} \right)^2 + \gamma_s^1 P_s^{ship} + \gamma_s^2 \right) \right]
\end{aligned}
\qquad (10.49)
$$

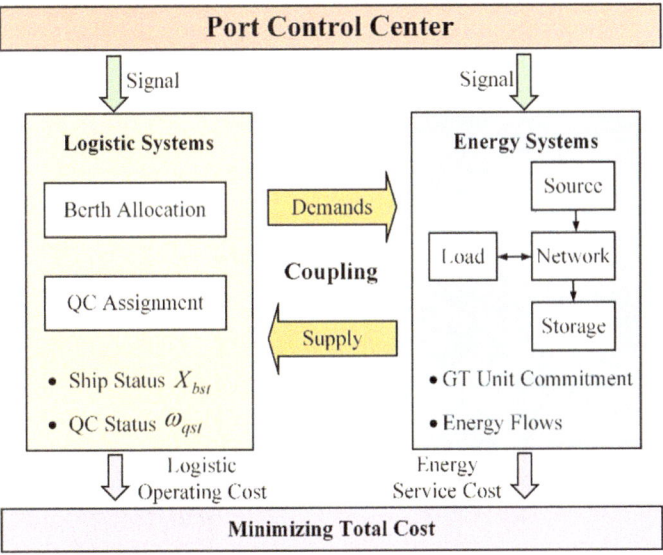

Fig. 10.3 Coordinated optimization framework of PIES

With the optimization goal and operating constraints, a coordinated optimization framework of PIES is proposed as Fig. 10.3 shows. In the framework, ships and QCs adjust their states under logistic-energy dual-constraints to achieve load temporal-spatial shifting and logistic operation. At the same time, energy dispatching schemes including GT unit commitment and energy flows are also determined in the coordination process.

10.4 Solution Methodology

Clearly, the proposed model is a nonlinear and non-convex optimization problem, which is challenging to solve. The solution method is stated in this section.

10.4.1 Linearizing Logistic Constraints

The nonlinearity of logistic operation constraints comes from Eqs. (10.1), (10.16–10.18), and can be linearized by big-M method.

Equation (10.1) can be transformed into Eqs. (10.50–10.54), which are equal to Eq. (10.1) logically.

$$bX_{bst} \leq B_s, \quad \forall b, s, t \tag{10.50}$$

$$bX_{bst} + M(1 - X_{bst}) \geq B_s, \quad \forall b, s, t \tag{10.51}$$

$$t X_{bst} \leq t_{2,s}, \quad \forall b, s, t \tag{10.52}$$

$$t X_{bst} + M(1 - X_{bst}) \geq t_{1,s}, \quad \forall b, s, t \tag{10.53}$$

$$\sum_{b \in B} \sum_{t \in T} X_{bst} = t_{2,s} - t_{1,s} + 1, \quad \forall s \tag{10.54}$$

Equation (10.54) indicates that berthing duration is the difference between departure time $t_{2,s}$ and berthing start time $t_{1,s}$, and it also equal to the sum of docking status X_{bst} in dimensions b and t.

Similarly, Eqs. (10.16), (10.17) and (10.18) are transformed to Eqs. (10.55–10.57), (10.58–10.59) and (10.60), respectively.

$$B_{qt} \leq B_s + M(1 - \omega_{qst}), \quad \forall q, s, t \tag{10.55}$$

$$B_{qt} + M(1 - \omega_{qst}) \geq B_s, \quad \forall q, s, t \tag{10.56}$$

$$B_{qt} \leq B^{max} \sum_{s \in S} \omega_{qst} \tag{10.57}$$

$$B_{qt} \geq \xi_{qt}, \quad \forall q, t \tag{10.58}$$

$$B_{qt} \leq B^{max} \xi_{qt} \tag{10.59}$$

$$B_{(q+1)t} + M(1 - \xi_{(q+1)t}) \geq B_{qt}, \quad \forall q, t \tag{10.60}$$

10.4.2   Convexifying the Energy Systems Equations

Replace $(V_i)^2$ with U_i and $(I_{ij})^2$ with i_{ij}, then Eq. (10.37) is convexified by SOC relaxations as follows

$$\left\| \begin{matrix} 2P_{ij} \\ 2Q_{ij} \\ i_{ij} - U_i \end{matrix} \right\|_2 \leq i_{ij} + U_i \tag{10.61}$$

Replace $(\pi_m)^2$ with $\hat{\pi}_m$ and introduce binary variables u^+/u^- to represent $\mathrm{sgn}(\pi_m - \pi_n)$ in Eq. (10.43), then Eq. (10.43) transforms to Eq. (10.62):

$$g_{mn}^2 = K_{mn}^2 \left(u_{mn}^+ - u_{mn}^- \right) \left(\hat{\pi}_m - \hat{\pi}_m \right) \tag{10.62}$$

Let $\prod_m$ be squares of nodal pressure bound and binary variables u^+/u^- can be calculated by:

$$\left(u_{mn}^+ - 1 \right) \underset{n}{\overline{\prod}} \le \hat{\pi}_m - \hat{\pi}_n \le \left(1 - u_{mn}^- \right) \underset{m}{\overline{\prod}} \tag{10.63}$$

$$\left(u_{mn}^+ - 1 \right) \underset{mn}{\overline{G}} \le g_{mn} \le \left(1 - u_{mn}^- \right) \underset{mn}{\overline{G}} \tag{10.64}$$

$$u_{mn}^+ + u_{mn}^- = 1 \tag{10.65}$$

Then, Eq. (10.62) is finally linearized and convexified as follows:

$$\psi_{mn} \ge \hat{\pi}_n - \hat{\pi}_m + \left(\underline{\prod}_m - \underset{n}{\overline{\prod}} \right) \left(u_{mn}^+ - u_{mn}^- + 1 \right) \tag{10.66}$$

$$\psi_{mn} \le \hat{\pi}_n - \hat{\pi}_m + \left(\underset{m}{\overline{\prod}} - \underline{\prod}_n \right) \left(u_{mn}^+ - u_{mn}^- + 1 \right) \tag{10.67}$$

$$\psi_{mn} \ge \hat{\pi}_m - \hat{\pi}_n + \left(\underset{m}{\overline{\prod}} - \underline{\prod}_n \right) \left(u_{mn}^+ - u_{mn}^- - 1 \right) \tag{10.68}$$

$$\psi_{mn} \le \hat{\pi}_m - \hat{\pi}_n + \left(\underline{\prod}_m - \underset{n}{\overline{\prod}} \right) \left(u_{mn}^+ - u_{mn}^- - 1 \right) \tag{10.69}$$

$$\left\| \begin{matrix} \frac{2g_{mn}}{K_{mn}} \\ \psi_{mn} - 1 \end{matrix} \right\|_2 \le \psi_{mn} + 1 \tag{10.70}$$

where the variable ψ_{mn} replaces $\left(u_{mn}^+ - u_{mn}^- \right) \left(\hat{\pi}_m - \hat{\pi}_n \right)$ in Eq. (10.62).

The convex quadratic term $h_0^{GT} \left(P_t^{GT} \right)^2$ in Eq. (10.31) can be convexified by an auxiliary variable δ_0 and the following SOC inequalities, which is defined as the epigraph form [34].

$$\delta_o + 1 \ge \left\| \begin{matrix} 2\sqrt{h_{0,o}^{GT}} \, P_{o,t}^{GT} \\ \delta_o - 1 \end{matrix} \right\| \tag{10.71}$$

10.4.3   The Final Optimization Formulation of PIES

Based on the linearization and relaxations discussed above, the following convex MISOCP problem is obtained:

$$\min F_L + F_E \qquad\qquad (10.72)$$

s.t.

Logistic system constraints (10.2–10.15), (10.50–10.60).

Energy system constraints (10.19–10.36), (10.38–10.42), (10.62–10.71).

Coupling constraints (10.44–10.46).

which is the final form of proposed coordinated optimization model.

10.5   Case Studies

In this section, the proposed model is applied to a typical port integrated energy system. It includes 6 shore-side power devices (6 berthing position), 18 available QCs, modified 33 electrical buses [35] and modified 7 natural gas nodes [36]. The topology of test system is shown in Fig. 10.4. Shore-side power devices, QCs and various energy technologies are connected to certain nodes in energy supply networks.

The electrical network and natural gas network are linked by two GT units and two P2Gs. In the electrical network, GT1 (5 MW) and GT2 (5 MW) are installed at bus 3 and bus 9, P2G1 (3 MW) and P2G2 (3 MW) are connected to bus 19 and bus 26, PV1 (5 MW), PV2 (2.4 MW) and PV3 (2 MW) are linked with bus 5, bus 19 and bus 26, respectively. There is also an ESS (3 MW) at bus 5. The electricity prices during valley (1:00–7:00, 24:00), normal (8:00–10:00, 16:00–18:00, 22:00–23:00) and peak (11:00–15:00, 19:00–21:00) hours are 0.13, 0.49, and 0.83¥/kWh respectively. The natural gas price is set as 5.41$/kcf. The carbon emission coefficient of ship auxiliary engines is set equally as 3.136 kg/MW^{2}h, 537.6 kg/MWh and 23.36 kg/h [13]. The carbon emission factors of main grid and gas wells are 712.9 kg/MWh and 64.0 kg/kcf. The carbon tax is set to be 6.192$/ton according to Shanghai Emission Trading System.

The operation horizon is set to be 48 h. Three kinds of ships are included: heavy ship, medium ship and small ship. The number of containers to be handled is 1500–3000 for heavy ships, 800–1500 for medium ships and 200–800 for small ships. The number of QCs that can be allocated for each ship is 3–6 for heavy ships, 2–5 for medium ships and 1–4 for small ships. The power demand of heavy ships, medium ship and small ships are 3 MW, 2 MW and 1 MW respectively. Each QC handle 35 containers per hour and its power is 0.3 MW. The logistic operating cost coefficients in Eq. (10.48) are 0.5, 1 and 1 times of the number of handled containers.

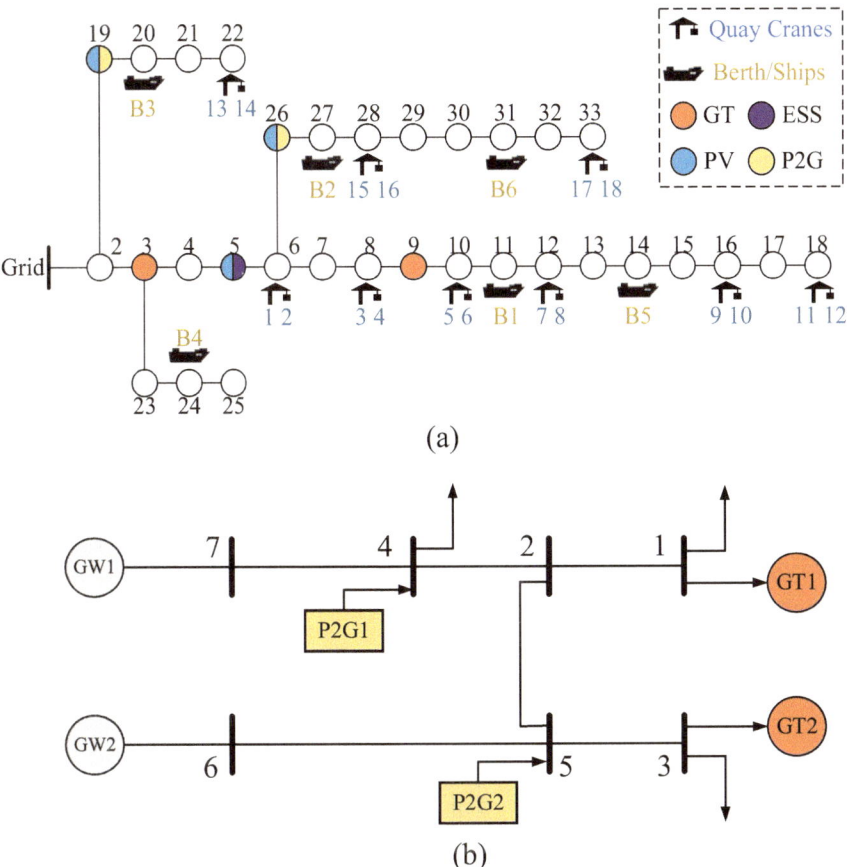

Fig. 10.4 Network topologies: **a** electrical network; **b** natural gas network

Four cases are set to demonstrate the advantages of proposed model. Case I is designed as an uncoordinated operating mode: the first stage is to minimize logistic operating cost F_L (48) under logistic operation constraints and obtain the status of ships and QCs. The second stage is to minimize energy service cost F_E (49) with obtained logistic parameters. Finally evaluate total cost $F_L + F_E$. Case II is coordinated optimization of PIES using proposed mothed. Cases III and Case IV decouple electricity networks and natural gas networks in Cases I and Case II.

All simulations are implemented by Gorubi 9.10 and Yalmip in MATLAB 2021a on a laptop with an 8-core Intel Core i7 processor and 16 GB memory.

10.5.1 Case of Sufficient Berths

In this section, ten arrival ships with fewer containers are considered so berths are not congested. Table 10.1 gives the information of 10 arrival ships. Table 10.2 shows the optimization results of four cases. Table 10.3 compares the berth allocation of Case I and Case II. The advantages of proposed method are analyzed from the following two aspects.

(1) Emission Reduction and Economic Benefits

In Case I, berths and QCs are pre-allocated. All ships immediately dock at preferred berths when arriving at seaport, then leaving as soon as possible. Considering that the determination of berth allocation and QC assignment is only under logistic transportation constraints without energy operating constraints, some ships cannot utilize multiple energy sources because their berthing position mismatches the multiple energy supply, e.g., the second ship docks at berth 3 at 3:00 but no PV are available at that time interval. Therefore, uncoordinated optimization cannot obtain

Table 10.1 Parameters of ten arrival ships

Ship	Arrival time (h)	Containers (TEU)	Preferred berth	Type
1	1	1600	1	Heavy
2	3	800	3	Medium
3	6	500	5	Small
4	10	1050	4	Medium
5	14	500	6	Small
6	18	900	3	Medium
7	22	1750	2	Heavy
8	25	650	5	Small
9	30	850	4	Medium
10	35	450	6	Small

Table 10.2 Optimization results in four cases

	Case I	Case II	Case III	Case IV
Berth deviation penalty (k$)	0	0.21	0	0.19
Berthing cost (k$)	8.94	8.94	8.94	9.94
Operating cost of GT (k$)	0.39	0.41	0	0
Electricity purchase (k$)	2.27	0.37	7.50	4.58
Natural gas purchase (k$)	4.60	5.03	3.11	3.11
Emission penalty (k$)	0.55	0.43	0.98	0.98
Total (k$)	16.75	15.39	20.53	18.79

Table 10.3 Berth allocation of ships in two cases

Ship	Actual berth		Anchoring time (h)		Docking time (h)	
	Case I	Case II	Case I	Case II	Case I	Case II
1	1	1	0	0	8	8
2	3	4	0	0	5	5
3	5	5	0	0	4	4
4	4	2	0	0	6	6
5	6	6	0	0	4	4
6	3	1	0	0	6	6
7	2	2	0	0	9	9
8	5	5	0	0	5	5
9	4	4	0	0	5	5
10	6	6	0	0	4	4

the optimal dispatching schemes of LS and ES. The carbon emission and total costs are higher than Case II.

In Case II, the emission penalty and total costs reduce by 21.82% and 8.12% compared with Case I. This reduction is at the expense of berth deviation penalty. The second ship deviates from preferred berth 3 to berth 4 linked with GT unit, so the electricity purchased from main grid is replaced by cheaper natural gas. The fourth and the sixth ship arrive at seaport during the day when PV sources are available. Under coordination, the fourth ship deviates from berth 4 to berth 2 in order to use PV sources. The sixth ship deviates from berth 3 and docks at berth 1 because berth 1 links to the child bus of PV1 that has higher capacity, so that more PV sources can be used if the sixth ship docks at berth 1. Overall, in the case of sufficient berths, coordinated optimization changes the spatial distribution of ship load. The loads are matched with the energy supply under logistic-energy dual constraints. As a result, energy purchase and carbon emission reduce.

Figure 10.5 compares the electricity balance of Case I and Case II. In Case I, surplus PV sources and cheaper natural gas are not fully utilized by ships and QCs. More electricity is purchased from main grid to support ships and QCs, while extra PV generation is transformed to natural gas via P2G to support natural gas loads. So that the energy service costs of Case I is higher than Case II. In Case II, the logistic operating costs slightly increase from 8.94 to 9.15 k$, while total energy service costs significantly decrease from 7.81 to 6.24 k$. Although some ships deviate from their preferred berths, the increase of berth deviation penalty is negligible compared with the decline of energy service costs. The reduction of electricity purchase is mainly due to the temporal-spatial match between ships/QCs loads and energy supply, which can deliver more photovoltaic and natural gas to ships and QCS to replace the expensive electricity of the main grid.

As for Case III and Case IV, both logistic operating and energy service costs increase compared with Case I and Case II. Due to the lack of support from the

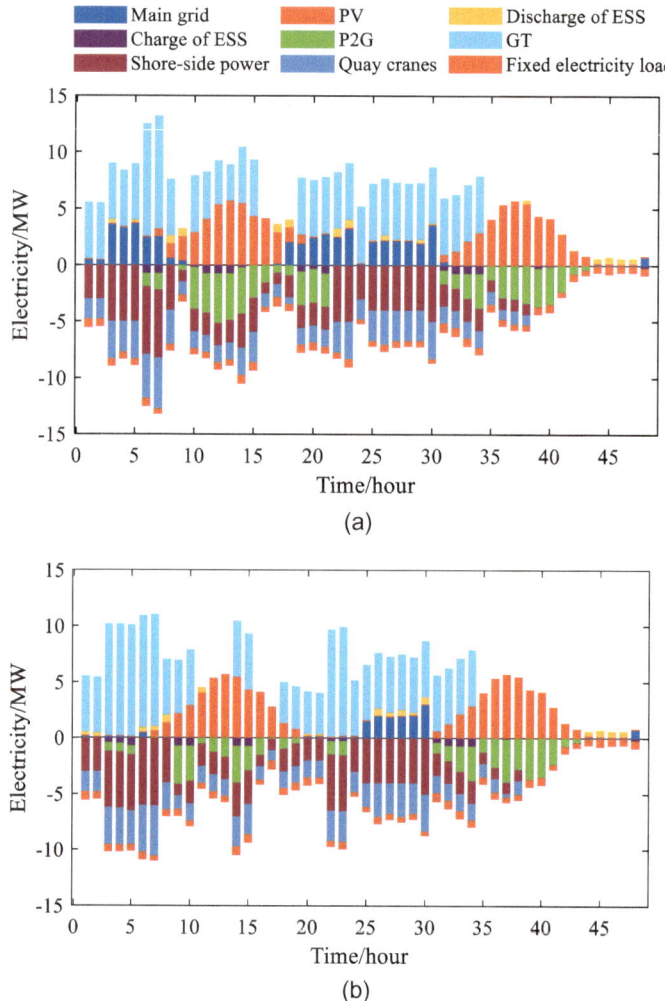

Fig. 10.5 Electricity balance in the coupled systems, **a** Case I; **b** Case II

natural gas network, seaports have to purchase expensive electricity from main grid to meet power demands, causing higher energy purchase costs. In addition, electricity price and PV generation become the main influencing factors of berth allocation and QC assignment. Some ships anchor to wait after arriving at seaports, and begin to dock when PV generation becomes larger or electricity price is relatively low, causing more berthing costs compared with Case I and Case II.

(2)   Sensitivity Analysis

The advantages of proposed coordinated optimization strategy may be affected by various parameters. A sensitivity analysis is performed on per-unit logistic operating costs and natural gas price to show the impact on total costs and operational decisions.

The per-unit logistic operating cost is set to be 10%, 50%, 100%, 150% and 200% of reference value in Case II. The results are shown in Tables 10.4 and 10.5 and Fig. 10.6. With the increasing of per-unit logistic operating cost, the economic benefits become less effective. Ships are more reluctant to change their preferred berth and try to minimize the sum of anchoring and docking time. Therefore, the results of Case I and Case II become more similar. In the cases of lower per-unit logistic operating cost (i.e., 10% and 50%), the logistic operating cost is negligible compared with energy service costs. Ships are more willing to deviate from preferred berths and adjust berthing time (see Table 10.4) in order to use multiple energy sources to reduce energy service costs. In this case, the operational decisions mainly depend on energy systems.

The natural gas price is set as 50%, 200% and 500% of reference value in Case II. Tables 10.6 and 10.7 show the results of berth allocation. With the decrease of natural gas price, the economic benefits become more obvious as Fig. 10.7 shows. This because when the price of natural gas become lower, the purchase cost of natural gas is closer to renewable energy (free to use). Ships are more willing to dock at berths that linked with GT units. When the price of natural gas reaches a certain high level, it is no longer an economical way to replace electricity with natural gas. As a result, in the case of 200% of natural gas, only the ninth ship docks at berth 4 that connected with child bus of GT unit, while there are three ships dock at berth 4 in the case of 50% natural gas price. The second and fourth ships dock at the berth 4 in the case of 50% natural gas price. However, the two ships dock at berth 3 and berth 2 when the price increase to 200%. When the price increases to 500%, only the nineth ship docks at berth 4 and is supported by natural gas. The other ships dock at other berths

Table 10.4 Berthing time of different per-unit logistic operating cost coefficients

Ship	Anchoring time (h)			Docking time (h)		
	10%	50%	150% 200%	10%	50%	150% 200%
1	0	0	0	8	8	8
2	0	0	0	5	5	5
3	0	0	0	4	4	4
4	0	0	0	6	6	6
5	2	0	0	4	4	4
6	0	0	0	6	6	6
7	0	0	0	9	9	9
8	1	0	0	7	5	5
9	1	0	0	5	5	5
10	1	0	0	4	4	4

Table 10.5 Berthing position of different per-unit logistic operating cost coefficients

Ship	Actual berth			
	10%	50%	150%	200%
1	1	1	1	1
2	4	4	3	3
3	5	5	5	5
4	2	2	2	4
5	1	6	6	6
6	4	1	1	1
7	1	2	2	2
8	6	5	5	5
9	2	1	1	4
10	6	6	6	6

Fig. 10.6 Cost analysis under different per-unit logistic operating costs. Every two slacks grouped together in each column represent the correlated costs in Case I and Case II, respectively. The percentage on the top of each stack group is the ratio of total cost reduction

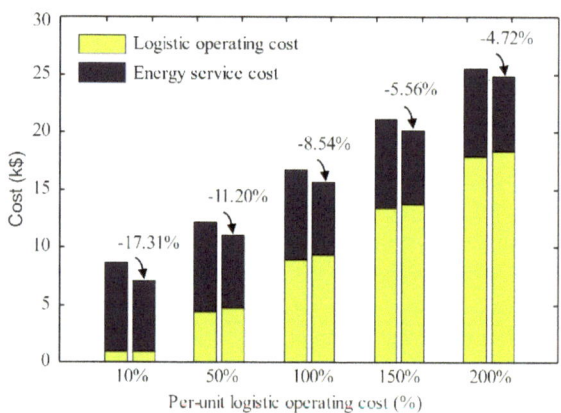

and are supplied by PV or electricity from power grid. In addition, berthing time is also influenced by natural gas price. The sixth ship anchors for four hours after arriving at seaport and begins to dock at 22:00 when the electricity price is low.

Figure 10.8 shows the energy supply structure in the cases of different natural gas price. When natural gas price is relatively low, the electrical loads are mainly supported by natural gas (67.53%) and PV (29.50%), while electricity from main grid only accounts for 2.96%. With the increase of natural gas price, the percentage of GT decreases significantly to 26.10% and the main grid increase to 41.94%. In the case of 500%, the natural gas is more expensive than electricity. 71.82% of the electricity supply comes from the main grid, and the proportion of gas turbines is negligible.

From the above discussion, the per-unit logistic operating cost and natural gas price have an obvious impact on operational decisions.

Table 10.6 Berthing time of different natural gas price

Ship	Anchoring time (h)			Docking time (h)		
	50%	200%	500%	50%	200%	500%
1	0	0	0	8	8	8
2	0	0	0	5	5	5
3	0	0	0	4	4	4
4	0	0	0	6	6	6
5	0	0	0	4	4	4
6	0	0	4	6	6	6
7	0	0	0	9	9	9
8	0	0	0	5	5	5
9	0	0	0	5	5	5
10	0	0	0	4	4	4

Table 10.7 Berthing position of different natural gas price

Ship	Actual berth		
	50%	200%	500%
1	1	1	1
2	4	3	3
3	5	5	5
4	4	2	2
5	6	6	3
6	1	1	1
7	2	2	2
8	5	5	5
9	4	4	4
10	6	6	6

Fig. 10.7 Cost analysis under different natural gas price. Every two slacks grouped together in each column represent the correlated costs in Case I and Case II, respectively. The percentage on the top of each stack group is the ratio of total cost reduction

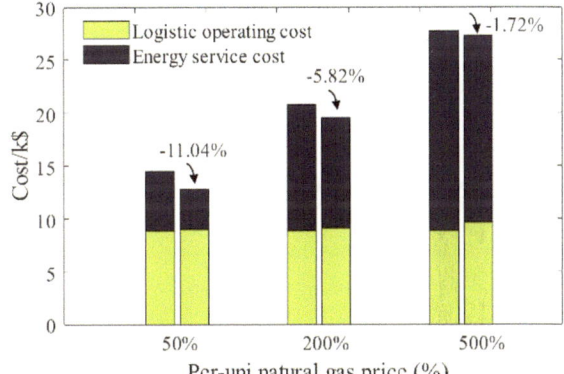

Fig. 10.8 The energy supply structure of the three cases

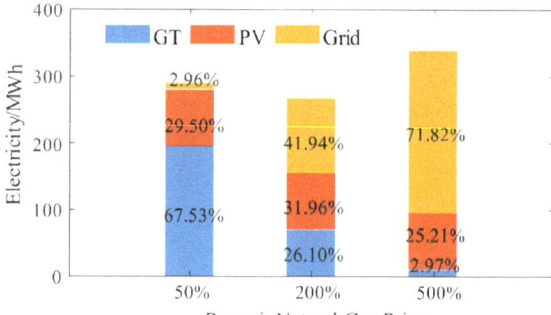

10.5.2   Case of Berth Congestion

This section focuses on the advantages of coordinated optimization in the case of berth congestion. Fifteen arrival ships with more containers are considered and the parameters of ships are shown in Table 10.8. The optimization results are given in Table 10.9, Figs. 10.9 and 10.10.

Figure 10.9 shows the results of berth allocation in Case I and Case II. In Case II, the fourteenth and the eighth ships dock at berth 1 in order to utilize cheaper natural gas. The berth deviation of the fourteenth ship makes berth 3 available, so that the six ship can begin to dock at berth 3 earlier than Case I to reduce anchoring costs. In this case, the berth deviation penalty of the fourteenth and the eighth ships

Table 10.8 Parameters of fifteen arrival ships

Ship	Arrival time (h)	Containers (TEU)	Preferred berth	Type
1	1	2900	1	Heavy
2	3	1500	3	Medium
3	6	1200	3	Medium
4	10	1400	4	Medium
5	14	800	6	Small
6	18	1250	3	Medium
7	22	2750	2	Heavy
8	25	700	5	Small
9	30	1350	4	Medium
10	35	650	6	Small
11	2	2750	1	Heavy
12	4	1250	4	Medium
13	5	800	5	Small
14	11	1450	3	Medium
15	13	750	6	Small

Table 10.9 Optimization results under berth congestion

	Case I	Case II
Average anchoring time (hour)	3.27	2.93
Average docking time (hour)	9.20	9.13
Logistic operating cost (k$)	44.59	45.40
Energy service cost (k$)	16.10	13.23
Total cost (k$)	60.69	58.63

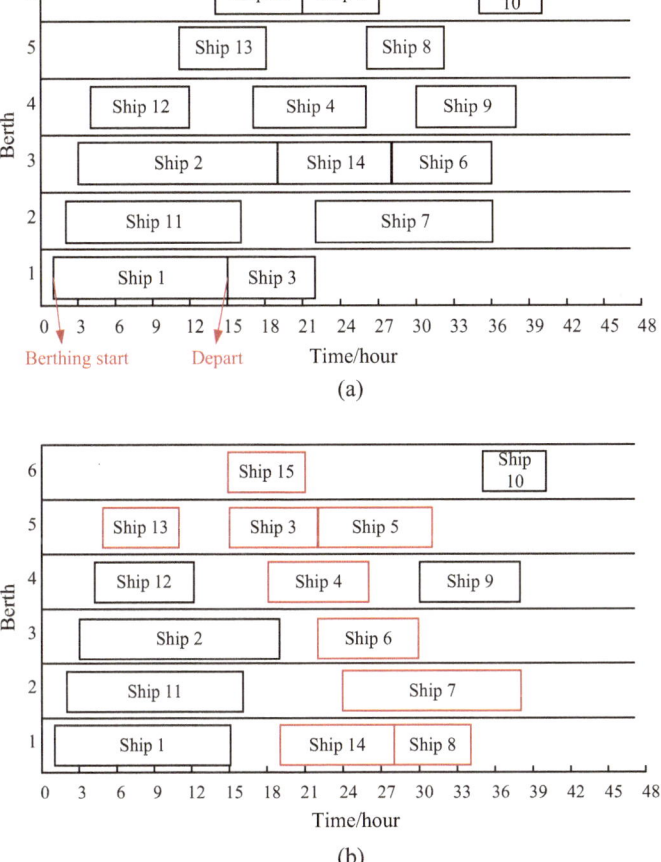

Fig. 10.9 The berthing time and position of ships: **a** Case I; **b** Case II. The red box represents the difference of Case II from Case I

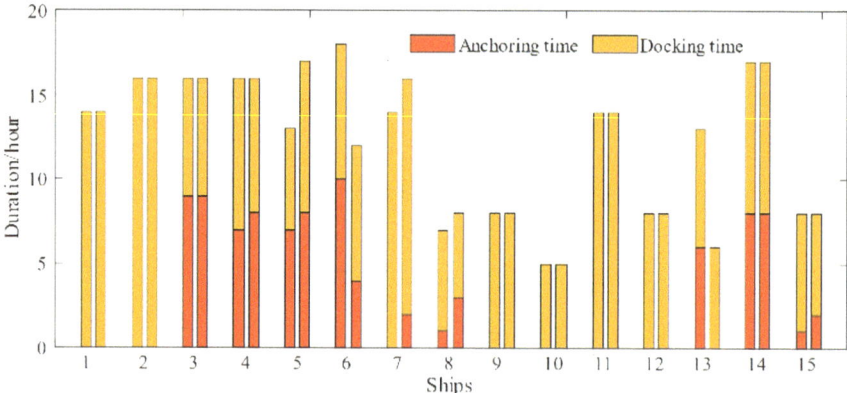

Fig. 10.10 Anchoring and docking time of the fifteen ships. The left-side bar represents the result of Case I and the right-side bar corresponds to Case II. The number above the bar is the berthing position of ships

is negligible compared with the reduction of energy purchase and anchoring time cost. For the fourth and the fifteenth ships, although the anchoring time of the two ships has increased by 1 h, the decrease of docking costs is more than the increase of anchoring costs. Similarly, the anchoring and docking time of the fifth ship increase because of less available QCs. But this ship deviates to berth 5 in order to reduce energy purchase cost.

It is worth mentioning that ships with higher power demands have higher priority to dock at berths that connect with more energy supply. For example, the third ship deviates from berth 1 to berth 5 and gives berth 1 to the fourteenth ship. This is because the fourteenth ship consumes more energy than the third ship and there is more energy supply at berth 1. Therefore, it is more economical to arrange the fourteenth at berth 1. As a result, higher power load from the fourteenth ship is supported by PV and natural gas, thereby reducing electricity purchase.

As for transportation efficiency, most of ships in Case II have the same or less anchoring and docking time in Case I (see Fig. 10.10). The average anchoring and docking time in Case II decrease by 0.34 and 0.07 h respectively compared with Case I. The increase of logistic operating cost is mainly caused by berth deviation penalty. Although some ships leave the seaports later than Case I, from an overall point of view, transportation efficiency is slightly affected. Therefore, even in the case of berth congestion, the proposed coordination optimization method can reduce total cost through changing berthing position/time of ships and the status of QCs, while ensuring transportation efficiency.

References

1. Fiadomor, R.: Assessment of alternative maritime power (Cold-ironing) and its impact on port management and operations. Ph.D. Dissertation. World Maritime University, Malmö (2009)
2. Yigit, K., Acarkan, B.: A new electrical energy management approach for ships using mixed energy sources to ensure sustainable port cities. Sustain. Cities Soc. **40**, 126–135 (2018)
3. Coppola, T., Fantauzzi, M., Lauria, D., et al.: A sustainable electrical interface to mitigate emissions due to power supply in ports. Renew. Sust. Energy Rev. **54**, 816–823 (2016)
4. Revised MARPOL annex VI: Regulations for the Prevention of Air Pollution from Ships (2008). Report Marine Environment Protection Committee, International Maritime Organization, London
5. Lu, S., Gu, W., Zhou, S., et al.: Pan adaptive robust dispatch of integrated energy system considering uncertainties of electricity and outdoor temperature. IEEE Trans. Industr. Inf. **16**(7), 4691–4702 (2020)
6. Massrur, H.R., Niknam, T., Fotuhi-Firuzabad, M.: Investigation of carrier demand response uncertainty on energy flow of renewable based integrated electricity-gas-heat systems. IEEE Trans. Industr. Inf. **14**(1), 5133–5142 (2018)
7. Yang, W., Liu, W., Chung, C.Y., et al.: Coordinated planning strategy for integrated energy systems in a district energy sector. IEEE Trans. Sustain. Energy **11**(3), 1807–1819 (2020)
8. Zhu, X., Yang, J., Liu, Y., et al.: Chen optimal scheduling method for a regional integrated energy system considering joint virtual energy storage. IEEE Access **7**, 138260–138272 (2019)
9. Xiang, X., Liu, C.: An expanded robust optimisation approach for the berth allocation problem considering uncertain operation time. Omega **103**(1), 102444 (2021)
10. Zhen, L.: Tactical berth allocation under uncertainty. Eur. J. Oper. Res. **247**(3), 928–944 (2015)
11. Liu, C., Xiang, X., Zheng, L.: A two-stage robust optimization approach for the berth allocation problem under uncertainty. Flex. Serv. Manuf. J. **32**(2), 425–452 (2019)
12. Xiang, X., Liu, C., Miao, L.: A bi-objective robust model for berth allocation scheduling under uncertainty. Transp. Res. Part E-logistics Transp. Rev. **106**, 294–319 (2017)
13. Schepler, X., Absi, N., Feillet, D., et al.: The stochastic discrete berth allocation problem. EURO J. Transp. Logis. **8**(4), 363–396 (2019)
14. Iris, Ç., Pacino, D., Ropke, S.: Improved formulations and an adaptive large neighborhood search heuristic for the integrated berth allocation and quay crane assignment problem. Transp. Res. Part E-logistics Transp. Rev. **105**, 123–147 (2017)
15. Malekahmadi, A., Alinaghian, M., Hejazi, S.R., et al.: Integrated continuous berth allocation and quay crane assignment and scheduling problem with time-dependent physical constraints in container terminals. Comput. Ind. Eng. **147**, 106672 (2020)
16. Agra, A., Oliveira, M.: MIP approaches for the integrated berth allocation and quay crane assignment and scheduling problem. Eur. J. Oper. Res. **264**(1), 138–148 (2018)
17. Tan, C., He, J.: Integrated proactive and reactive strategies for sustainable berth allocation and quay crane assignment under uncertainty. Ann. Oper. Res. **3**, 1–32 (2021)
18. Bouzekri, H., Alpan, G., Giard, V.: Integrated laycan and berth allocation and time-invariant quay crane assignment problem in tidal ports with multiple quays. Eur. J. Oper. Res. **293**(3), 892–909 (2021)
19. Nourmohammadzadeh, A., Voß, S.: A robust multiobjective model for the integrated berth and quay crane scheduling problem at seaside container terminals. Ann. Math. Artif. Intell., 1–23 (2021)
20. Parise, G., Parise, L., Martirano, L., et al.: Wise port and business energy management: port facilities, electrical power distribution. IEEE Trans. Ind. Appl. **52**(1), 18–24 (2016)
21. Misra, A., Venkataramani, G., Gowrishankar, S., et al.: Renewable energy based smart microgrids—a pathway to green port development. Strat. Planning Energy Environ. **37**(2), 17–32 (2017)
22. Duin, J., Geerlings, H., Verbraeck, A., et al.: Cooling down: a simulation approach to reduce energy peaks of reefers at terminals. J. Clean. Prod. **193**, 72–86 (2018)

23. Parise, G., Parise, H., Malerba, A., et al.: Comprehensive peak-shaving solutions for port cranes. IEEE Trans. Ind. Appl. **53**(3), 1799–1806 (2017)
24. Parise, G., Honorati, A.: Port cranes with energy balanced drive. In: 2014 AEIT Annual Conference-From Research to Industry: The Need for a More Effective Technology Transfer, Trieste, pp. 1–5 (2014)
25. Kanellos, F.D., Olanis, E.M.V., Hatziargyriou, N.D.: Power management method for large ports with multi-agent systems. IEEE Trans. Smart Grid **10**(2), 1259–1268 (2019)
26. Kanellos, F.D.: Multiagent-system-based operation scheduling of large ports' power systems with emissions limitation. IEEE Syst. J. **13**(2), 1831–1840 (2019)
27. Gennitsaris, S.G., Kanellos, F.D.: Emissions-aware and costeffective distributed demand response system for extensively electrified large ports. IEEE Trans. Power Syst. **34**(6), 4341–4351 (2019)
28. Kanellos, F.D.: Real-time control based on multi-agent systems for the operation of large ports as prosumer microgrids. IEEE Access **5**, 9439–9452 (2017)
29. Song, T., Li, Y., Zhang, X., et al.: Integrated port energy system considering integrated demand response and energy interconnection. Int. J. Electr. Power Energy Syst. **117**, 105654 (2020)
30. Pu, Y., Chen, W., Zhang, R., et al.: Optimal operation strategy of port integrated energy system considering demand response. In: 2020 IEEE 4th Conference on Energy Internet and Energy System Integration, Wuhan, pp. 518–523 (2020)
31. Fang, S., Wang, Y., Gou, B., et al.: Toward future green maritime transportation: an overview of seaport microgrids and all electric ships. IEEE Trans. Veh. Technol. **69**(1), 207–219 (2019)
32. Peng, Y., Dong, M., Li, X., et al.: Cooperative optimization of shore power allocation and berth allocation: a balance between cost and environmental benefit. J. Clean. Prod. **279**, 123816 (2020)
33. Baran, M.E., Wu, F.: Network reconfiguration in distribution systems for loss reduction and load balancing. IEEE Trans. Power Deliv. **4**(2), 1401–1407 (1989)
34. Boyd, S., Andenberghe, L.V.: Convex Optimization. Cambridge University Press (2004)
35. Wu, L., He, C., Dai, C., et al.: Robust network hardening strategy for enhancing resilience of integrated electricity and natural gas distribution systems against natural disasters. IEEE Trans. Power Syst. **33**(5), 5787–5798 (2018)
36. Liu, C., Shahidehpour, M., Fu, Y., et al.: Security-constrained unit commitment with natural gas transmission constraints. IEEE Trans. Power Syst. **24**(3), 1523–1536 (2009)

Chapter 11
Joint Scheduling of Power Flow and Berth Allocation in Port Microgrids

11.1 Introduction

As mentioned in the last chapter, the maritime transportation accounted for almost 90% world's trades [1]. However, due to the high reliance on fossil fuels, the shipping industry has caused enormous gas emission and correspondingly made ports one of the major sources of shipping-related air pollution in recent decade [2, 3]. To solve the urgent environmental problems, the Marine Environment Protection Committee implemented an ambitious emissions reduction mission in 2018, targeting a 50–70% reduction in greenhouse gas emissions from the shipping industry by 2050 [4]. Under this context, maritime electrification technologies, including renewable energy-based seaport microgrids and all-electric ships, come into existence and become an inevitable trend toward sustainable development of seaports.

Conventionally, seaports only provide logistic services to berthing ships, including berth allocation [5–7] and quay crane (QC) assignment for handling cargos on ships [8–10], which are the emphasis of most existing literatures in maritime community. However, these studies only focus on transportation efficiency while completely ignoring energy management associated with these industrial processes. As a results, seaports consume a large amount of fossil energy, leading to noise and air pollutions on the harbor territory.

To reduce gas emission within harbor territory, seaport microgrids become a representative and promising technology toward future sustainable maritime transportation [11]. With the development of green technologies, many measures have been implemented in seaports recently. For example, cold-ironing technology can provide onshore electrical power to berthing ships [12]. Ships are powered by onshore electricity while completely shutting down auxiliary diesel generators. Besides, electricity-driven equipment [13], renewable energy [14], and integrated energy system [15, 16], have also be applied into seaports. In this sense, the connections between the seaport and ships are no longer limited in logistic-side, but also expanded

© The Author(s) 2023

W. Huang et al., *Energy Management of Integrated Energy System in Large Ports*, Springer Series on Naval Architecture, Marine Engineering, Shipbuilding and Shipping 18, https://doi.org/10.1007/978-981-99-8795-5_11

to electric-side [17]. Better energy management strategies should be implemented in future seaports to improve energy efficiency and economic benefits.

Last decades have witnessed many results on seaport energy management. Literature [18] achieved the demand response of port reefers through adjusting their charging time. Literature [19] optimized the energy absorptions of port cranes in an evaluated drive system. They also shaved the peak power demands of port cranes by installing ultracapacitors and flywheels into port cranes [20]. In Literature [21–24], a multiagent-based hierarchical framework was proposed to optimize a large number of port flexible loads. Various power loads, including electric vehicles, reefers and all electric ship (AES), are aggregated into upper-level agents. Such method reduces the scale of optimization model, thereby enhancing the efficiency of seaport energy management. Although the above studies have made remarkable progresses in seaport energy management, most of them only optimized the curve of power demands that given in advance. The most distinctive features of seaports (e.g., berth allocation) are completely ignored.

In fact, since the power load distributions of AESs and QCs totally depend on berth allocation schemes, the berth allocation process has a significant impact on the power dispatch of microgrids. Therefore, the interdependency between berth allocation and the power dispatch should be considered in seaport microgrid operations. However, there is a limited number of existing studies on this topic. Literature [17] conceptually proposed the benefits of the coordination between berth allocation and power dispatch. Literature [25] optimized the electric and thermal demands of port integrated energy systems by considering flexible berth allocation, which provided adequate flexibility to facilitate port operation. Literature [26] adjusted ship's berthing duration and the number of assigned cranes to reduce the operational costs of seaport microgrids under uncertain renewable generation.

Although the above literature has initially explored this area, and some of them considered the uncertainty of renewable energy generation, the impact of uncertainty from berth allocation, e.g., AES arrival time, on the seaport microgrid operations has not been fully discussed. In fact, the uncertain AES arrival directly affects berth allocation, and thus having an impact on power load distribution and power dispatch of seaport microgrids. Moreover, in literatures [25, 26], the berthing duration of AES are determined day-ahead. However, the uncertainty of AES arrival is revealed intra-day. The day-ahead decisions may be infeasible when actual arrival time of AES is later than planned berthing start time. From this perspective, considering the uncertainty of AES arrival is crucial for the operation of seaport microgrid. Besides, since the benefits of coordination between power dispatch and berth allocation highly depends on the available renewable generation [26], the uncertain renewable generation will also influence berth allocation process in turn. Therefore, the multiple uncertainties of both AES arrival and renewable generation will simultaneously post challenges on the operation of seaport microgrids, and thus should be considered comprehensively. However, to our best knowledge, there is no existing literatures considering this issue.

Based on the above discussion, it can be found that there may lack a coordination between berth allocation and power dispatch in seaport microgrids considering the multiple uncertainties of AES arrival and renewable generation. To fill the existing research gaps, this chapter proposes an optimal joint scheduling strategy to coordinate power dispatch and berth allocation in a uniform framework under the mentioned multiple uncertainties.

11.2 Deterministic Joint Scheduling Model

11.2.1 Problem Description

This study aims to jointly schedule berth allocation process of AES and the power dispatch of green port microgrids to improve energy efficiency and economic benefits. Figure 11.1 illustrates a typical structure of port microgrids. On the shore-side, a renewable energy-based microgrid combining onsite photovoltaic (PV), battery energy storage system (BESS), dispatchable distributed generator (DG) and substation connecting to the main grid provides electricity for power loads on both shore-side and ship-side. On the ship-side, AESs anchor to wait firstly when arrive the seaport. Then the seaport allocates berths to the anchoring AESs. At the same time, certain number of QCs are assigned for berthing AESs for cargo handling tasks.

Due to the application of cold-ironing and electrification technologies into seaports, AESs and QCs are both supported by onshore electricity. The berth allocation schemes will have a significant impact on the power load distribution of AESs and QCs, thereby influencing the power flow of shore-side microgrids. However, the objective of shore-side microgrids is to minimize electricity supply costs, while the ship-side aims to minimize the total services time of AESs. The two different objectives may conflict with each other due to the mutual impact between shore-side and ship-side. Therefore, the berth allocation and the power dispatch should be implemented in a coordinated framework to achieve a better synergy and trade-off between electricity supply costs and AES service efficiency.

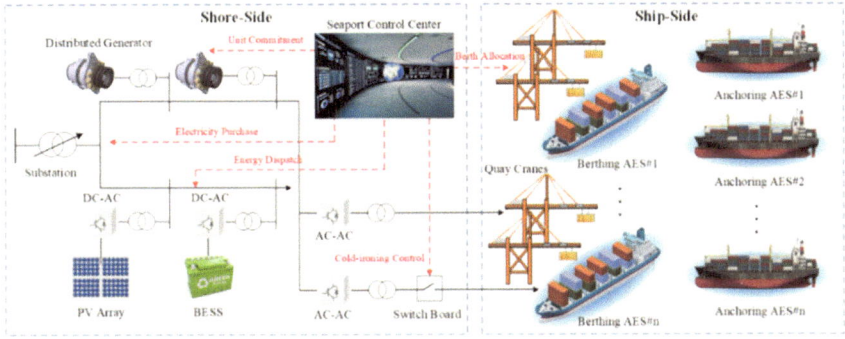

Fig. 11.1 Schematic diagram of port microgrids

The coordination between berth allocation and power dispatch is executed by seaport control center shown in Fig. 11.1. The microgrid determines unit commitment of DGs, charging and discharging power of BESS, power output of PV array and power flow in electrical networks. The decisions of ship-side includes the berthing position and duration of each AES, and the number of QCs assigned for each AES at each time slot. By jointly optimizing the decisions of microgrids and ship-side under the uniform management of seaport control center, an optimal joint scheduling scheme that can achieve the balance between electricity supply costs and AESs service efficiency can be obtained.

11.2.2 Objective Function

The objective of ship-side is to minimize the total service time of AES, which is measured differently from the electricity supply costs of microgrids. To compare the benefits between power dispatch and berth allocation, berthing related cost coefficients are introduced to convert service time of AES into economic costs. Then, the objective function of deterministic joint optimization model can be formulated to minimize the total costs of microgrid operations and AES berth allocation services as follows:

$$
\begin{aligned}
F = \sum_{t \in T} \sum_{g \in N_{DG}} \left(u_{g,t} S_g^{on} + v_{g,t} S_g^{off} \right) \\
+ \sum_{t \in T} \left[c_t^{grid} P_t^{grid} + \sum_{g=1}^{N_{DG}} \left[\begin{matrix} h_{g,2} \left(P_{g,t}^{DG} \right)^2 \\ + h_{g,1} P_{g,t}^{DG} + h_{g,0} \end{matrix} \right] \right] \\
+ \sum_{s \in S} \left[c_s^w \left(t_s^a - t_s^b \right) + c_s^b \left(t_s^d - t_s^b \right) \right]
\end{aligned} \tag{11.1}
$$

where T, S, N_{DG} are the set of dispatch period, AESs and DG units, t is the indice of time slot, g is the indice of DG unit, s is the indice of AES, $u_{g,t}$ is the binary variable indicating whether DG unit g start-up, $v_{g,t}$ is the binary variable indicating whether DG unit g shut-down. S_g^{on} and S_g^{off} are the start-up/shut-down cost of DG unit g respectively. c_t^{grid} is the unit purchasing price of electricity from the grid, P_t^{grid} is the purchased electricity from the grid, $P_{g,t}^{DG}$ is the power output of DG unit g, $h_{g,2}$, $h_{g,1}$ and $h_{g,0}$ are the generation cost coefficient of DG unit g, c_s^w, c_s^b are the cost coefficients related to anchoring and berthing of AES respectively, t_s^a is the arrival time of AES s, t_s^b is the berthing start time of AES s, t_s^d is the end of berthing time of AESs.

The first and second terms are electricity supply costs of microgrids, which includes the start-up and shut-down cost of dispatchable DGs, electricity purchase cost from the main grid, and generation cost of dispatchable DGs. The third term is the equivalent economic costs of AES berth allocation services, which includes the waiting and berthing costs of AES.

11.2.3 Constraints

1. Modeling of Ship-Side

(1) *Berth Allocation of AES*

The berth allocation aims to determine the berthing position and berthing duration of each AES. Conventionally, berth allocation process is formulated by a binary variable φ_{bsk} that represents the berthing status of ships [27]. The binary variable φ_{bsk} is indexed by berthing position b, ship number s, and service order k, meaning that ship s is served at berth b as the kth ship if $\varphi_{bsk} = 1$. Such model formulation is applicable in terms of independent berth allocation problems. However, since the binary variable φ_{bsk} is irrevent to time slot t, the power demands of AES at each time slot cannot be expressed by the binary variable φ_{bsk}. Therefore, if adopting the traditional AES service order-based model, berth allocation and power dispatch cannot be operated in a uniform time scale.

To establish the bridge between berth allocation and power dispatch, we improve the traditional AES service order-based berth allocation model by replacing the binary variable φ_{bsk} with a new time-indexed binary variable φ_{bst}. The binary variable φ_{bst} represents ship s is served at berth b at time slot t if $\varphi_{bsk} = 1$. In this way, the power demands of AES can be expressed mathematically. Meanwhile, the time-indexed model can still formulate the process of berth allocation. Therefore, the time-index model mathematically couples berth allocation and power dispatch. The time-indexed berth allocation model is formulated as follows:

Firstly, the binary variable φ_{bst} is a binary variable indicating if AES s is served at berth b at timeslot t, which is associated with berthing position B_s, berthing start time t_s^b, and berthing end time t_s^d by Eq. (11.2):

$$\varphi_{bst} = \begin{cases} 1 & \text{if } b = B_s \text{ and } t \in \left(t_s^b, t_s^d\right) \\ 0 & \text{otherwise} \end{cases}, \quad \forall s \tag{11.2}$$

Equation (11.2) can be linearized by big-M method as follows:

$$b\varphi_{bst} \leq B_s, \quad \forall b, s, t \tag{11.3}$$

$$b\varphi_{bst} + M(1 - x_{bst}) \geq B_s, \quad \forall b, s, t \tag{11.4}$$

$$t\varphi_{bst} \leq t_{2,s}, \quad \forall b, s, t \tag{11.5}$$

$$t\varphi_{bst} + M(1 - \varphi_{bst}) \geq t_s^b, \quad \forall b, s, t \tag{11.6}$$

$$\sum_{b \in B} \sum_{t \in T} \varphi_{bst} = t_s^d - t_s^b + 1, \quad \forall s \tag{11.7}$$

where B is the set of berths.

Constraints (11.3) and (11.4) ensure that the AES cannot change its berthing position once it starts berthing. Constraints (11.5) and (11.6) ensure that the binary variable φ_{bst} is equal to 1 only when the AES s is served at berth.

$$t_s^a \leq t_s^b, \quad \forall s \tag{11.8}$$

$$t_s^b + 1 \leq t_s^d \leq t_s^{latest}, \quad \forall s \tag{11.9}$$

$$1 \leq B_s \leq B_{\max}, \quad \forall s \tag{11.10}$$

$$\sum_{b \in B} \varphi_{bst} \leq 1, \quad \forall s, t \tag{11.11}$$

$$\sum_{s \in S} \varphi_{bst} \leq 1, \quad \forall b, t \tag{11.12}$$

where t_s^{latest} is the latest departure time of AES s, $B_{\max}$ is the number of available berths.

Constraints (11.8) and (11.9) present the relationship between arrival time, berthing start time, and departure time. The berthing start time should be greater than or equal to arrival time. The berthing end time should be greater than berthing start time and less than or equal to latest departure time. Constraint (11.10) limits the berthing position of each AES within an allowable range. Constraint (11.11) ensures that each AES can only be assigned to one berth, and constraint (11.12) restricts that each berth can only serve at most one AES at a time.

(2) QC Assignment for AES

In berth allocation process, the actual berthing duration of AES relies on the cargo handling speed, which depends on the number of assigned QCs for each AES. The QCs assignment schemes not only directly influence the berthing duration of AES, but also have an impact on the power demands of QCs. Therefore, the QC assignment should also be considered in berth allocation process, which are formulated as follows:

$$\sum_{s \in S} \omega_{qst} \leq 1, \quad \forall q, t \tag{11.13}$$

$$Q_{st} = \sum_{q \in Q} \omega_{qst}, \quad \forall s, t \tag{11.14}$$

$$\sum_{s \in S} Q_{st} \leq Q^{\max}, \quad \forall t \tag{11.15}$$

$$Q_s^{\min} \sum_{b \in B} \varphi_{bst} \leq Q_{st} \leq Q_s^{\max} \sum_{b \in B} \varphi_{bst}, \quad \forall s, t \tag{11.16}$$

$$\sum_t \eta Q_{st} \geq TEU_s, \quad \forall s, t \tag{11.17}$$

where Q is the set of QCs, ω_{qst} is the binary variable to indicate if QC q is serving AES, ω_{qst} equals to 1 when the QC q serves for the AES s at time slot t. Q_{st} is the number of QCs serving for AES s at time slot t, $Q^{\max}$ is the number of available QCs, $Q_s^{\min}$ and $Q_s^{\max}$ is the minimum and maximum number of QC hat can be assigned for AES, η is the cargo handling efficiency of QC, TEU_s is the number of carges on AES s.

Constraint (11.13) ensures that each QC can serve for at most one AES at each time slot. The number of assigned QCs for AES s is calculated by Eq. (11.14). Constraint (11.15) presents that the total number of working QCs cannot be larger than the total number of available QCs. Each AES has the minimum and maximum number of QCs that can be assigned, which is presented as constraint (11.16). Constraint (11.17) ensures that enough QCs should be assigned for AES to finish cargo handling tasks before AES departs the seaport.

(3) Power demands of AES and QC.

Based on the binary variables φ_{bst} and ω_{qst}, the power demands of AES $P_{i,t}^{AES}$ and QC $P_{i,t}^{QC}$ at electrical bus i at time slot t can be formulated respectively as follows:

$$P_{i,t}^{AES} = \sum_{s \in S} \varphi_{bst} P_s^{AES}, b \in \Gamma(i) \tag{11.18}$$

$$P_{i,t}^{QC} = \sum_{q \in \Phi(i)} \sum_{s \in S} \omega_{qst} P_q^{QC} \tag{11.19}$$

where $\Gamma(i)$ and $\Phi(i)$ are the set of berths and QCs that are linked with the electrical bus i, P_s^{AES} and P_q^{QC} are the rated power demands of AES s and QC q.

It can be found that the power demands of AES and QC should be restricted by berth allocation and QC assignment constraints since they are formulated by the binary variables φ_{bsk} and ω_{qst}. Meanwhile, AES and QC are the power loads of microgrids, thus they should also be limited by operational constraints of microgrids. From this perspective, Eqs. (11.18) and (11.19) establish the interface between microgrids and ship-side.

2. Modeling of Microgrids

(1) Power Flow Balance

The electrical network of seaport microgrids operates a radial network. The SOC-based DistFlow model [28] is employed to represent power flows in seaport microgrids:

$$\sum_{i\in\Pi(j)} P_{ij,t} + P_{j,t}^{in} - r_{ij}i_{ij,t} = \sum_{k\in\Theta(j)} P_{jk,t} + P_{j,t}^{L} \tag{11.20}$$

$$\sum_{i\in\Pi(j)} Q_{ij,t} + Q_{j,t}^{in} - x_{ij}i_{ij,t} = \sum_{k\in\Theta(j)} Q_{jk,t} + Q_{j,t}^{L} \tag{11.21}$$

$$U_{j,t} = U_{i,t} - 2(r_{ij}P_{ij,t} + x_{ij}Q_{ij,t}) + \left(r_{ij}^2 + x_{ij}^2\right)i_{ij,t} \tag{11.22}$$

$$2P_{ij,t}2Q_{ij,t}i_{ij,t} - U_{i,t2} \leq i_{ij,t} + U_{i,t} \tag{11.23}$$

$$i_{ij,t} \leq i_{ij,t}^{max}, U_{j,t}^{min} \leq U_{j,t} \leq U_{j,t}^{max} \tag{11.24}$$

$$P_{j,t}^{in} = P_{j,t}^{RE} + P_{j,t}^{dis} + P_{j,t}^{DG} \tag{11.25}$$

$$P_{j,t}^{L} = P_{j,t}^{d} + P_{j,t}^{ch} + P_{j,t}^{AES} + P_{j,t}^{QC} \#(11.26) \tag{11.26}$$

where $\Pi(j)$ and $\Theta(j)$ are the sets of upstream/downstream buses of bus j, $P_{ij,t}$ and $Q_{ij,t}$ are the active and reactive power flow from bus i to j at time slot t, similarly, $P_{j,t}^{in}$ and $Q_{j,t}^{in}$ are the injected active/reactive power of bus j at time slot t, $P_{jk,t}$ and $Q_{jk,t}$ are the active and reactive power flow from bus j to k at time slot t. r_{ij} and x_{ij} are the resistance and inductance of the line ij, $P_{j,t}^{L}$ and $Q_{j,t}^{L}$ are the active and reactive power of the load at bus j at time slot t, $i_{ij,t}$ is the amplitude square of current from bus i to j, $U_{i,t}$ is the voltage at bus i at time slot t, $i_{ij,t}^{max}$ is the maximum current allowed for line ij at time slot t, $U_{j,t}^{min}$ and $U_{j,t}^{max}$ are the lower and upper bound of voltage of bus j at time slot t, $P_{j,t}^{RE}$, $P_{j,t}^{DG}$, $P_{j,t}^{QC}$ are the power of renewable generation, DG and QC of bus j at time slot t, $P_{j,t}^{dis}$ and $P_{j,t}^{ch}$ are the discharging/charging power of BESS, $P_{j,t}^{d}$ is the conventional load power of bus j at time slot t.

Constraints (11.20) and (11.21) present the nodal active and reactive power balance. The line power flow $P_{ij,t} = P_{t}^{grid}$ and $Q_{ij,t} = Q_{t}^{grid}$ if j is the substation bus. Constraints (11.22) presents the voltage drop at each electrical bus. Constraints (11.23) is the SOC relaxation of the limit $P_{ij,t}^2 + Q_{ij,t}^2 = i_{ij,t}U_{i,t}$. Constraint (11.24) imposes bounds on line currents and nodal voltages. Constraints (11.25) and (11.26) show the nodal power injections and loads, where AES and QC are regarded as the part of nodal power loads.

(2) Energy Technologies

Energy technologies, including dispatchable DG units, BESS and renewable generators, are implemented in seaport microgrids. The operational constraints of these devices are presented as follows:

$$I_{g,t} P_g^{\min} \leq P_{g,t}^{DG} \leq I_{g,t} P_g^{\max} \tag{11.27}$$

$$-D_g^{\max} \leq P_{g,t}^{DG} - P_{g,t-1}^{DG} \leq U_g^{\max} \tag{11.28}$$

$$T_g^{on}\left(I_{g,t}^{DG} - I_{g,t-1}^{DG}\right) + \sum_{\tau=t-T_g^{on}}^{t-1} I_{g,\tau} \geq 0 \tag{11.29}$$

$$T_g^{off}\left(I_{g,t-1}^{DG} - I_{g,t}^{DG}\right) + \sum_{\tau=t-T_g^{off}}^{t-1} I_{g,\tau} \geq 0 \tag{11.30}$$

$$u_{g,t} + v_{g,t} = I_{g,t} - I_{g,t-1} \tag{11.31}$$

$$u_{g,t} + v_{g,t} \leq 1 \tag{11.32}$$

$$SOC_{e,t}^{BESS} = SOC_{e,t-1}^{BESS} + \eta_e^{ch} P_{e,t}^{ch} - \frac{P_{e,t}^{dis}}{\eta_e^{dis}} \tag{11.33}$$

$$SOC_{e,\min}^{BESS} \leq SOC_{e,t}^{BESS} \leq SOC_{e,\max}^{BESS} \tag{11.34}$$

$$P_{e,\min}^{\frac{ch}{dis}} \leq P_{e,t}^{\frac{ch}{dis}} \leq P_{e,\max}^{\frac{ch}{dis}} \tag{11.35}$$

$$P_{r,t}^{RE} \leq P_{r,t}^{RE,\max} \tag{11.36}$$

where $I_{g,t}$ is the binary variable indicating whether DG unit g is on, $P_g^{\min}$ and $P_g^{\max}$ are the lower and upper bound of power of DG unit g, $D_g^{\max}$ and $U_g^{\max}$ are the ramp-up/ ramp down limit of DG unit g, T_g^{on} and T_g^{off} are the minimum up/down time of DG unit g, $SOC_{e,t}^{BESS}$ is the SOC of the BESS at time slot t, $SOC_{e,\min}^{BESS}$ and $SOC_{e,\max}^{BESS}$ are the lower and upper bound of the SOC of the BESS, η_e^{ch} and η_e^{dis} are the charging/ discharging efficiency of the BESS, $P_{e,t}^{ch}$, $P_{e,t}^{dis}$ are the charging/discharging power of the BESS e at time slot t, $P_{e,\min}^{ch}$, $P_{e,\max}^{ch}$ are the lower and upper bound of the charging power of the BESS, $P_{e,\min}^{dis}$ and $P_{e,\max}^{dis}$ are the lower and upper bound of the discharging power of the BESS, $P_{r,t}^{RE}$ is the power output of the renewable generator r at time slot t, $P_{r,t}^{RE,\max}$ is the upper limit of the power output of the renewable generator r at time slot t.

Constraint (11.27) imposes the lower and upper bound on the power output of DG unit. Constraint (11.28) presents the up and down ramp limit. Constraints (11.29) and (11.30) limit the minimum up- and down-time of DG unit. Constraints (11.31) and (11.32) present the logical relation between the binary variables that indicate whether DG unit is on, start-up and shut-down. Constraints (11.33–11.35) present the operational restrictions of BESS, which include the limits of charging status

(11.33 and 11.34) and charging/discharging power (11.35). Constraint (11.36) limits the available renewable generation.

11.3 DRO-Based Joint Scheduling Model Under Multiple Uncertainties

11.3.1 Joint Scheduling Framework

In Sect. 11.2, the deterministic joint optimization model is formulated with objective function and operational constraints. To hedge with the multiple uncertainties from both AES arrival and renewable energy output, this section proposes a two-stage joint scheduling model based on DRO approach for seaport microgrids.

Different from traditional power load uncertainty that is revealed in the second stage and influences the decisions of the second stage, the uncertainty of AES arrival time is revealed in the second stage but influences the berthing allocation decision in the first stage. Specifically, in literature [25, 26], the berth allocation decision is determined in the first stage, ignoring the uncertainty of AES arrival time. However, the day-ahead berth allocation schemes may become infeasible if actual arrival time is later than planned berthing start time.

To this end, a novel two-stage joint scheduling framework is proposed as Fig. 11.2 shows. The berthing position of AES and the unit commitment of dispatchable DGs are determined in the first stage. The berthing start and departure time of AES, the power dispatch of microgrids and the QC assignment schemes are adjusted according to the actual AES arrival and renewable generations observed in the second stage. In this way, the feasibility of berth allocation scheme can be guaranteed regardless of future uncertainty realizations.

11.3.2 Two-Stage Joint Scheduling Model Based on DRO Method

(1) *Ambiguity Set Construction*

In this section, data-driven distributionally robust approaches are adopted to describe the uncertainties. In literatures, the probability distribution is generally established based on moment [29], Wasserstein metric [30] and so on. The moment and Wasserstein metric-based method will lead to a semi-infinite subproblem in the second stage, thus strong dual theory is required for model solution. However, in this study, various binary and integer variables are included in berth allocation and QC assignment model, which does not meet strong dual theory and cannot directly adopt the

> **First Stage: Optimal berthing position of AESs and unit commitment of DGs**
>
> *Objective:* Minimizing start up/shut down costs of DG units
>
> *Subject to:* Berthing position and unit commitment constraints

Realization of multiple uncertainties

> **Second Stage: Optimal berthing duration of AESs, QC assignment and power dispatch of microgrids**
>
> *Objective:* Minimizing berthing costs of AES and electricity supply costs of microgrids
>
> *Subject to:* Berthing duration, QC assignment, power flow and device operation constrains

Fig. 11.2 The proposed two-stage joint scheduling framework

moment and Wasserstein metric-based method. To this end, a discrete scenarios-based DRO method [31] is adopted to modeling the multiple uncertainties. The subproblem in each scenario is independent to the scene probability distribution and thus can be solved independently without complex transformation.

Based on the historical data samples K, typical scenarios set N are obtained. The historical data includes renewable energy output and the difference between actual and estimated arrival time at each historical AES visiting. Then, the arrival time of AES in each typical scenario is calculated as the sum of estimated arrival time $t_s^{a,e}$ in the next operational day and the historical data of arrival time deviation Δt_s^a, i.e., $t_s^a = t_s^{a,e} + \Delta t_s^a$.

Theoretically, the probability distribution of independent scenarios can be any value. However, to make the scene probability distribution more appropriate to the actual data, a probability distribution set is constructed based on the initial probability value of each scenario. It includes the 1-norm and ∞-norm constraints:

$$\Omega = \left\{ \{p_n\} \middle| \begin{array}{l} p_n > 0, n = 1, 2, \ldots, N \\ \sum_{n=1}^{N} p_n = 1 \\ p_n - p_{n1}^0 \leq \theta_1 \\ p_n - p_{n\infty}^0 \leq \theta_\infty \end{array} \right. \tag{11.37}$$

where p_n^0 is the nominal probability distribution obtained from the historical data, p_n is the probability distribution, θ_1 and θ_∞ are the allowable deviation limit of possibility. They are relevant with the confidence of scenario probability α_1 and α_∞:

$$\begin{cases} \theta_1 = \frac{N}{2K} ln \frac{2N}{1-\alpha_1} \\ \theta_\infty = \frac{1}{2K} ln \frac{2N}{1-\alpha_\infty} \end{cases}$$

(11.38)

(2) DRO Model Construction

The proposed two-stage joint scheduling model intends to minimize the total costs of both microgrid operations and AES services under the worst-case probability distribution. The objective function (11.1) is rewritten as follows:

$$\min_{x \in X} F_1 + \max_{p \in \Omega} \sum_{n=1}^{N} \left(p_n \min_{y_n \in Y} F_2^n \right)$$

(11.39)

where $\mathbf{x} = \{B_s, I_{g,t}, u_{g,t}, v_{g,t}\}$ are the decision variables of the first stage, and F_1 is the first stage costs shown as follows:

$$F_1 = \sum_{t \in T} \sum_{g \in N_{DG}} \left(u_{g,t} S_g^{on} + v_{g,t} S_g^{off} \right)$$

(11.40)

The second stage determines the berthing duration of each AES, the QC assignment schemes and the power dispatch of microgrids. The decision variables of the second stage are represented by variable $\mathbf{y}$. The operational costs of the second stage in each scenario is stated as follows:

$$F_2 = \sum_{t \in T} \left[c_t^{grid} P_t^{grid} + \sum_{g=1}^{N_{DG}} \left[h_{g,2} \left(P_{g,t}^{DG} \right)^2 + h_{g,1} P_{g,t}^{DG} + h_{g,0} \right] \right]$$
$$+ \sum_{s \in S} \left[c_s^w \left(t_s^a - t_s^b \right) + c_s^b \left(t_s^d - t_s^b \right) \right]$$

(11.41)

The above model is written in compact form as follows:

$$\min_{x \in X} a^T x + \max_{p \in \Omega} \sum_{n=1}^{N} \left(p_n \min_{y_n \in Y} \left(b^T y_n + c^T \xi_n \right) \right)$$

(11.42)

s.t.

$$\begin{cases} A_1 x \leq d_1 \\ A_2 x = d_2 \end{cases}$$

(11.43)

$$Bx + Cy_n \leq e$$

(11.44)

$$\begin{cases} \boldsymbol{D}_1 \boldsymbol{y}_n \leq \boldsymbol{f}_1 \\ \boldsymbol{D}_2 \boldsymbol{y}_n = \boldsymbol{f}_2 \\ \left\| \boldsymbol{D}_3 \boldsymbol{y}_n + \boldsymbol{f}_3 \right\|_2 \leq \boldsymbol{D}_4 \boldsymbol{y}_n + \boldsymbol{f}_4 \end{cases} \tag{11.45}$$

$$\boldsymbol{E} \boldsymbol{y}_n + \boldsymbol{F} \boldsymbol{\xi}_n \leq \boldsymbol{g} \tag{11.46}$$

where $\boldsymbol{A}_1, \boldsymbol{A}_2, \boldsymbol{B}, \boldsymbol{C}, \boldsymbol{D}_1 - \boldsymbol{D}_4$ are the constant matrixes, $\boldsymbol{\xi}_n = \left\{ \Delta t_{s,n}^a, P_{r,t,n}^{RE,\max} \right\}$ is arrival time deviation of AES and available renewable generation in scenario n.

11.3.3 Solution Methodology

The proposed two-stage joint scheduling model can be solved by Column and Constraint Generation (C&CG) algorithm, and decomposed into main problem (MP) and subproblem (SP) as follows.

MP finds the decisions of the first stage under a given worst-case probability distribution:

$$\begin{array}{c} \min \quad \boldsymbol{a}^T \boldsymbol{x} + W \\ \boldsymbol{x} \in X, W \\ \boldsymbol{y}_n^{(l)} \in Y(\boldsymbol{x}, \boldsymbol{\xi}_n) \end{array} \tag{11.47}$$

s.t.

$$W \geq \sum_{n=1}^{N} \left(p_n^{(l)} \left(\boldsymbol{b}^T \boldsymbol{y}_n^{(l)} + \boldsymbol{c}^T \boldsymbol{\xi}_n \right) \right) \tag{11.48}$$

$$\begin{cases} \boldsymbol{A}_1 \boldsymbol{x} \leq \boldsymbol{d}_1 \\ \boldsymbol{A}_2 \boldsymbol{x} = \boldsymbol{d}_2 \end{cases} \tag{11.49}$$

$$\boldsymbol{B} \boldsymbol{x} + \boldsymbol{C} \boldsymbol{y}_n^{(l)} \leq \boldsymbol{e} \tag{11.50}$$

$$\begin{cases} \boldsymbol{D}_1 \boldsymbol{y}_n^{(l)} \leq \boldsymbol{f}_1 \\ \boldsymbol{D}_2 \boldsymbol{y}_n^{(l)} = \boldsymbol{f}_2 \\ \left\| \boldsymbol{D}_3 \boldsymbol{y}_n^{(l)} + \boldsymbol{f}_3 \right\|_2 \leq \boldsymbol{D}_4 \boldsymbol{y}_n^{(l)} + \boldsymbol{f}_4 \end{cases} \tag{11.51}$$

$$\boldsymbol{E} \boldsymbol{y}_n^{(l)} + \boldsymbol{F} \boldsymbol{\xi}_n \leq \boldsymbol{g} \tag{11.52}$$

where l represents the current iteration number, $y_n^{(l)}$ and $p_n^{(l)}$ are second stage variables and worst-case distribution at l^{th} iteration.

SP finds the worst-case-scenario distribution according to the first stage decision variables x^* obtained in the MP:

$$W(x^*) = \max_{p_n \in \Omega} \sum_{n=1}^{N} \left(p_n \min_{y_n \in Y(x^*, \xi_n)} \left(b^T y_n + c^T \xi_n \right) \right) \#11.53 \tag{11.53}$$

s.t.

$$Bx^* + Cy_n \le e \tag{11.54}$$

$$\begin{cases} D_1 y_n \le f_1 \\ D_2 y_n = f_2 \\ \left\| D_3 y_n + f_3 \right\|_2 \le D_4 y_n + f_4 \end{cases} \tag{11.55}$$

$$E y_n + F \xi_n \le g \tag{11.56}$$

Then, the worst-case distribution can be easily obtained by solving the following optimization model.

$$W(x^*) = \max_{p_n \in \Omega} \sum_{n=1}^{N} p_n \left(b^T y_n^* + c^T \xi_n \right) \tag{11.57}$$

Based on the above MP and SP, the C&CG algorithm can be implemented by the following steps:

Step 1: Set initial value of the worst-case distribution p_n, lower bound $LB = -\infty$, upper bound $UB = +\infty$, iteration number $l = 0$, convergence criterion ε.

Step 2: Solve the MP and obtain the optimal decisions of the first stage x^* under the given worst-case distribution $p_n^{(l)}$. Update $LB^{(l+1)} = \max\left(LB^{(l)}, W^*\right)$.

Step 3: Solve the inner minimization problem in SP based on the obtained x^* and obtain the optimal value $b^T y_n^* + c^T \xi_n$ of each scenario. Based on that, the worst-case distribution $p_n^{(l+1)}$ can be obtained by solving the optimization problem. Update $UB^{(l+1)} = \max\left(UB^{(l)}, W^*(x^*)\right)$.

Step 4: Terminate the solution if the convergence criterion $UB^{(l+1)} - LB^{(l+1)} \le \varepsilon$ is satisfied. Otherwise, generate variable $y_n^{(l+1)}$ and constraints to the MP. Set $l = l + 1$ and repeat the above steps until the convergence criterion is satisfied

$$\begin{cases} W \ge \sum_{n=1}^{N} \left(p_n^{(l+1)} \left(b^T y_n^{(l+1)} + c^T \xi_n \right) \right) \\ Bx + Cy_n^{(l+1)} \le e \\ D_1 y_n^{(l+1)} \le f_1 \\ D_2 y_n^{(l+1)} = f_2 \\ \left\| D_3 y_n^{(l+1)} + f_3 \right\|_2 \le D_4 y_n^{(l+1)} + f_4 \\ E y_n^{(l+1)} + F \xi_n \le g \end{cases} \tag{11.58}$$

11.4 Case Studies

11.4.1 Case Description

The effectiveness of the proposed model is verified through numerical experiments on a modified IEEE-33 bus seaport microgrids shown in Fig. 11.3. The operation horizon is 48 h. There are 6 berths, 12 deployable QCs, 2 dispatchable DG units, 2 PV units and 1 BESS. The parameters of these devices are shown in Table 11.1. Ten ships are considered in the operational days. The parameters of the ten ships are shown in Table 11.2. Each QC can handle 35 TEU cargo per hour and its rated power demand is 0.3 MW. The electricity purchase price of main grid during valley (1:00–7:00, 24:00), normal (8:00–10:00, 16:00–18:00, 22:00–23:00) and peak (11:00–15:00, 19:00–21:00) hours are 0.067, 0.139, and 0.216\$/kWh respectively. The berthing related cost coefficient c_s^w and c_s^b in Eq. (11.1) are set as 15.6\$/h. The number of the number of historical data samples K is 100, the number of independent scenarios N is 10, and the confidence levels α_1 and α_∞ are 0.2 and 0.5. The uncertainty realizations of AES arrival and PV generation in the second stage are shown in Table 11.2 and Fig. 11.4 respectively.

Three cases are set to demonstrate the advantages of proposed model:

Case I: Two-stage non-joint scheduling model considering multiple uncertainties, where the berth allocation of AES and the QC assignment are determined independently under AES arrival uncertainty according to berth allocation and QC assignment constraints. Then power dispatch of microgrid is determined under renewable energy uncertainty based on the obtained berth allocation and QC assignment schemes.

Case II: Two-stage joint scheduling model without considering AES arrival uncertainty. The berthing position and berthing duration of AES, and the unit commitment of DG units are determined in the first stage. The QC assignment and the power dispatch of microgrids are decided in the second stage based on uncertainty realizations.

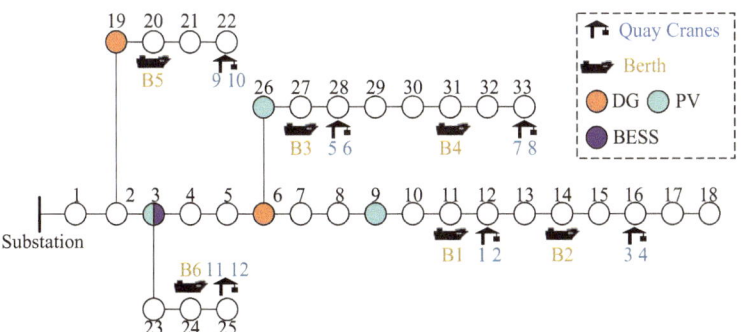

Fig. 11.3 The topology of 33 bus-based seaport microgrid

Table 11.1 Parameters of devices

Device	Parameter	Value	Electrical bus
PV 1/2/3	Capacity (MWh)	3	3/9/26
DG 1/2	Power output limit (MW)	[0.5, 5]	6/19
	Ramp limit	2.5	
	Minimum up/down time (h)	2	
BESS	Capacity (MWh)	5	3
	Charging/discharging power limit (MW)	1.25	
	Charging/discharging efficiency	0.95	
	Limit of SOC	[0.2, 0.8]	

Table 11.2 Parameters of the ten ships

Ship	Forecast arrival time (h)	Actual arrival time (h)	Latest departure time (h)	Cargo (TEU)
1	1	3	12	750
2	3	5	15	800
3	6	4	17	550
4	10	12	21	750
5	14	17	27	500
6	18	14	35	800
7	22	16	37	750
8	25	28	40	650
9	30	34	45	550
10	35	30	48	450

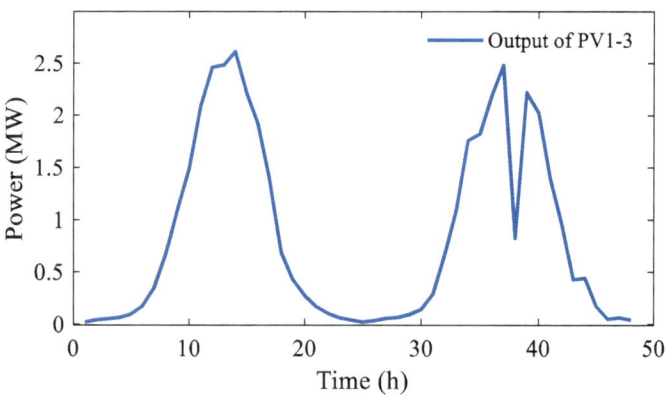

Fig. 11.4 The uncertainty realization of PV output

Case III: The proposed two-stage joint scheduling model, considering multiple uncertainties. The berthing position of AES and the unit commitment of DG units are determined in the first stage. The berthing duration of AES, QC assignment and the power dispatch of microgrids are decided in the second stage based on uncertainty realizations.

All simulations are implemented by Gurobi 9.5 and Python 3.8 on a laptop with an 8-core Intel Core i7 processor and 16 GB memory.

11.4.2 Comparison of Different Scheduling Model

Based on the day-ahead decisions obtained in the first stage and the uncertainty realizations of AES arrival and PV generation observed in the second stage, the optimization results are shown in Table 11.3. The power balance in microgrid and the berth allocation of AES are illustrated in Figs. 11.5 and 11.6 respectively.

In Case I, the berth allocation of AES and the power dispatch of microgrids are determined independently. Berths and QCs are pre-allocated before power dispatch with the highest AES service efficiency. All AESs immediately dock at allocated berths after arriving the seaport, and then departing as soon as possible. Despite the waiting and berthing cost of AES in Case I is lowest in the three cases, because the berth allocation of AES is determined without considering power dispatch constraints, the electricity supply costs are much higher than Case III. Therefore, joint operation of berth allocation and power dispatch is necessary for energy efficiency and economic benefits of seaport microgrids.

In Case II, the berth allocation of AES and the unit commitment are determined without considering uncertain arrival. However, in the second stage, the actual arrival time of AES deviates from the planned arrival time, causing the infeasibility of the first stage decisions. That means for some AESs, the decision of berthing duration in the first stage cannot be implemented because the actual arrival time in the second stage is later than planned berthing start time (i.e., constraint (4.1) is violated). From this perspective, making day-ahead berth allocation schemes without considering AES arrival uncertainty may provide an infeasible solution when the actual arrival time of AES deviating from expectations. To obtain the scheduling results of operational day, we deliberately re-schedule the AESs whose the first stage decisions

Table 11.3 Optimization results in different cases

	Case I	Case II	Case III
DG unit start-up/shut-down cost ($)	80.3	89.8	80.3
Waiting and berthing cost of AES ($)	866.8	1284.8	1331.2
Electricity purchase cost from main grid ($)	15,539.1	12,739.8	13,166.2
DG unit generation costs ($)	11,248.6	10,347.4	9219.6
Total Cost ($)	28,734.8	24,461.8	23,797.2

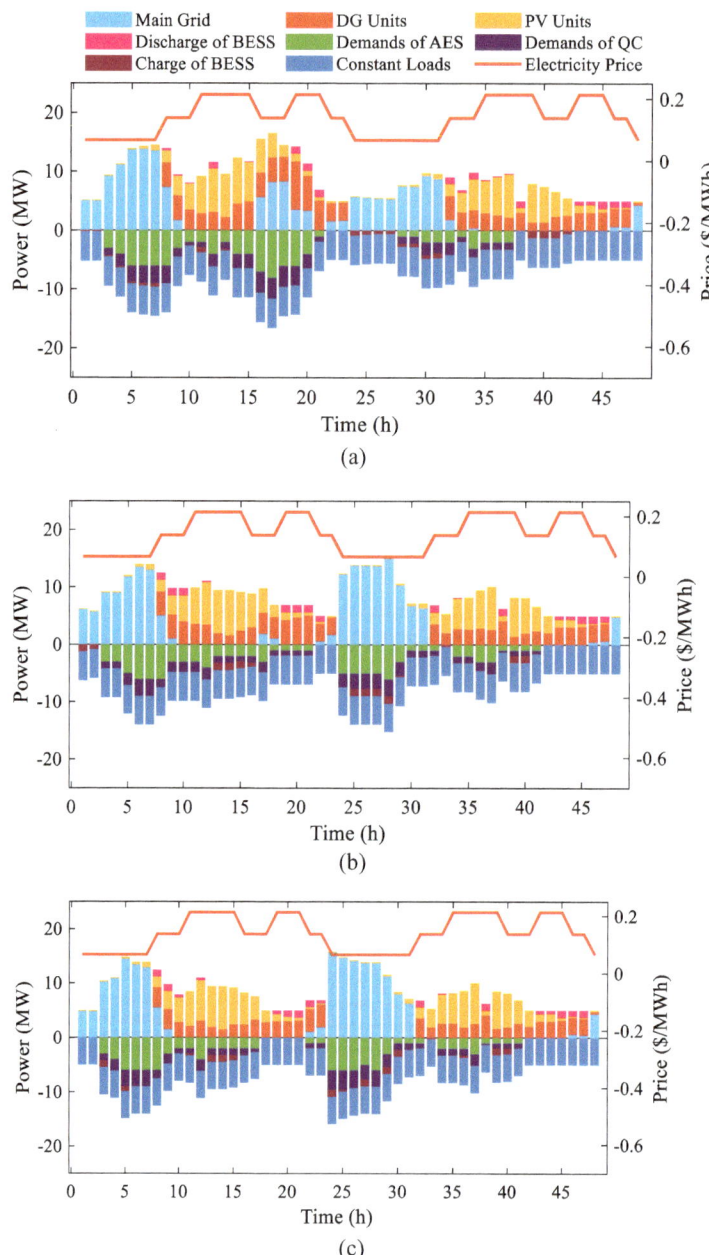

Fig. 11.5 The power balance in microgrids: **a** Case I; **b** Case II; **c** Case III

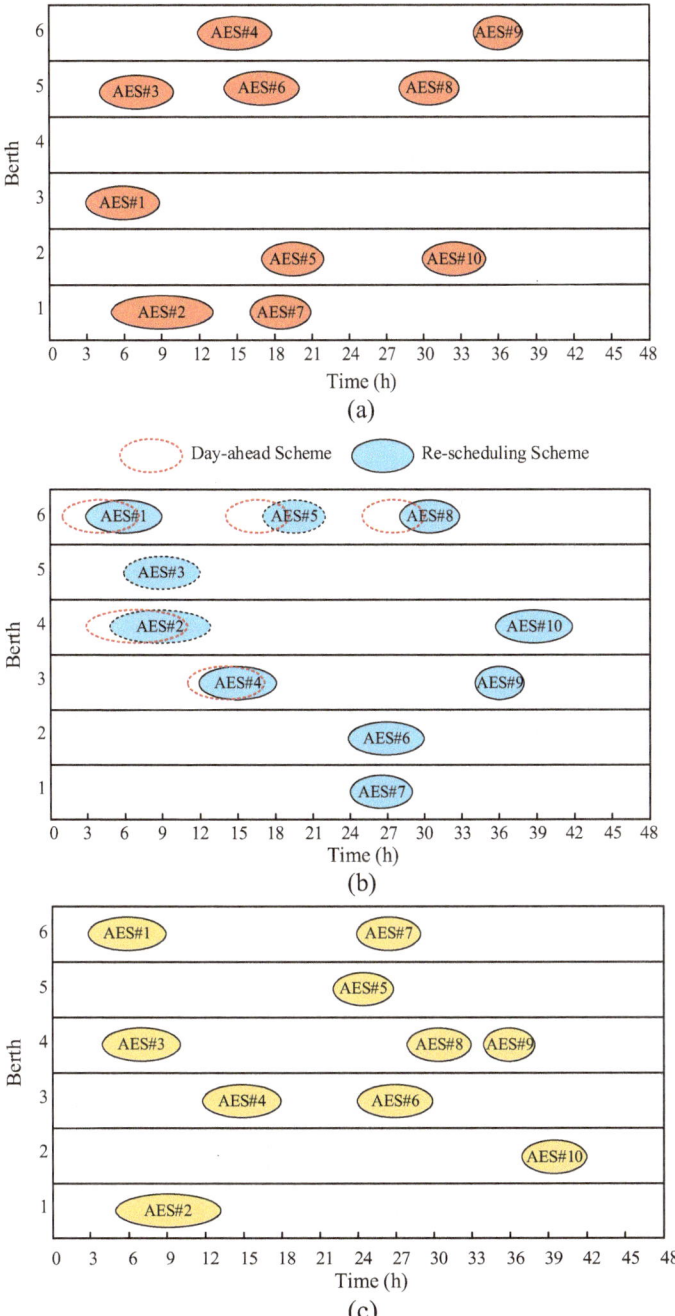

Fig. 11.6 The intra-day berth allocation scheme of AES based on uncertainty realizations: **a** Case I; **b** Case II; **c** Case III

are infeasible, as Fig. 11.6b shows. Additionally, because uncertain AES arrival is ignored, the decisions in the first stage are not optimal for the actual arrival situation in the second stage. Correspondingly, the total costs of Case II are higher than Case III.

In Case III, the berth allocation of AES and the power dispatch of microgrids are jointly optimized considering multiple uncertainties. Compare with Case I, the berth allocation and QC assignment schemes can be adjusted according to electricity price of main grid and available PV generation. Correspondingly, the power loads of AES and QC shifts to the time-period with lower unit supply cost, thus the total costs reduce by 4937.6$. Compare with Case II, the feasibility of first stage decisions in Case III can be guaranteed. Besides, more AES and QC loads shift to time-period with lower supply cost since the AES arrival uncertainty is considered. As a result, the total cost reduces by 664.6$. Therefore, the proposed two-stage joint scheduling model not only can guarantee the feasibility of the first stage decisions, but also can hedge with the multiple uncertainties well.

Figure 11.5 illustrates the power balance in the second stage under uncertainty realizations. In Case I, AES and QC are non-dispatchable. The power loads of AES and QC distribute at the time-period with high unit supply cost, such as 15:00–20:00, leading to higher electricity purchase and DG generation costs. In Case II, AES and QC are scheduled jointly with power dispatch. The total costs reduce compared with Case I. However, the uncertainty of AES arrival is not considered. After re-scheduling of infeasible AES berthing duration, the electricity supply costs are still higher than Case III. Because the first stage decisions are made without considering uncertain AES arrival, which is not optimal under AES arrival uncertainty realization. In Case III, it can be found that some demands of AES and QC further shift to the time-period with lower unit supply costs. For example, the electricity purchase increases compared with Case II during 1:00–9:00 and 24:00–30:00. Correspondingly, the DG generation decreases during 10:00–24:00. Because the unit DG generation cost is higher than electricity price during valley, the increase of electricity purchase costs is lower than the decrease of DG generation costs. Therefore, the total costs of Case III are lower than Case II.

The berth allocation of AES in the second stage under uncertainty realizations is shown in Fig. 11.6. In Case I, the berths and QCs are allocated with highest AES service efficiency. All AES start berthing immediately after arriving the seaport. In Case II, the red dotted line represents the day-ahead berth allocation scheme of AES. Because the berthing duration of AES is determined in the first stage without considering uncertain arrival time, the decisions become infeasible. For example, the AES#1 is arranged to start berthing at 1:00 in the first stage. However, the AES#1 arrives the seaport at 3:00 in the second stage. The actual arrival time is later than planned berthing start time, causing that the first stage decision cannot be implemented. Thus, the AES#1 should be re-scheduled, and the actual berthing start time of AES#1 delays to 3:00. From this perspective, it can be found that the uncertain AES arrival influences not only the economic benefits, but also the feasibility of scheduling schemes. In Case III, the berthing duration of AES are determined in the second stage based on the uncertainty realizations. The total costs further reduce due to flexible adjustment of AES and QC. Moreover, the scheduling schemes in the first stage can be implemented in the second stage without infeasibility. Therefore, compared with Case I and Case II, the proposed method can guarantee both economic benefits and the feasibility of scheduling schemes.

11.4.3 Sensitivity Analysis of System Parameters

The advantages of the proposed joint scheduling model may be affected by various parameters. This section investigates the impact of system parameters on the optimization results.

The berthing related cost coefficients in Eq. (11.1) affects the proportion of berth allocation costs in total costs, thus further influencing the economic benefits of the proposed method. We set the cost coefficients as 10, 50, 100, 150 and 200% of basic value. Figure 11.7 shows the total cost and cost reduction ratio in different cost coefficients.

For Case I and Case III, when the cost coefficients are low (e.g., 10% and 50), the waiting and berthing costs are negligible compared to electricity supply costs. The results are dominated by the power dispatch of microgrids. AESs are willing to adjust berthing position and berthing duration to reduce electricity costs. However, with the increasing of cost coefficients, the priority of berth allocation is growing. AESs are more sensitive to service efficiency and try to reduce the total waiting and berthing duration, which may lead to higher electricity supply costs. Therefore, a trade-off between the service efficiency of AES and the economic benefits of microgrids can be found by adjusting the cost coefficients.

For Case II and Case III, considering that Case II and Case III both jointly optimize berth allocation and power dispatch, there is no an obvious unidirectional trend of cost reduction. Therefore, we can see that the berthing-related cost coefficients mainly

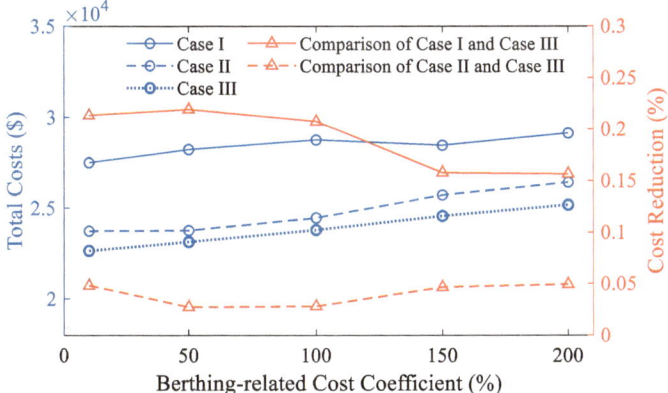

Fig. 11.7 Sensitivity analysis of berthing-related cost coefficient

Table 11.4 Optimization results under different historical data samples

Number of historical data samples	Expected value of total costs ($)
100	23,736.2
500	23,724.7
1000	23,718.8
2000	23,704.1
5000	23,690.6

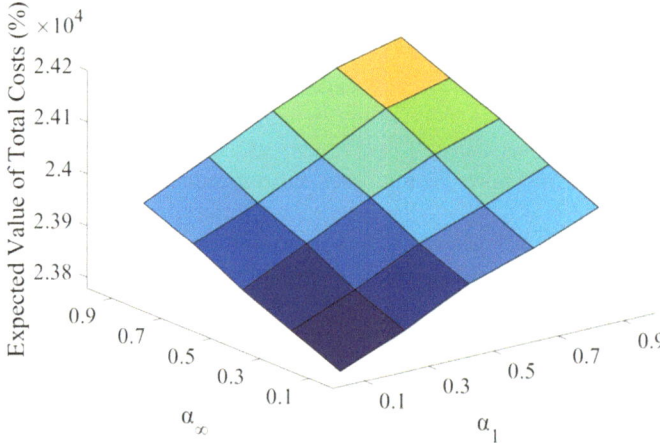

Fig. 11.8 Optimization results under different confidence level

influence the trade-off between berth allocation service efficiency and electricity supply costs.

Moreover, we further investigate the impact of the number of historical data samples and the confidence level on the optimization results, which are shown in Table 11.4 and Fig. 11.8 respectively.

With the increase of confidence level, the expected value of total operational costs increases. This is because the higher confidence level will lead to a wider volatility of probability distribution, causing stronger uncertainties. As a result, more resources need to be scheduled to hedge with the stronger uncertainties.

Similarly, the increase of historical data samples will decrease the value of θ_1 and θ_∞ in Eq. (11.38). Less volatility of probability distribution is allowed. Therefore, the expected value of total operational costs can reduce due to more realistic probability distribution.

References

1. Siemens: An environmentally friendly, economic power supply for berthed ships (2013). http://www.siemens.com/siharbo
2. Friese, H.S., Poulsen, R.T., Nowinska, A.U., et al.: What drives ports around the world to adopt air emissions abatement measures. Transp. Res. Part D: Transp. Environ. **90**, 102644 (2021)
3. Molavi, A., Lim, G.J., Shi, J.: Stimulating sustainable energy at maritime ports by hybrid economic incentives: a bilevel optimization approach. Appl. Energy **272**, 115188 (2020)
4. International Maritime Organization: Reducing greenhouse gas emissions from ships (2020). https://www.imo.org/en/MediaCentre/HotTopics/Pages
5. Xiang, X., Liu, C.: An expanded robust optimisation approach for the berth allocation problem considering uncertain operation time. Omega **103**(1), 102444 (2021)
6. Liu, C., Xiang, X., Zheng, L.: A two-stage robust optimization approach for the berth allocation problem under uncertainty. Flex. Serv. Manuf. J. **32**(2), 425–452 (2019)
7. Schepler, X., Absi, N., Feillet, D.: The stochastic discrete berth allocation problem. EURO J. Transp. Logist. **8**(4), 363–396 (2019)
8. Tan, C., He, J.: Integrated proactive and reactive strategies for sustainable berth allocation and quay crane assignment under uncertainty. Ann. Oper. Res. **3**, 1–32 (2021)
9. Agra, A., Oliveira, M.: MIP approaches for the integrated berth allocation and quay crane assignment and scheduling problem. Eur. J. Oper. Res. **264**(1), 138–148 (2018)
10. Bouzekri, H., Alpan, G., Giard, V.: Integrated laycan and berth allocation and time-invariant quay crane assignment problem in tidal ports with multiple quays. Eur. J. Oper. Res. **293**(3), 892–909 (2021)
11. Ahamad, N.B.B., Guerrero, J.M., Su, C.: Microgrids technologies in future seaports. In: 2018 IEEE International Conference on Environment and Electrical Engineering and 2018 IEEE Industrial and Commercial Power Systems Europe, Palermo, pp. 1–6 (2018)
12. Kumar, J., Kumpulainen, L., Kauhaniemi, K.: Technical design aspects of harbour area grid for shore to ship power: state of the art and future solution. Int. J. Electr. Power Energy Syst. **104**, 840–852 (2019)
13. Parise, G., Parise, L., Martirano, L., et al.: Wise port and business energy management: port facilities, electrical power distribution. IEEE Trans. Ind. Appl. **52**(1), 18–24 (2016)
14. Misra, A., Venkataramani, G., Gowrishanka, S., et al.: Renewable energy based smart microgrids—a pathway to green port development. Strategic Plann. Energy Environ. **37**(2), 17–32 (2017)
15. Song, T., Li, Y., Zhang, X., et al.: Integrated port energy system considering integrated demand response and energy interconnection. Int. J. Electr. Power Energy Syst. **117**, 105654 (2020)
16. Pu, Y., Chen, W., Zhang, R., et al.: Optimal operation strategy of port integrated energy system considering demand response. In: 2020 IEEE 4th Conference on Energy Internet and Energy System Integration, Wuhan, pp. 518–523 (2020)
17. Fang, S., Gou, B., Wang, Y., et al.: Towards future green maritime transportation: an overview of seaport microgrids and all-electric ships. IEEE Trans. Veh. Technol. **69**(1), 207–219 (2020)
18. Duin, J., Geerlings, H., Verbraeck, A., et al.: Cooling down: a simulation approach to reduce energy peaks of reefers at terminals. J. Clean. Prod. **193**, 72–86 (2018)
19. Parise, G., Honorati, A.: Port cranes with energy balanced drive. In: 2014 AEIT Annual Conference—From Research to Industry: The Need for a More Effective Technology Transfer, Trieste, pp. 1–5 (2014)
20. Parise, G., Parise, H., Malerba, A., et al.: Comprehensive peak-shaving solutions for port cranes. IEEE Trans. Ind. Appl. **53**(3), 1799–1806 (2017)
21. Kanellos, F.D., Volanis, E.M., Hatziargyriou, N.D.: Power management method for large ports with multi-agent systems. IEEE Trans. Smart Grid **10**(2), 1259–1268 (2019)
22. Kanellos: Multiagent-system-based operation scheduling of large ports' power systems with emissions limitation. IEEE Syst. J. **13**(2), 1831–1840 (2019)

23. Gennitsaris, S.G., Kanellos, F.D.: Emissions-aware and costeffective distributed de-mand response system for extensively electrified large ports. IEEE Trans. Power Syst. **34**(6), 4341–4351 (2019)
24. Kanellos, F.D.: Real-time control based on multi-agent systems for the operation of large ports as prosumer microgrids. IEEE Access **5**, 9439–9452 (2017)
25. Mao, A., Yu, T., Ding, Z., et al.: Optimal scheduling for seaport integrated energy system considering flexible berth allocation. Appl. Energy **308**, 118386 (2022)
26. Iris, C., Lam, J.S.L.: Optimal energy management and operations planning in sea-ports with smart grid while harnessing renewable energy under uncertainty. Omega **103**(3), 102445 (2021)
27. Kramer, A., Lalla-Ruiz, E., Iori, M., et al.: Novel formulations and modeling enhancements for the dynamic berth allocation problem. Eur. J. Oper. Res. **278**(1), 170–185 (2019)
28. Gan, L., Li, N., Topcu, U., et al.: Exact convex relaxation of optimal power flow in radial networks. IEEE Trans. Autom. Control **60**(1), 72–87 (2015)
29. He, C., Zhang, X., Liu, T., et al.: Distributionally robust scheduling of integrated gas–electricity systems with demand response. IEEE Trans. Power Syst. **34**(5), 3791–3803 (2019)
30. Wang, C., Gao, R., Wei, W., et al.: Risk-based distributionally robust optimal gas-power flow with Wasserstein distance. IEEE Trans. Power Syst. **34**(3), 2190–2204 (2019)
31. Gao, H., Wang, J., Liu, Y., et al.: An improved ADMM-based distributed optimal operation model of AC/DC hybrid distribution network considering wind power uncertainties. IEEE Syst. J. **15**(2), 2201–2211 (2021)

Chapter 12
Port Electrification and Integrated Energy Cases

12.1 Port Carbon Allowance Projections

Carbon emission quota refers to a certain quota of a given enterprise in the state. During the annual settlement, if the emissions exceed the allocated quota, it becomes necessary to purchase additional allowances from other enterprises to compensate. And if the emissions do not surpass the quota, the surplus carbon emission allowances can be sold. Therefore, in order to achieve the goal of carbon neutrality in green port, it is necessary to make the emissions equal to or less than the carbon quota. The difference between existing emissions and carbon allowances is the amount needed to be reduced each year.

According to data from the National Bureau of Statistics, China's carbon emissions in 2021 are 10 billion tons, and the country's GDP is 117 trillion yuan. Based on the recent Q-Series report titled "From 10 billion tons to zero: How can China achieve its carbon neutrality goal" by the UBS research team and the latest data released by the Tsinghua University Institute of Climate Change and Sustainable Development, it is estimated that the country's carbon emissions will be around 1.9 billion tons in 2060. According to the forecasts of Tsinghua University, the Business and Finance Research Center of Waseda University and other institutions, the national GDP in 2060 will be about 400 to 600 trillion yuan, and the median value in this report is 500 trillion yuan.

Let's use the example of Rizhao Port in Shandong Province to illustrate carbon emissions. In 2021, Rizhao Port's carbon emissions are 162,166 tons, and the throughput is 463.08 million tons. The forecast for Rizhao Port's carbon emission quota is determined based on the contribution of port throughput to GDP and the proportion of Rizhao Port's carbon emissions in relation to the national carbon emissions. And the specific calculation formula is as follows:

W. Huang et al., *Energy Management of Integrated Energy System in Large Ports*,
Springer Series on Naval Architecture, Marine Engineering, Shipbuilding and Shipping 18, https://doi.org/10.1007/978-981-99-8795-5_12

$$C_{60}^r = \frac{\frac{T_{60}^r}{G_{60}^c}}{\frac{T_{21}^r}{G_{21}^c}} \times \frac{C_{21}^r}{C_{21}^c} \times C_{60}^c \tag{12.1}$$

where C_{60}^r is the 2060 Rizhao Port carbon emission quota, C_{60}^c is the national carbon emission quota in 2060, C_{20}^r is the 2021 Rizhao Port carbon emission quota, and C_{20}^c is the national carbon emission in 2021. T_{60}^r is the throughput of Rizhao Port in 2060, and T_{20}^r is the throughput of Rizhao Port in 2021. G_{60}^c is the national GDP in 2060, G_{20}^c and for the national GDP in 2020.

According to the above formula, Rizhao Port's carbon emission quota from 2022 to 2060 can be determined by China's projected total GDP and national carbon emissions. The comparison of carbon emission quotas and carbon emission forecasts without emission reduction measures is shown in Fig. 12.1 and Table 12.1 is shown. It can be seen from the figure that the carbon emissions of Rizhao Port will increase year by year without taking emission reduction measures, and the carbon emission quota will be reduced year by year with the gradual strictness of carbon emission requirements. And the difference between the two is the carbon emissions that Rizhao Port needs to reduce in the carbon neutrality action plan.

According to the existing practice, when counting the total national carbon emissions, indirect carbon emissions have actually been calculated in the power generation and heat industries, and should not be included in the total national carbon emissions in order to avoid double counting. Secondly, the heavy oil consumption of some ports accounts for 20%, but it is mainly used for liner transportation consumption, and the carbon emission site is outside the port area and is not included in the port carbon emissions.

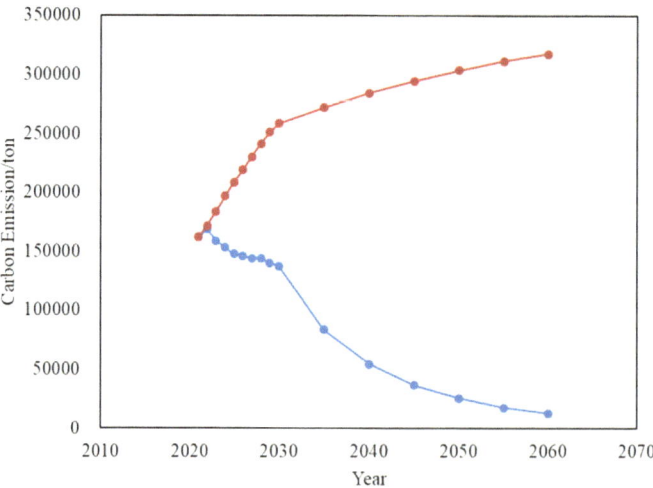

Fig. 12.1 Carbon emission forecast results under Rizhao Port's carbon emission quota (blue curve) and no emission reduction measures (orange curve)

Table 12.1 Rizhao Port carbon emission quota and carbon emission forecast result table under no emission reduction measures

Annual	CO$_2$ emission allowances (10,000 tons)	Carbon emission forecast without emission reduction measures (10,000 tons)
2021	162,166	162,166
2022	168,659	171,896
2023	159,160	183,929
2024	153,374	196,804
2025	148,007	208,612
2026	145,806	219,043
2027	144,308	229,995
2028	144,136	241,495
2029	140,111	251,154
2030	137,655	258,689
2035	83,746	271,885
2040	54,401	284,342
2045	36,556	294,434
2050	25,554	303,374
2055	17,775	311,034
2060	12,966	317,305

The measurement items of direct carbon emissions of ports include: (1) Carbon emissions from combustion generated by the production of port production equipment; (2) Transportation carbon emissions generated by port foreign vehicles in port activities; (3) Carbon emissions from combustion during the anchorage period of ships in port; (4) Carbon emissions generated by other energy-consuming facilities or capacity facilities in the port area. The indirect carbon emissions of Rizhao Port include: (1) Carbon emissions from electricity use; (2) Carbon emissions from heat or steam use.

Taking Rizhao Port as an example, the total direct carbon emissions of Rizhao Port are approximately 162,200 tons. This includes around 101,200 tons from the carbon emissions of port production equipment, 24,400 tons from ships calling at the port, approximately 2.53 × 10,000 tons from foreign vehicles, and roughly 11,300 tons from domestic carbon emissions generated by non-production units like communication companies, information companies, and hospitals. The main carbon emission sources of Rizhao Port in 2021 are shown in Fig. 12.2.

(1)   Carbon Emissions of Production Equipment in Rizhao Port

Rizhao Port production equipment mainly includes: container loading and unloading equipment, including quay bridges, yard bridges, rail cranes and other equipment. Dry bulk loading and unloading equipment, including stacker/reclaimer, belt machine, door machine, ship unloader, ship loader and other equipment. Loaders,

Fig. 12.2 Proportion of
major carbon emission
sources in Rizhao Port in
2021

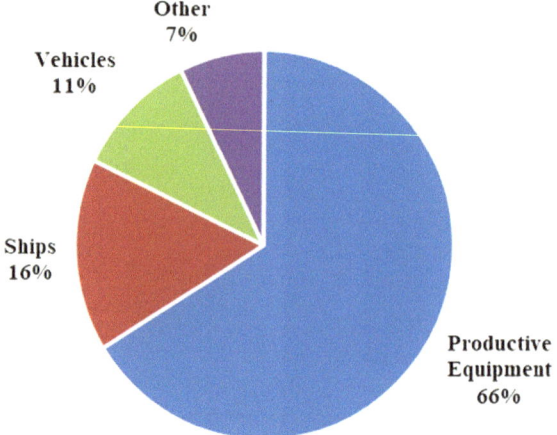

excavators and other mobile machinery, as well as ships, locomotives and other equipment.

The main way to calculate the carbon emissions of port handling tools is to convert it into carbon emissions by calculating the amount of fuel consumed during production equipment operations. For production equipment that uses electricity, based on the existing commonly used carbon emission calculation rules in the world, electricity is included in the carbon emissions of the power plant when it is generated, so on the consumption side, the equipment using electricity can be considered to not produce carbon emissions. The main production equipment of Rizhao Port oil includes loaders, locomotives, excavators, environmental protection vehicles, jib cranes, dump trucks, reach stackers, forklifts, etc. The estimation of diesel consumption and carbon emissions of major production equipment in 2021 is shown in Table 12.2.

(2) Carbon Emissions of Ships Calling at Rizhao Port

Ships are means of transportation that can sail or moor in water for transportation or operations. Maritime transportation by ships holds significant importance in international logistics, serving as the primary mode of transport and also plays a vital role in establishing ports as essential logistics nodes. The emission of the ship is mainly determined by the emission of each engine and boiler, and each engine and boiler has different emission characteristics in different sailing conditions of the ship.

The emission of the ship is mainly determined by the emission of each engine and boiler, and each engine and boiler has different emission characteristics in different sailing conditions of the ship.

Navigation status: sailing in open water, such as sailing in the sea;

Operational status: navigation in restricted waters, such as sailing from the warning area to the breakwater;

Berth status: the ship is docked at the berth, generally in the loading and unloading operation;

Table 12.2 Estimation of diesel consumption and carbon emissions of main production equipment in Rizhao Port

Device name	Quantity	Diesel consumption (t)	Carbon emissions (t)
Loader	208	11,373	35,938.68
shipping	27	8733	27,596.28
locomotive	16	4166	13,164.56
excavator	75	2343	7403.88
Eco-friendly vehicles	155	2041	6449.56
Curved boom crane	24	939	2967.24
Dump truck	22	767	2423.72
Reach stacker	10	479	1513.64
Forklift	70	430	1358.80
Other	106	753	2379.48
Total	713	32,024	101,195.84

Anchorage status: the ship docks at the anchorage;

Transfer status: The state in which a ship is transferred from a berth or anchorage to another berth or anchorage.

Carbon emissions other than berthing and anchorage are not included in the carbon emissions of Rizhao Port Area, so they are not considered in this chapter. According to the above principles, the calculation principle of ship carbon emissions is shown in Fig. 12.3.

In the above state, in addition to the navigation state and operation state, the ship needs greater power to sail and start the main engine, and the other states basically only use the auxiliary engine and boiler to provide power or energy to maintain the power required by the ship in the non-sailing state and operating state. This book studies the carbon emissions of port integrated logistics. It can be considered that the status of ships sailing in development waters and some restricted waters is not affected by the port integrated logistics system, but is only affected after the ship enters the entry and exit channels. Therefore, the navigation state and most of the operating states do not belong to the carbon emission measurement scope of the port. The calculation method of carbon emissions of ships calling at Rizhao Port can be found in the appendix.

(3) Carbon emissions From Foreign Vehicles in the Port

Foreign vehicles in the port mainly include collection and distribution of inbound and outbound trucks. Studies have shown that the driving characteristics of motor vehicles have a significant impact on their emissions, with speed having the most significant impact on emissions. The emission model based on average speed corrects the emission factor by using the average speed as a parameter, and then multiplies the modified factor by the number of vehicle kilometers to obtain the total emissions, which is generally suitable for macro and meso scales. A speed-based emissions

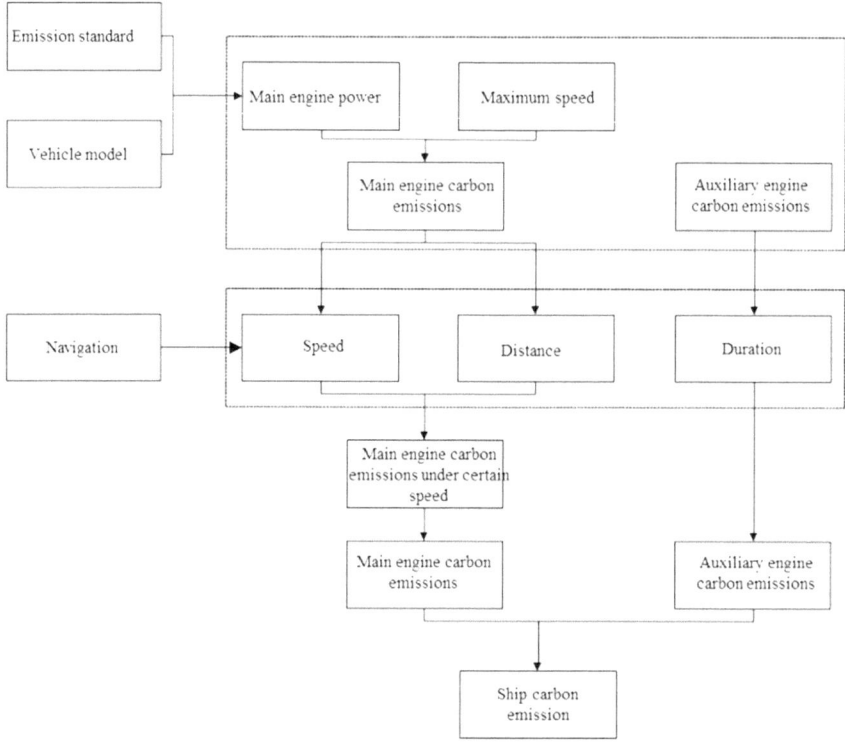

Fig. 12.3 Calculation of carbon emissions from ships

model can be represented by the following mathematical formula:

$$E = L \times e \tag{12.2}$$

where E is the total carbon emission of the vehicle, L is the mileage driven, and e is the carbon emission factor that takes into account the speed correction. According to the data, the carbon emission factors of large and medium-sized trucks (trucks) are shown in the following table 12.3.

12.2   Port Energy Consumption and Carbon Emission Projections

China's major decision to reach carbon peak and carbon neutrality highlights China's strategic determination and responsibility as a major country in the construction of ecological civilization, and releases a positive signal to the world that China firmly follows the path of green and low-carbon development and leads the construction of

Table 12.3 Carbon emission factors of large and medium-sized trucks (trucks)

Speed (km/h)	Truck/truck carbon emission factor (g/km)
10	1188
20	725
30	487
40	388
50	329
60	289
70	261
80	240
90	223
100	210

global ecological civilization and a beautiful world. In order to help achieve the dual carbon goals, this project will first make a forecast of Rizhao Port's energy consumption from 2022 to 2060. Based on the throughput data, energy consumption data and carbon emission data of Rizhao Port from 2016 to 2021, the future throughput, energy consumption and carbon emissions of Rizhao Port are forecasted. In this process, throughput prediction is greatly affected by world economic development and international political and economic situation, and has high uncertainties. At the same time, the types of port energy use are also affected by technological development and policy factors, and have great uncertainties.

First, the preliminary analysis of the problem is carried out, and the throughput forecast is based on the global maritime market demand forecast, the national port transportation development and the 14th Five-Year Plan of Rizhao Port released by the China Industry Research Report Network. Rizhao Port handles eight types of goods: iron ore, containers, crude oil, coal, wood chips, steel, soybeans and bauxite.

It is predicted that the transportation volume of oil, gas and products will maintain slow growth until 2030, and will gradually decrease after 30 years due to the impact of energy consumption depletion, energy conservation and emission reduction. Iron ore and steel throughput maintain rapid growth before 2030, but slow down after 2030. The throughput of coal and its products, timber and grain maintain a steady growth trend, but subject to the slowdown in globalization after 30 years. Other cargo types are mainly transported in the form of containers, and the throughput growth is relatively fast, and it will maintain a growth rate of about 1.5% in 2050. The prediction results of various cargo throughput are shown in Fig. 12.4, where blue curve represents historical data and orange curve represents forecast data.

Based on the above cargo types and global development situation of Rizhao Port, the throughput forecast results of Rizhao Port from 2022 to 2060 (annual forecast from 2022 to 2030, and every five years from 2030 to 2060) are obtained, as shown in Fig. 12.5 and Table 12.4. It can be seen that from 2006 to2030, the throughput of Rizhao Port has been in a stage of rapid growth. From 2030 to 2060, throughput growth slows due to the slowdown in globalization and possible

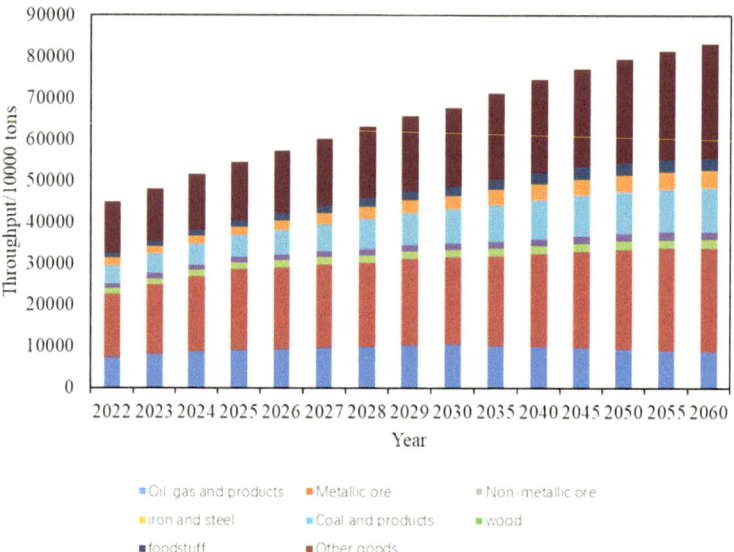

Fig. 12.4   Forecast of various cargo throughput in Rizhao Port

geographical conflicts, and the final throughput forecast for 2060 is around 830 million tonnes.

On the basis of Rizhao Port throughput forecast, taking the energy structure and energy consumption data of Rizhao Port in 2021 as a reference, the total energy consumption of Rizhao Port from 2022 to 2060 is obtained by the equal proportion

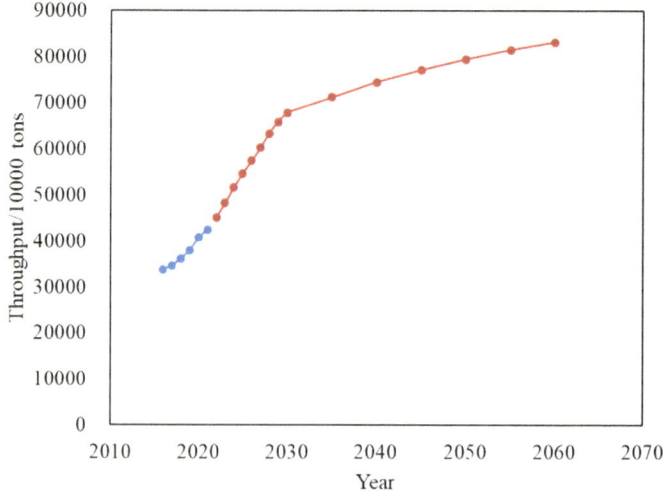

Fig. 12.5   Overall throughput of Rizhao Port

Table 12.4 Summary table of the overall throughput of Rizhao Port

Annual	Throughput historical data (10,000 tons)	Annual	Throughput prediction results (10,000 tons)
2016	33,792	2026	57,421
2017	34,639	2027	60,292
2018	36,092	2028	63,306
2019	37,885	2030	67,814
2020	40,739	2035	71,273
2021	42,511	2040	74,539
2022	45,062	2045	77,184
2023	48,216	2050	79,528
2024	51,591	2055	81,536
2025	54,686	2060	83,180

method, that is, the relationship between various energy consumption and the total throughput of Rizhao Port is considered to be a positive proportional relationship. As the throughput increases, the energy consumption increases in equal proportion. The specific formula of the method is shown in the appendix.

The specific energy consumption of each sub-item is shown in Table 12.5.

According to the energy consumption forecast results and the carbon emission statistics of ships calling at Rizhao Port, 3.33 tons of carbon dioxide are produced

Table 12.5 Energy consumption by sub-item

Year	Throughput (10,000 tons)	Gasoline/ diesel (t)	Electricity (million kWh)	Natural gas (t)	Discounted coal (t)
2022	45,062	35,518	34,945	2449	99,063
2023	48,216	38,004	37,391	2620	105,998
2024	51,591	40,664	40,008	2803	113,418
2025	54,686	43,104	42,409	2972	120,223
2026	57,421	45,259	44,529	3120	126,234
2027	60,292	47,522	46,756	3276	132,546
2028	63,306	49,898	49,094	3440	139,173
2029	65,839	51,894	51,057	3578	144,740
2030	67,814	53,451	52,589	3685	149,082
2035	71,273	56,177	55,272	3873	156,687
2040	74,539	58,751	57,804	4050	163,866
2045	77,184	60,837	59,856	4194	169,682
2050	79,528	62,684	61,673	4322	174,834
2055	81,536	64,267	63,230	4431	179,249
2060	83,180	65,562	64,505	4520	182,862

by the complete combustion of 1 ton of gasoline/diesel, and 2.85 tons of carbon dioxide is produced by the complete combustion of 1 ton of natural gas. The carbon emissions of Rizhao Port from 2022 to 2060 are obtained, as shown in Fig. 12.6 and Table 12.6. Under the existing energy ratio, based on the prospect analysis of the stable development of Rizhao Port, if carbon neutrality is not adopted, the carbon emissions of Rizhao Port in 2060 will be 1.5 times of the current level. The huge increase in carbon emissions will prevent Rizhao Port from achieving its carbon neutrality goal.

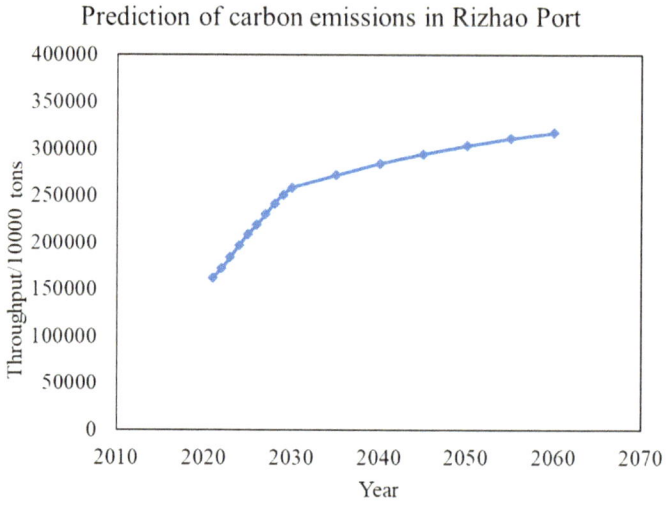

Fig. 12.6   Carbon emission projections without emission reduction measures

Table 12.6   Carbon emission forecast table without emission reduction measures

Annual	CO_2 emissions (10,000 tons)	Annual	CO_2 emissions (10,000 tons)
2022	171,896	2030	258,689
2023	183,929	2035	271,885
2024	196,804	2040	284,342
2025	208,612	2045	294,434
2026	219,043	2050	303,374
2027	229,995	2055	311,034
2028	241,495	2060	317,305
2029	251,154	–	–

12.3 Port Low-Carbon Planning

According to the requirements of carbon peaking and carbon neutrality, as well as comprehensive consideration of the 14th Five-Year Plan for Port Action and national policies, the action plan for ports to achieve the dual carbon goal is divided as follows.

12.3.1 Action Plan for Electrification Equipment

From 2022 to 2030, the preparatory period for achieving carbon peak and carbon neutrality will be deeply optimized, the proportion of electrification equipment will be further increased, and electric energy substitution technology will be fully applied in transportation vehicles and mobile machinery. By 2025, the technical level of electrification equipment such as electric trucks, electric tugboat ships, and electric flow machines will be further developed, and it will have a relatively stable market supply capacity. Rizhao Port will phase out diesel equipment, including trucks, locomotives, loaders, excavators, forklifts, reach stackers, environmental protection vehicles, dump trucks, etc., and replace them with electrified equipment to achieve a high degree of electrification.

By 2035, the technology of electrification equipment is expected to reach a mature stage, with a gradual decrease in economic costs. Additionally, the marketization level will see further improvement, resulting in a 100% adoption rate of electric drive equipment, and there will be rapid development in electric energy replacement technology. 2030–2035 is the adaptation period of energy transformation to achieve carbon peak and carbon neutrality. The number of electrification equipment continues to increase, and the number of vehicle electrification equipment with mature electric drive technologies such as trucks and tractors will grow rapidly.

From 2035 to 2045, the energy transition to achieve carbon peak and carbon neutrality is gradual, and the port has the ability to implement large-scale electrification transformation, which is the golden period for the port to carry out large-scale electrification transformation. The number of electrified equipment doubles. By 2045, there will be 2.5 times as many electrified trucks as in 2025, and the rest of the electrified equipment will be more than three times that of 2025.

From 2045 to 2055, the energy transition to achieve carbon peak and carbon neutrality will accelerate, the cost of electrification equipment will be rapidly reduced, and all types of electrification equipment will grow explosively. By 2055, there will be three times as many electrified trucks as in 2022, and more than four times as many electrified devices as in 2022.

2055–2060 is the energy transition period to achieve carbon peak and carbon neutrality, and the completion of the carbon neutrality goal will be completed. Carbon emissions will reach the lowest point since 2022 in 2060, the number of electrification equipment will increase by about 10% per year, and the quality level will be deeply optimized.

12.3.2 Action Plan for New Energy

New energy construction plans include shore power, wind turbines, photovoltaics and hydrogen energy equipment.

(1) Construction Plan of Shore Power Berths

For shore power, the ship uses oil-fired power generation auxiliaries to supply power to the load during arrival and docking, which will produce a large amount of fuel exhaust emissions, bring serious air and water pollution problems to the port, and put greater pressure on environmental protection and sustainable development. According to statistics, 70% of the world's ship waste emissions occur in the near port area, of which 60%-90% occur during berthing, and the ship call in port will produce a large amount of carbon emissions and cause serious pollution impacts on the area near the port. The shore power system refers to the power supply system that replaces the ship's power generation auxiliary during the ship's arrival and port call, and provides the power required for the ship's pumps, ventilation, lighting, communications and other loads during the ship's arrival and docking. The use of shore power for energy supply by ships in port is of great significance to reduce regional environmental pollution, reduce the energy cost of ships in port and improve energy efficiency, and is one of the effective ways for port energy conservation and emission reduction.

In combination with the current use of shore power in the port, the shore power construction action plan is formulated as follows. (1) Increase the construction of shore power equipment. According to the overall plan of carbon peak and carbon neutrality of Rizhao Port, it is recommended that Rizhao Port maintain the construction of new shore power piles every year after 2030, and the shore power berth coverage rate will reach 9 5 by 2055% or more. (2) Improve shore power service capacity. After the completion of shore power pile construction, the shore power cable intelligent management system, ship-shore quick connection device and automatic synchronous merging device should be gradually completed, so as to improve the efficiency of shore power access to ships and reduce labor costs. At the same time, a shore power management system and operation service platform should be established to fully integrate shore power facility information, and provide ship users with services such as power connection guidance, ship monitoring, and shore power operation monitoring, so as to improve the convenience of shore power use and increase the use of shore power. (3) Improve the shore power use mechanism, improve the shore power service fee pricing mechanism, and establish a dynamic service fee pricing mechanism on a long-term scale based on price information such as oil prices, transformation costs and government subsidies to achieve mutual benefit and win–win results for both ships and ports, so as to guide more ships to connect to shore power.

(2) Construction Plan of Wind Turbine

Breakwater fans and shoreline fans are planned to be installed in green ports to increase the proportion of clean energy generation. Based on the current application status of offshore wind power generation equipment and the application situation of other ports, this action plan selects 3–5 MW wind turbines to be installed on the shoreline and waveproof lifting of the port. According to the overall planning and step-by-step implementation of the principle of design, it is expected that the total proportion of wind turbines and photovoltaics will be more than 10% in 2030, and the wind turbines will reach more than 150 MW in 2060.

(3) **Construction Plan of Distributed PV**

Distributed photovoltaic is a more suitable new energy power generation method in green ports. It can effectively solve the problem of consumption, because the distributed photovoltaic power supply is on the user side, the power generation is supplied to the local use to achieve nearby consumption, which can effectively reduce the dependence on grid power supply and reduce line loss. The photovoltaic operation mode adopts the form of spontaneous self-consumption, and the power generation surplus can be connected to the Internet. Considering the economy and the needs of port operations, the installation location of photovoltaic equipment should be mainly located on the roof of warehouses and processing workshops. The roof PV equipment of different buildings should be independent of each other and adopt the azimuth and inclination angle that is most suitable for the type of roof.

(4) **Construction Plan of Hydrogen Energy Facility**

With the constant development of new energy technology and the continuous maturity of hydrogen energy technology, Rizhao Port can introduce hydrogen-powered equipment in the later stage to enrich the path of carbon neutrality. Hydrogen-powered equipments have been put into use in some ports at home and abroad, such as hydrogen-powered trucks, hydrogen-powered forklifts, hydrogen-powered bridge cranes, etc. According to the technical development and cost changes, hydrogen fuel equipment can be considered to replace electric equipment by referring to the hydrogen power equipment input scheme of domestic and foreign ports.

In addition, as an emerging technical solution, hydrogen-powered tugboats can be used as one of the feasible solutions for hydrogen energy replacement in Rizhao Port. At present, hydrogen-powered tugboats are already in service in Germany, and in January 2022, the German "Elektra" will serve as a demonstration of zero-emission vessels, the power system of which is designed for all kinds of inland and offshore vessels. In addition to propulsion, the ship's electricity will also provide for the crew's living and working needs. The waste heat generated by the ship's fuel cells can be reused by heating continuous cold water. The vessel will use fuel cell technology to provide the basic energy for the drivetrain and electrical systems on board. At peak loads, the battery provides additional energy. The hydrogen supplied to the fuel cell is produced by electrolysis from green electricity generated by wind power. It is expected that after 2040, hydrogen-powered ship technology will become more mature, and its cost will be lower than that of electric tugboat ships, so it can be put into use.

12.3.3   Optimization Plan for Collection and Distribution System

(1)   **Further Promotion of the Port Function Optimization, Transformation and Upgrading**

The intensive and large-scale development of ports will be promoted. The overall management and efficient utilization of shoreline resources should be strengthened, and the zoning and centralized arrangement of goods and types will be implemented. The level of port specialization and process needs to be improved. By Realizing the optimization of port stock resources and carrying out port specialization and process transformation, the terminal loading and unloading efficiency and production energy efficiency will be improved, so as to essentially essentially improve the operating environment of the port area. Focusing on the southern area of Shijiu Port Area, professional, process-oriented and intelligent transformation will be implemented to realize the entire process transportation and interconnection of ore, coal, grain and other berths and rear storage and transportation systems, minimize the short fall of automobiles, and create a leading smart green bulk cargo functional area in China.

The dock will be green planned and its design level will also be improved. To expand the high-quality incremental supply of the terminal, the whole life cycle of the design, construction and operation of the new terminal will be integrated into the concept of green development. This is specifically manifested in strengthening the energy-saving evaluation and process review of new engineering projects, strictly controlling the energy consumption of unit products of the terminal, and strengthening the management of pollution sources in construction and operation.

(2)   **Optimization of the Structure and Organization Mode of Logistics and Transportation**

The proportion of collection and distribution of railways, waterways and pipelines needs to be increased. In accordance with the principle of "suitable iron is iron, suitable water is water, and suitable is public is public", overall plans for the transportation of bulk goods "from road to iron and road to water" will be made. Bulk raw materials and industrial products such as coal and ore will be explored for transportation through railways, sealed corridors, pipelines, and other modes.

"Scattered revisions" and "piecemeal recollections" need to be vigorously developed. The special integrated service business such as "scattered reform collection", "scattered collection" and railway open-top box evacuation will continue to be implemented. Bulk goods such as ore, coal, and coke will be converted into container railway collection and distribution ports, and the proportion of "piece recollection" of fertilizer, grain and other types of goods will be increased. Cooperation between Hong Kong and Hong Kong and port and shipping need to be strengthened, and measures such as preferential freight rates, space guarantees, and priority in trunking and entry and exit stations will be adopted to promote the rapid growth of "scattered and regrouped" cargo volume on waterways.

12.3.4 Operation Process Transformation Plan

In the operation process, by using three-dimensional laser scanning technology, simulation processing technology, high-precision positioning technology, PLC control technology, computer software, etc., based on the realization of the automatic unloading operation control of grab type ship unloader, automatic loading operation control of ship loader, automatic loading operation control of loading machine and digital dock, the equipment chain control model and ore terminal operation process full automatic control system are established according to the loading and unloading process operation line. They carry out the overall control of the operation line process, and achieve the whole process automation of unloading process, mixing process and loading process, which realizes the automation and centralized control of the whole process of dry bulk cargo terminal operation and improves the level of terminal management.

The application of tubular belt conveyor technology in ports is planned to increase. The "tubular belt conveyor" process replaces the traditional dump truck transfer method to realize the "oil to electricity" of bulk cargo transportation. The transformation of bulk cargo operation process will be strengthened. Pilot study will be carried out on the application of full-process technology for conditional professional dry bulk terminals. Promoting the full-process system project of berths in the south operation area of Shijiu Port Area will realize the full process transportation and interconnection of ore, coal, grain and other berths and rear storage and transportation systems, minimize the short fall of cars, create a new model of port transformation and upgrading, and establish a "new benchmark of smart and green dry bulk cargo operations". The automation reconstruction project for the loading and unloading process system of the yard behind the bulk cargo berth will be continued, and the construction of the automated yard and the stacker, reclaimer and belt transportation system will be carried out step by step, realizing the specialization and automation of the ore loading and unloading process. The general bulk cargo terminal gives priority to the use of energy-saving mobile hopper loading method for unloading operations to reduce energy consumption in shore loading and unloading.

12.3.5 Action Plan for the Construction of Management Mechanisms

(1) **Further Advancing the Management of Low-Carbon and Green Development in Ports**

Using advanced science and technology and modern systems, the overall planning for the low-carbon and green development of the port industry are made. In terms of port production equipment management, the corresponding inspection plan should be formulated, and the corresponding management system should be strictly implemented in the production and operation process, so as to realize the port's low-carbon green development goal.

In addition, in order to make all employees participate in the low-carbon and green construction of the port, a certain reward and punishment mechanism should be formulated for employees, encourage employees to carry out technological innovation, and actively contribute to the low-carbon and green development of the port.

(2) **Strict Limitations on the Entry and Exit of Equipment with High Fossil Energy Consumption**

In the production and operation of the port, the most important source of carbon emissions is some production machinery and equipment in the port, which generally need to consume large fossil energy and emit a large quantity of polluting gases. In view of this phenomenon, port enterprises should strictly manage it, establish a strict assessment mechanism for the use and withdrawal of equipment, set energy consumption and carbon emission limit standards for the equipment used, and replace some old, high-energy, polluting machinery and equipment in time.

(3) **Promoting Energy-Saving Technologies and Advocating for Low-Carbon Practices**

Publicity and education activities on low-carbon green development within the port industry should be carried out actively, so that the concept of low-carbon green development of the port is deeply rooted in the hearts of the people. This will promote the employees of port enterprises to actively participate in the low-carbon green development of the port, consciously practice various relevant policies and systems, set an example, and make due contributions to the low-carbon and green development of the port. It is also necessary to actively develop the active role of employees in the construction of low-carbon green ports, encourage innovation, and strengthen the training of various energy-saving and environmental protection technologies.

(4) **Establishing the Port Smart Energy Management Platform to Facilitate Regional Low-Carbon Operations**

A smart energy ecological management platform for Rizhao Port will be built to monitor and manage the comprehensive energy system of Rizhao Port in real time, integrate 5G communication, carbon footprint tracking, Internet of Things and other technologies to improve Rizhao Port's energy ecological management capabilities and low-carbon service capabilities. Through the smart energy ecological management platform, Rizhao Port's comprehensive energy system is digitally intelligently supervised, managed and served, and carbon reduction services are created from source to terminal aggregation, ensuring energy supply and safe and efficient management in the port area, mastering the overall energy consumption, realizing refined management, helping energy conservation and consumption reduction, escorting safe production, and realizing green safety in the port area.

In summary, it can be seen that as long as the port through the above measures, it can achieve carbon peak in 2030 and carbon neutrality in 2060. After the above measures, the port can achieve deep optimization of energy structure, and achieve remarkable results in energy conservation and carbon reduction; The transportation structure is significantly optimized, and the logistics system is clean and efficient; The port structure has been greatly optimized, and the industrial structure has been comprehensively transformed; The low-carbon management means have been innovated and the green and intelligent coordinated development has been realized; The environmental governance mode has been innovated, and the synergy of pollution reduction and carbon reduction has been achieved. A world-class marine port will be built, including the world's leading smart green port, the world's leading logistics hub port, the world's leading industry-city integration port, the world's leading financial and trade port, and the world's leading cruise cultural tourism port.